Perspectives on Biodiversity:

Case Studies of Genetic Resource Conservation and Development

Perspectives on Biodiversity:

Case Studies of Genetic Resource Conservation and Development

Christopher S. Potter
Joel I. Cohen
Dianne Janczewski
editors

AAAS PRESS A publishing division of the American Association for the Advancement of Science

Library of Congress Cataloging-in-Publication Data

Perspectives on biodiversity : case studies of genetic resource conservation and development / edited by Christopher S. Potter, Joel I. Cohen, Dianne Janczewski.
p. cm. — (AAAS publication ; 93-10S)
Includes biographical references and index.
ISBN 0-87168-512-4 (alk. paper)
I. Potter, Christopher S. II. Cohen, Joel I. III. Janczewski, Dianne. IV. Series.
Q181.A1A68 no. 93-01S
[QH75]
333.095'16—dc20 93-10544
CIP

The findings, conclusions, and opinions stated or implied in this publication are those of the editors. They do not necessarily reflect the views of the Board of Directors, Council, or membership of the American Association for the Advancement of Science.

Cover photographs — © Uniphoto/Tim Ribar
Design — Julie A. Cherry

AAAS Publication: 93-10S
International Standard Book Number: 0-87168-512-4
Printed in the United States of America on acid-free recycled paper

Table of Contents

Fisheries and Wildlife

Managed Forest Ecosystems

Conservation and Regional Development

Contributors List

Paul R. Ehrlich, Stanford University, Stanford, CA

Edward O. Wilson, Harvard University, Museum of Comparative Zoology, Cambridge, MA

Joel I. Cohen, Bureau for Research and Development, Office of Agriculture, Agency for International Development, Washington, DC

Christopher S. Potter, NASA-Ames Research Center, Moffett Field, CA

Michael E. Soulé, Board of Environmental Studies, University of California, Santa Cruz, Santa Cruz, CA

Terry L. Erwin, National Museum of Natural History, Smithsonian Institution, Washington, DC

Harold J. Morowitz, George Mason University, Fairfax, VA

J.T. Williams, International Fund for Agricultural Research, Arlington, VA

Miguel A. Altieri, Division of Biological Control, University of California, Berkeley, CA

Camila Montecinos, Centro de Educacion y Tecnología, Santiago, Chile

Stephen R. Gliessman, Agroecology Program, Environmental Studies, University of California, Santa Cruz, Santa Cruz, CA

Garrison Wilkes, Biology Department, University of Massachusetts at Boston, Boston, MA

James S. Diana, School of Natural Resources, University of Michigan, Ann Arbor, MI

Robert R. Twilley, Department of Biology, University of Southwestern Louisiana, Lafayette, LA

Alejandro Bodero, Proyecto de Manejo de Recursos Costeras, Guayaquil, Ecuador

Donald Robadue, Coastal Resources Center, University of Rhode Island, Narragansett Bay, RI

William Weber, Wildlife Conservation Internationl, New York Zoological Society, Bronx, NY

Gary S. Hartshorn, World Wildlife Fund, Washington, DC

William Pariona A., Fundación de Conservación de la Naturaleza, Lima, Peru

Christine Padoch, Institute of Economic Botany, The New York Botanical Garden, Bronx, NY

Charles Peters, Institute of Economic Botany, The New York Botanical Garden, Bronx, NY

David S. Wilcove, Environmental Defense Fund, Washington, DC

Jeffrey T. Olson, The Wilderness Society, Washington, DC

David A. Neill, Missouri Botanical Garden, St. Louis, MI and Museo Ecuatoriano de Ciencias Naturales, Quito, Ecuador

Miguel Pinedo-Vasquez, Yale School of Forestry and Environmental Studies, Greeley Memorial Laboratory, New Haven, CT

Christopher Uhl, Department of Biology, Pennsylvania State University, University Park, PA, and EMBRAPA/CPATU, and IMAZON, Belém, Pará, Brazil

Oswaldo Bezerra, Instituto do Homem e do Meio Ambiente da Amazônia, Bélem, Pará, Brazil

Adriana Martini, Instituto do Homem e do Meio Ambiente da Amazônia, Belém, Pará, Brazil

Prologue

Biodiversity Studies: Science and Policy*

Paul R. Ehrlich and Edward O. Wilson

Biodiversity studies comprise the systematic examination of the full array of different kinds of organisms together with the technology by which the diversity can be maintained and used for the benefit of humanity. Current basic research at the species level focuses on the process of species formation, the standing levels of species numbers in various higher taxonomic categories, and the phenomena of hyperdiversity and extinction proneness. The major practical concern is the massive extinction rate now caused by human activity, which threatens losses in the esthetic quality of the world, in economic opportunity, and in vital ecosystem services.

From Linnaeus to Darwin to the present era of cladograms and molecular evolution, a central theme of biology has always been the diversity of life. A new urgency now impels the study of this subject for its own sake: just as the importance of all life forms for human welfare becomes most clear, the extinction of wild species and ecosystems is seen to be accelerating through human action (1). The dilemma has resulted in the rise of biodiversity studies: the systematic examination of the full array of organisms and the origin of this diversity, together with the methods by which diversity can be maintained and used for the benefit of humanity. Biodiversity studies thus combine elements of evolutionary biology and ecology with those of applied biology and public policy. They are based in organismic and evolutionary biology in the same manner that biomedical studies are based in molecular and cellular biology. They include the newly emergent discipline of conservation biology but are even more eclectic, subsuming pure systematic research and the practical applications of such research that accrue to medicine, forestry, and agriculture, as well as research on policies that maximize the preservation and use of biodiversity. In biodiversity studies, the systematist meets the economist and

---

* This chapter was originally published in *Science*, vol. 253(1991), pp. 758–62. It is reprinted here with permission.

political scientist. In this article we will present some of the key issues that newly link these two principal domains.

SPECIES FORMATION

A rich medley of models has been constructed to account for the origin of species by reproductive isolation. Two broad categories have been substantiated by empirical evidence. The first is polyploidy, the multiplication of entire chromosome numbers within individual species or within hybrids of species, a process that isolates the new breed from its ancestor in one step. This instantaneous mode has generated 40% of contemporaneous plant species and a much smaller number of animal species (2). Of comparable importance is geographic (or allopatric) speciation, the origin of intrinsic isolating mechanisms in two or more daughter populations while they are isolated by a geographic barrier, such as a sea strait, desert basin, or mountain range. Evidence of this two-step process, which occurs widely in plants and animals, has been documented minutely, often to the level of the gene, in birds, mammals, and a few groups of insects such as drosophilid flies and butterflies (3).

The diversification processes of polyploidy and geographic isolation are generally appreciated because they follow an easily traced pathway of measurable steps. Other modes of speciation are more difficult to conceive and test, but this does not mean they do not occur widely. Perhaps the most common is nonpolyploid sympatric speciation, in which new species emerge from the midst of parental species even when individuals of both populations are close enough to intermingle during part of their life cycles. The dominant process of this category, at least the one most persuasively modeled and documented, is by intermediate host races. Members of the parental species feed upon and mate in the vicinity of one kind of plant; they give rise to an alternate host race that shifts to a second species of host plant growing nearby; the two races, thus isolated by their microhabitat differences, diverge further in other traits that reinforce reproductive isolation. Sympatric speciation may play a key role in the origin of the vast numbers of insects and other invertebrates specialized on hosts or other types of microhabitats. The early stages are difficult to detect, however, and few studies have been initiated in the invertebrate groups most likely to display them (4).

Certain forms of speciation can thus occur very rapidly, within one to several generations. And when species meet, they can displace one another genetically within ten or fewer generations, reducing competition and the likelihood of hybridization (5). A question of central importance is the impact of high speciation rates on standing diversity. Although the probability of extinction of species within a particular group at a particular place (say, the anole lizards of Cuba) eventually rises with the number of species, the number of species should increase with greater speciation rates at all levels up to equilibrium. But does it really? And if so, in which groups and to what degree?

CURRENT LEVELS OF BIODIVERSITY

Also in an early stage, and surprisingly so, is the elementary taxonomic description of the world biota. At the present time approximately 1.4 million species of plants, animals, and microorganisms have been given scientific names (1, 6). Terrestrial and freshwater species diversity is greater than marine diversity. The overwhelming elements are the flowering plants (220,000 species) and their coevolutionary partners, the insects (750,000 species). The reverse is the case at the highest taxonomic levels, with all of the 33 living animal phyla present in the sea and only 17, or half, present on land and in fresh water (7).

Known species diversity is only a small fraction of actual species diversity, especially in the invertebrates and microorganisms. In this century the class Insecta has always been considered the most speciose group at the class level. As early as 1952, Sabrosky estimated that the number of living species is as high as 10 million (8). In 1982, Erwin found that beetle diversity in Neotropical trees, revealed in samples knocked down by insecticidal fogs, suggest far higher levels of insect and other arthropod diversity in tropical rain forests than had previously been estimated for the entire world fauna and flora (9). His figure, 30 million, was reached by extrapolating from counts of beetle species (1200) in a Panamanian tree species through estimates of total arthropod diversity per tree species to the percentages of species limited to each tree species to the total of tree species in tropical rain forests. Stork (10) reassessed this bold extrapolation, and in essence agreed with it, adding data of his own from Indonesian forests to produce a possible range of 10 to 80 million tropical forest arthropods. The most sensitive parameter remains the degree to which species of beetles and other arthropods are found uniquely on individual tree species.

In fact, because the life of the planet remains mostly unexplored at the species and infraspecies levels, systematists do not know the species diversity of the total world fauna and flora to the nearest order of magnitude. It is easily possible that the true number of species is closer to 10^8 than 10^7. Relatively little effort has been expended on nematodes, mites, or fungi, each highly diverse and containing undescribed species that could easily range into the hundreds of thousands or millions. Bacteria, with only about 4000 described species, remain a terra vitae incognita because of the astonishingly small amount of research devoted to their diversity, as opposed to the genetics and molecular biology of select species.

HYPERDIVERSITY

Certain taxa are hyperdiverse, that is, they contain more species, genera, or higher ranked groups within them than expected by a null model of random assortment (11). Examples include arthropods among animal phyla, insects among arthropods, rodents among mammalian orders, orchids among monocotyledonous plant families, *Sciurus* among the genera of Sciuridae (squirrels), and so forth. It can be expected in a Darwinian world, where chance and opportunism prevail, that production of great diversity depends to substantial degree on special adaptations allowing penetration of multiple niches, such that each hyperdiverse group has its own magic key. For example, the ants appear to have expanded by virtue of fungistatic secretions, series-parallel work operations, and a highly altruistic worker caste (12). But recent research has also begun to identify properties possessed by many groups: small size, permitting fine niche subdivision (7, 13, 14); phytophagy and parasitism with specialization on hosts (15); specialized life stages that allow species to occupy multiple niches; entry into new geographic areas with subsequent adaptive radiation and preemption; and greater dispersal ability, promoting the colonization of empty areas. Southwood has neatly summarized the likely causes of the extreme hyperdiversity of insects as "size, metamorphism, and wings" (13).

Hyperdiversity also occurs in certain habitats and geographical areas. The strongest trend worldwide is the latitudinal diversity gradient, with group after group reaching its maximum richness in the tropics and most particularly in the tropical rain forests and coral reefs. (Exceptions include conifers, salamanders, and aphids.) The hyperdiversity of continental rain forests is legendary. Gentry found about 300 tree species in single-hectare plots in Peru (16), to be compared with 700 native tree species in all of North America. A single tree in the same area yielded 43 species of ants in 26 genera, about equal to the ant fauna of the entire British isles (17). Explaining the latitudinal diversity gradient has proven an intractable problem. But clues exist which when pieced together suggest the possibility of a general explanation, involving climatic stability and extreme biological specialization and niche division (18).

NATURAL EXTINCTION

One of the qualities reducing diversity in particular groups is extinction proneness, which renders populations vulnerable to environmental change and reduces taxonomic groups to one or a very few threatened species. A threatened or endangered species (the two grades commonly employed by conservationists) is one with a high probability of extinction during the next few years or decades. The principal demographic properties contributing to the status are a low maximum breeding population size and a high coefficient of variation in that size (19). When the breeding size drops to a hundred or less, the likelihood of extinction is enhanced still further by inbreeding depression (20).

The overall natural extinction rate (at times other than mass extinction episodes) estimated from fossil data to the nearest order of magnitude is 10^{-7} species per species year (21). This estimate refers to true extinction, from the origin of a species to the extinction of that species and any species descended from it (altogether, called the clade) and excludes "pseudoextinction," the evolution of one species into another. Wide variation exists among major taxonomic groups in the longevity of clades. Mesozoic ammonoid and Silurian graptolite clades lasted only 1 million to 2 million years, whereas most other Paleozoic and Mesozoic invertebrate clades lasted closer to 10 million years (21). In general, planktonic and sessile marine animals, including corals and brachiopods, have had higher extinction rates than mobile benthic animals such as gastropods and bivalves (22). Using anatomical evidence from fossils and comparisons with related living species, paleobiologists have begun to infer the determinants of clade longevity by relating the adaptations of the organisms to maximum population size, population fluctuation, and dispersal ability (23).

HUMAN-CAUSED EXTINCTION

Biodiversity reduction is accelerating today largely through the destruction of natural habitats (1). Because of the latitudinal diversity gradient, the greatest loss occurs in tropical moist forests (rain forests) and coral reefs. The rate of loss of rain forests, down to approximately 55% of their original cover, was in 1989 almost double that in 1979. Roughly 1.8% of the remaining forests are disappearing per year (24). By the most conservative estimate from island biogeographic data, 0.2 to 0.3% of all species in the forests are extinguished or doomed each year (25). If two million species are confined to the forests, surely also a very conservative estimate, then extinction due to tropical deforestation alone must be responsible for the loss of at least 4000 species annually.

But there may well be 20 million or more species in the forests, raising the loss tenfold. Also, many species are very local and subject to immediate extinction from the clearing of a single habitat isolate, such as a mountain ridge or woodland patch (26). The absolute extinction rate thus may well be two to three orders of magnitude greater than the area-based estimates given above. If current rates of clearing are continued, one-quarter or more of the species of organisms on Earth could be eliminated within 50 years—and even that pessimistic estimate might be conservative (25). Moreover, for the first time in geological history, plants are being extinguished in large numbers (27).

Another data set illuminating the urgency of dealing with the extinction problem measures the human impact on global net primary productivity (NPP) (28); global NPP is roughly the total food supply of all animals and decomposers. Almost 40% of all NPP generated on land is now directly used, coopted, or forgone because of the activities of just one animal species—*Homo sapiens*.

Since the overwhelming majority (possibly more than 90%) of species now exists on land, the 40% human appropriation there alone shows why there is an extinction crisis. Furthermore, the human population is projected to double in the next half-century or so—to more than 10 billion people. Most ominous of all, the widely admired Brundtland Report speaks of a five- to tenfold increase in global economic activity needed during that period to meet the demands and aspirations of that exploding population (29). If anything remotely resembling that population-economic growth scenario is played out, with an acceleration of habitat destruction, most of the world's biodiversity seems destined to disappear.

WHY SHOULD WE CARE?

The loss of biodiversity should be of concern to everyone for three basic reasons (1, 30). The first is ethical and esthetic. Because *Homo sapiens* is the dominant species on Earth, we and many others think that people have an absolute moral responsibility to protect what are our only known living companions in the universe. Human responsibility in this respect is deep, beyond measure, beyond conventional science for the moment, but urgent nonetheless. The popularity of ecotourism, bird-watching, wildlife films, pet-keeping, and gardening attest that human beings gain great esthetic rewards from those companions (and generate substantial economic activity in the process).

The second reason is that humanity has already obtained enormous direct economic benefits from biodiversity in the form of foods, medicines, and industrial products, and has the potential for gaining many more. Wheat, rice, and corn (maize) were unimpressive wild grasses before they were "borrowed" from the library and developed by selective breeding into the productive crops that have become the feeding base of humanity. All other crops, as well as domestic animals, have their origins in the genetic library, as do many medicines and various industrial products, including a wide variety of timbers (1, 30). Throughout the world almost a quarter of all medical prescriptions are either for chemical compounds from plants or microorganisms, or for synthetic versions or derivatives of them (31). One plant compound, quinine, is still a mainstay of humanity's defense against its most important disease, malaria.

Biodiversity is a precious "genetic library" maintained by natural ecosystems. But the potential of the library to supply such benefits has barely been tapped.

Only a tiny portion of plant species has been screened for possible value as providers of medicines (31), and although human beings have used about 7000 plant species for food, at least several times that number are reported to have edible parts (1).

The third reason, perhaps the most poorly evaluated to date, is the array of essential services provided by natural ecosystems, of which diverse species are the key working parts. Ecosystem services include maintenance of the gaseous composition of the atmosphere, preventing changes in the mix of gases from being too rapid for the biota to adjust. In Earth's early history, photosynthesizing organisms in the seas gradually made Earth's atmosphere rich in oxygen. Until there was enough oxygen for an ozone shield to form, the land surface was bathed in ultraviolet-B radiation. Up to some 450 million years ago life was confined to the seas. Only with the protection of the ozone shield were plants, arthropods, and amphibians able to colonize the land.

Significant alteration of the atmosphere has signaled the arrival over the past few decades of *Homo sapiens* as a global force, one capable of destroying most of biodiversity. As a result of human activities (32), the ozone shield has thinned by as much as 5% over Europe and North America (33), and there is some evidence that the surface intensity of ultraviolet-B radiation has increased there (34). Each spring the shield is now reduced over the Antarctic by approximately 50%. The global impact of the human economy is even more evident in the prospect of climatic change in response to increasing concentrations of greenhouse gases (35).

The organisms in natural ecosystems influence the climate in ways other than the role they play in regulating atmospheric gases. The vast rain forests of Amazonia to a large degree create the moist conditions that are required for their own survival by recycling rainfall. But as the forest shrinks under human assault, many biologists speculate that there will be a critical threshold beyond which the remaining forest will no longer be able to maintain the climate necessary for its own persistence (36). Deforestation and the subsequent drying of the climate could have serious regional effects in Brazil outside of Amazonia, conceivably reducing rainfall in important agricultural areas to the south. There also appear to be regional effects on climate when semi-arid regions are desertified (37), but their extent remains unknown.

The generation and maintenance of soils is another crucial service supplied most efficiently by natural ecosystems. Soils are much more than fragmented rock; they are themselves complex ecosystems with a rich biota (38). The elements of biodiversity in soil ecosystems are crucial to their fertility—to their ability to support crops and forests.

Many green plants enter into intimate relationships with mycorrhizal fungi in the soil. The plants nourish the fungi, which in turn transfer essential nutrients into the roots of the plant. In some forests where trees appear to be the dominant organisms, the existence of the trees is dependent upon the functioning of these fungi. On farms, other microorganisms play similar critical roles in transferring nutrients to crops such as spring wheat.

Organisms are very much involved in the production of soils, which starts with the weathering of the underlying rock. Plant roots can fracture rocks and thus help generate particles that are a major physical component of soils; plants and animals also contribute CO_2 and organic acids that contribute to the weathering of parent rock. More importantly, many species of small organisms, especially bacteria, decompose organic matter (shed leaves, animal droppings, dead organisms, and so on), releasing carbon dioxide and water into the soil and leaving a residue of humus, or tiny organic particles. These are resistant to further decomposition, help maintain soil texture and retain water, and play a critical role in soil chemistry, permitting the retention of nutrients essential for plant growth.

Soil ecosystems themselves are the main providers on land of two more essential ecosystem services: disposal of wastes and cycling of nutrients. Decomposers break wastes down into nutrients that are essential to the growth of green plants. In some cases, the nutrients are taken up more or less directly by plants near where the decomposers did their work. In others, the products of decomposition circulate through vast biogeochemical cycles before being reincorporated into living plants.

Another critical service provided by natural ecosystems is the control of the vast majority of species that can attack crops or domestic animals. Most of those potential pests are herbivorous insects, and the control is provided primarily by numerous species of predacious and parasitic insects that naturally feed upon them.

While natural ecosystems are providing crop plants with stable climates, water, soils, and nutrients, and protecting them from pests, they also often pollinate them. Although honeybees, essentially domesticated organisms, pollinate many crops, numerous other crops depend on the services of pollinators from natural ecosystems. One such crop is alfalfa, which is most efficiently pollinated in cooler areas by wild bees.

The biodiversity in natural ecosystems also supplies people with food directly—most notably with a critical portion of their dietary protein from fishes and other marine animals. This service is provided by oceanic ecosystems in conjunction with coastal wetland habitats that serve as crucial nurseries for marine life.

The ecosystem services in which biodiversity plays the critical role are provided on such a grand scale and in a manner so intricate that there is usually no real possibility of substituting for them, even in cases where scientists have the requisite knowledge. In fact, one could conclude that virtually all human attempts at large-scale inorganic substitution for ecosystem services are ultimately unsuccessful, whether it be introductions of synthetic pesticides for natural pest control, inorganic fertilizer for natural soil maintenance, chlorination for natural water purification, dams for flood and drought control, or air-conditioning of overheated environments. Generally, the substitutes require a large energy subsidy, thereby adding to humanity's general impact on the environment, and are not completely satisfactory in even the short run (39).

It is important to note that in supplying ecosystem services the species and genic diversity of natural systems is critical. One might assume that one grass or tree species can function as well as any other in helping control the hydrologic cycle in a watershed, or that one predator will be as good as another in controlling a potential pest. But, of course, organisms are generally highly adapted to specific physical and biotic environments—and organic substitutions, like inorganic ones, are likely to prove unsatisfactory.

In sum, much of biodiversity and the quality of ecosystem services generated by it will be lost if the epidemic of extinctions now under way is allowed to continue unabated.

PUBLIC POLICY

Many steps can be taken to preserve biodiversity, if the political will is generated. Perhaps the first step, which would be seen as especially extreme by Americans, would be to cease "developing" any more relatively undisturbed land. Every new shopping center built in the California chaparral, every hectare of tropical forest cut and burned, every swamp converted into a rice paddy or shrimp farm means less biodiversity.

In rich countries, stopping the more destructive forms of "development" is relatively simple in principle. Age structures are such that population shrinkage in most rich nations could be achieved with little effort (a few are already in that desirable mode). When new facilities are needed, they should replace deteriorating old ones. Forestry should be placed on a sustainable basis with careful attention to the conservation of precious reserves of old growth. And much more scientific effort and public support should go into biodiversity studies, including the cataloging of the genetic library and national biological inventories (1, 31).

In poor nations, the task is both more urgent and vastly more difficult. It cannot be accomplished immediately, and will not be accomplished at all without massive assistance from the rich. For instance, stopping the expansion of cropland and pasture into virgin areas cannot be accomplished unless birth rates can be dramatically lowered and the development of sustainable high-yield agricultural systems is backed by land reform and a sound agricultural infrastructure and economy. In many cases, new social and economic systems must be developed in which preservation of biodiversity and its sustainable exploitation go hand in hand. The social, political, economic, and scientific barriers to achieving the goal are so formidable that nothing less than the kind of commitments so recently invested in the Cold War could possibly suffice to accomplish it. And we are 45 years late in starting.

But ending direct human incursions into remaining relatively undisturbed habitats would be only a start. Simultaneously, global cooperative efforts to reduce anthropogenic impacts on ecosystems, such as those directed at a reduction of emissions of greenhouse gases and ozone-destroying compounds, must be greatly enhanced. They are much more likely to be successful if population growth can be halted and the cessation of forest destruction can be achieved.

Finally, because humanity already occupies so much of Earth's surface, substantial effort should be directed at making areas already used by people more hospitable to other organisms. Those efforts can range from the substitution of game ranching for cattle and sheep ranching in many areas to the substitution of native vegetation for European-style lawns in desert cities.

If there is to be any chance of abating the loss of biodiversity, action must be taken immediately. The essential tactics of conservation are being developed within conservation biology, as a subdiscipline of biodiversity studies. The indispensable strategy for saving our fellow living creatures and ourselves in the long run, is, as the evidence compellingly shows, to reduce the scale of human activities. The task of accomplishing this goal will involve a cooperative worldwide effort unprecedented in history. Unless humanity can move determinedly in that direction, all of the efforts now going into in situ conservation will eventually lead nowhere, and our descendents' future will be at risk.

P. R. Ehrlich is Bing Professor of Population Studies and Professor of Biological Sciences at Stanford University, Stanford, CA 94305. E. O. Wilson is Baird Professor of Science and Curator in Entomology at Harvard University, Museum of Comparative Zoology, Cambridge, MA 02138–2902. This article is based on their Crafoord Prize lectures given at the Royal Swedish Academy of Sciences, Stockholm, 26 September 1990.

REFERENCES AND NOTES

1. See, for example, P. R. Ehrlich and A. H. Ehrlich, *Extinction: The Causes and Consequences of the Disappearance of Species* (Random House, New York, 1981); E. O. Wilson and F. M. Peter, Eds., *Biodiversity* (National Academy Press, Washington, DC, 1988); C. C. Black *et al.*, *Loss of Biological Diversity: A Global Crisis Requiring International Solutions* (National Science Foundation, Washington, DC, 1989); W. V. Reid and K. R. Miller, *Keeping Options Alive: The Scientific Basis for Conserving Biodiversity* (World Resources Institute, Washington, DC, 1989).
2. That 40% of living plant species originated by polyploidy is the estimate widely accepted by plant systematists; the figure is likely to be higher if the earlier origins of stocks giving rise to living species are also considered.
3. D. J. Futuyma, *Evolutionary Biology* (Sinauer, Sunderland, MA, ed. 2, 1986); D. Otte and J. A. Endler, Eds., *Speciation and Its Consequences* (Sinauer, Sunderland, MA, 1989).
4. G. L. Bush, *Annu. Rev. Ecol. Syst.* 6, 399 (1975); C. A. Tauber and M. J. Tauber, in *Speciation and Its Consequences* D. Otte and J. A. Endler, Eds. (Sinauer, Sunderland, MA, 1989), pp. 307–344.
5. W. H. Bossert and E. O. Wilson, in *The Genetics of Colonizing Species,* H. G. Baker and G. L. Stebbins, Eds. (Academic Press, New York, 1964), pp. 7–24; J. Diamond, S. L. Pimm, M. E. Gilpin, M. LeCroy, *Am. Nat.* 134, 675 (1989).
6. E. O. Wilson, *Issues Sci. Technol.* 2 (Fall), 20 (1985).
7. R. M. May, *Science* 241, 1441 (1988).
8. C. W. Sabrosky, *Insects, The Year of Agriculture*, 1952 (U.S. Department of Agriculture, Washington, DC, 1952), pp. 1–7.
9. T. L. Erwin, *Coleopt. Bull.* 36, 74 (1982); *Bull. Ent. Soc. Am.* 30, 14 (1983).
10. N. E. Stork, *Biol. J. Linn. Soc.* 35, 321 (1988).
11. K. P. Dial and J. M. Marzluff, *Syst. Zool.* 38, 26 (1989).
12. E. O. Wilson, *Success and Dominance in Ecosystems: The Case of the Social Insects* (Ecology Institute, Oldendorf-Luhe, Germany, 1990).
13. T. R. E. Southwood, in *Diversity of Insect Faunas*, L. A. Mound and N. Waloff, Eds. (Blackwell, London, 1978), pp. 17–40.
14. K. P. Dial and J. M. Marzluff, *Ecology* 69, 1620 (1988).
15. P. R. Ehrlich and P. H. Raven, *Evolution* 18, 586 (1965); C. Mitter, B. Farrell, B. Wiegmann, *Am. Nat.* 132, 107 (1988).
16. A. H. Gentry, *Proc. Natl. Acad. Sci. U.S.A.* 85, 156 (1988).
17. E. O. Wilson, *Biotropica* 19, 245 (1987).
18. H. L. Sanders, *Am. Nat.* 102, 243 (1968); F. Grassle, *Trends Ecol. Evol.* 4, 12 (1989); G. C. Stevens, *Am. Nat.* 133, 240 (1989); E. R. Pianka, *Trends Ecol. Evol.* 4, 223 (1989); H. R. Pulliam, *Am. Nat.* 132, 652 (1988).
19. S. L. Pimm, H. L. Jones, J. M. Diamond, *Am. Nat.* 132, 757 (1988).
20. K. Ralls, J. D. Ballou, A. Templeton, *Conserv. Biol.* 2, 185 (1988).
21. D. M. Raup, in *Biodiversity*, E. O. Wilson, Ed. (National Academy Press, Washington, DC, 1988), pp. 51–57.
22. M. L. McKinney, *Nature* 325, 143 (1987).
23. S. M. Stanley, *Paleontology* 20, 869 (1977); *Bull. Am. Acad. Arts. Sci.* 40, 29 (1987); P. D. Taylor, *Hist. Biol.* 1, 45 (1988).
24. N. Myers, *Deforestation Rates in Tropical Forests and Their Climatic Implications* (Friends of the Earth, London, 1989).
25. E. O. Wilson, *Sci. Am.* 261, 108 (September 1989). Other, higher estimates are reviewed by W. V. Reid and K. R. Miller [*Keeping Options Alive: The Scientific*

Basis for Conserving Diversity (World Resources Institute, Washington, DC, 1989)].

26. J. M. Diamond, *Discover* 11, 55 (1990).
27. A. H. Knoll, in *Extinctions*, M. H. Nitecki, Ed. (Univ. of Chicago Press, Chicago, 1984), pp. 21–68.
28. P. M. Vitousek, P. R. Ehrlich, A. H. Ehrlich, P. A. Matson, *BioScience* 36, 368 (1986).
29. World Commission on Environment and Development, *Our Common Future* (Oxford Univ. Press, New York, 1987).
30. N. Myers, *A Wealth of Wild Species* (Westview Press, Boulder, CO, 1983).
31. T. Eisner, *Issues Sci. Technol.* 6 (Winter), 31 (1989).
32. See summary in S. Roan, *Ozone Crisis: The 15 Year Evolution of a Sudden Global Emergency* (Wiley, New York, 1989).
33. R. A. Kerr, *Science* 247, 1297 (1989).
34. M. Blumthaler and W. Ambach, *ibid.* 248, 206 (1990).
35. S. H. Schneider, *Global Warming* (Sierra Club Books, San Francisco, 1989); D. Abrahamson, Ed., *The Challenge of Global Warming* (Island Press, Washington, DC, 1989); P. R. Ehrlich and A. H. Ehrlich, *Healing the Planet* (Addison-Wesley, Reading, MA, in press).
36. J. Shukla, C. Nobre, P. Sellers, *Science* 247, 1322 (1990).
37. J. Charney, P. H. Stone, W. J. Quirk, *ibid.* 187, 434 (1975).
38. For more details on soils and technical citations on what follows, see P. R. Ehrlich, A. H. Ehrlich, J. P. Holdren, *Ecoscience: Population, Resources, Environment* (Freeman, San Francisco, 1977).
39. P. R. Ehrlich and H. A. Mooney, *BioScience* 33, 248 (1983).
40. This article is dedicated to some of the key individuals who have influenced our thinking on biodiversity: W. L. Brown, E. Mayr, C. D. Michener, and R. R. Sokal; and the late J. H. Camin, R. W. Holm, and R. H. MacArthur. For helpful comments on the manuscript we are indebted to G. C. Daily, J. M. Diamond, A. H. Ehrlich, T. Erwin, J. Harte, M. E. Harte, C. E. Holdren, J. P. Holdren, T. Lovejoy, J. P. Meyers, S. Conway Morris, D. M. Raup, P. H. Raven, and M. E. Soulé.

Introduction

Conservation of Biodiversity in Natural Habitats and the Concept of Genetic Potential

J.I. Cohen and C.S. Potter

Over the past several decades a development paradigm has emerged that seeks to link environmental quality and human welfare. Warnings of the adverse effects of deforestation, pollution, soil erosion, climate warming, and species extinction have focused worldwide attention on threatened natural habitats and species. Consequently, while the world's remaining biological diversity is subject to massive loss and destruction, it is simultaneously being explored as a primary resource base for sustainable development.

Recognizing the need to conserve biological diversity reflects increasing awareness of the importance of diversity—genetic, species, ecosystem—to sustaining the biological and ecological processes upon which production systems, economic development, and human welfare and survival depend. Biologists have long argued that species diversity is fundamental to the maintenance of critical evolutionary processes, as well as to the healthy, sustained functioning of virtually all ecological systems. The focus of this book is on one component of this diversity, that of genetic resources, meaning the economic, scientific, or societal value of heritable materials contained within and among species (1), and their relation to natural environments.

HIERARCHICAL LEVELS OF DIVERSITY

A recognized hierarchy of diversity exists between whole ecosystems, at the broadest level, and discreet pieces of genetic information at the molecular level. Soulé has identified five levels within this hierarchy (2). Ecosystems remaining intact at the broadest level contain an array of interactions between these five levels and among the three components of biological diversity: ecosystem, species, and genetic. These three components are described briefly before examining the role of genetic resources in detail.

Ecosystem diversity refers to the diversity of natural habitats of assemblages of species. Ecosystems provide critical life-support functions necessary for all life on earth—aiding in nutrient cycling, water filtration and purification, and climatic stability. Unfortunately, we know relatively little about the diversity of ecosystems, their resiliency in the face of human perturbation, or how critical a particular species is for the health and vitality of a particular system.

Species diversity refers to the variety of living organisms on earth. The number of species is unknown, but has been variously estimated to be between 5 and 50 million (3). Considering that only 1.4 million species have been described, the task ahead is to understand the number and kind of organisms that exist, particularly those in poorly known habitats, such as tropical rain forests, marine environments, and soil communities, and of poorly known taxa such as invertebrates and microorganisms.

Of the three levels that comprise global biological diversity, genetic diversity has received the greatest attention within the agricultural community. By genetic diversity, we refer to the total genetic information contained in an individual of a species, each containing a unique assembly of genes constituting its evolutionary heritage. his diversity begins at the molecular level, is carried as sequences of instructions on chromosomes, and provides a foundation for environmental adaptation and ultimately for the evolution of species. Genetic diversity can be characterized at the molecular level, the population level, the species level, and at the ecosystem level.

GENETIC RESOURCES

This book specifically addresses issues, strategies, and case studies involving the conservation and use of genetic resources. This component of diversity has particular relevance to international development since it encompasses economically useful and heritable characteristics. As such, they form a subset of total diversity available for domestication, improvement, or direct use by humanity. This diversity is represented by gene pools that cannot be created artificially, except on a limited scale through mutation breeding or genetic engineering (4).

Biodiversity is of immense importance to agriculture, food production, and consumption. Indeed, diversity at the genetic level is the basis for agronomic advancements. The pool of diversity referred to as plant genetic resources exists because of a number of factors: First, the great age of cultivation within centers of diversity; second, the wide range of habitats and farming practices found in them; and third, natural selection caused by the presence of many different pathotypes of pests and diseases and by variable climatic and edaphic factors (4).

GENETIC POTENTIALITY

It is the premise of this book that the loss of diversity—genetic, species, ecosystem—increasingly constrains local development options as well as threatens the biological and ecological processes upon which human welfare and survival depend. Reducing the storehouse of genetic resources negates development options by reducing the genetic potential of available biodiversity. We offer the term "genetic potential" to explain the likelihood that a species or a portion of its gene pool will contribute directly or indirectly to goods and services that support sustainable development (5).

The concept of genetic potential focuses attention on the option value of biodiversity. The option value of diversity is one of five values recognized (6) and is based upon the idea that loss of additional species reduces options for nature and people to respond to changing conditions by limiting the number of gene pools available for adaptation. However, genetic potential goes beyond options alone. It should be analyzed within the context of multiple conservation and utilization strategies, including user groups that either positively or negatively affect genetic potential, and technology inputs, such as biotechnology, that can affect broader conservation, use, and dissemination of valuable gene pools (7). Genetic potential should provide a means to analyze the hypothesis that reduction of diversity through loss of species, habitats, and nonsustainable utilization practices will reduce development options.

Increasing genetic potential must come from positive indications of the many variables that together increase the likelihood of particular genetic sequences, genomes, or even gene pools, all of which can contribute to sustainable development. Genetic potential should also increase when genetic sequences, genomes, or gene pools spread over larger areas. For example, a new species isolated from a reserve, entering local and then global production, has increased its genetic potential many times. To maximize this effect, appropriate conservation measures must be encouraged as an integral part of development. Values obtained as a measure of genetic potential should decrease if multiple strategies for conservation are ignored or if sustainable use is not demonstrated.

Variation at the genetic level has sustained mankind. If such contributions are to continue, each species, or a portion of its genetic wealth, must have the opportunity to increase its genetic potential. With regard to conservation, opportunities will be best protected through integrated conservation approaches (8). Such strategies must take into account ex situ conservation programs, particularly with regard to developing countries, as well as *in situ* efforts (9) if greater genetic potential is to be realized for species of economic importance. Case studies in this book provide illustrations of how such an analysis might begin. Ideally, genetic potential should merge integrated conservation measures with innovative measures to stimulate the utility of the resource conserved.

EXAMPLES OF GENETIC RESOURCE CONSERVATION

In this book, we examine strategies or case studies in the context of international development that optimize species or genetic resource potential or examine the factors leading to a decline in their potential. These studies represent a treatment of genetic resource conservation in a broad context, including examples of plant and animal conservation at both global and local levels. All case studies are expressly set against the background of development—community, national, and international—and the competing interests of corresponding stakeholders. Our intention was to bring together experiences from active field researchers in the area of genetic resource conservation and development and to look for common themes running through their various examples that might provide lessons for students, policy makers, and the scientific community at large.

In casting a broad net, however, we have intentionally created a painstaking process of synthesis. Our goal was no less than an integrated test of the concept that conservation of biodiversity enhances the potential for sustainable development and hence increases the genetic potential of the materials being conserved.

herefore, to facilitate integration of experiences, all case study authors were asked to address three common issues: (i) documentation of species loss and genetic erosion in natural habitats, (ii) presentation of management approaches that sought to meet the needs of both biodiversity conservation and sustainable development, with emphasis wherever possible on genetic resources, and (iii) evaluation of potential applicability of a given management approach in similar environments outside the region of immediate interest.

The reader will find what we believe is a unique collection of examples of development approaches that emphasize *in situ* biodiversity conservation as a critical element. The book is organized into two major sections: Themes in Species and Genetic Resource Conservation, and Case Studies of Biodiversity Conservation in Natural Habitats. The former section identifies contemporary issues in biodiversity conservation within the socioeconomic setting of development. In the latter section, chapters are grouped for convenience under the headings *Agroecosystems*, *Fisheries and Wildlife*, *Managed Forest Ecosystems*, and *Conservation and Regional Development*. There is, nonetheless, a good deal of crossover between disciplines and habitats among the case studies, some of which is addressed in a final chapter on synthesizing the potential of biodiversity to enhance development.

While we seek to highlight the opportunities provided by genetic resource conservation, we also realize the daunting challenges that lie ahead. Given the current rates of human population growth and global levels of landlessness and poverty, we anticipate that the possibility of setting aside increasing numbers of undisturbed natural habitats solely for conservation of genetic resources is exceedingly slim. Nevertheless, this book focuses on the potential benefits to be derived from protecting and using biodiversity wherever it can be found. The volume is chiefly about options, the critical element for sustainable development in a changing world.

The editors thank AAAS Press, particularly Patricia Morgan and Louise Goines, for their steadfast support in the preparation of this book. Our gratitude also to the participants in the original symposium at the 1991 Annual Meeting of the AAAS, which emphasized many of the issues addressed in the book: Peter Raven, Daniel Janzen, Christine Padoch, William Weber, David Wilcove, and Richard Bissell. Thanks also to Merin McDonell for assistance in the preparation of figures.

REFERENCES AND NOTES

1. Food and Agricultural Organization (FAO). 1989. *Plant Genetic Resources*. Rome.
2. Soulé, M.E. 1991. Conservation: Tactics for a Constant Crisis. *Science* 253:744.
3. See, for example, Erwin, T.L. 1982. *Colept. Bull.* 36:74; *Bull. Ent. Soc. Am.* 1983. 30:14; Stork, N.E. 1988. *Biol. J. Linn.* 35:321; Myers, N.L. 1990. *Palaeog. Palaeoclin, Paleoecol.* 82:175.
4. Hawkes, J.G. 1991. *Biol. J. Linn.* 43:3.
5. The authors interpret sustainable development to mean the progressive transformations in societies and economies that are designed to meet human needs of the present without compromising the ability of future generations to meet their own needs. Sustainable development takes into consideration access to resources, degradation of the environment, and distribution of costs and benefits.
6. McNeely, J.A. et al., 1990. *Conserving the World's Biological Diversity*. International Union for the Conservation of Nature and Natural Resources (IUCN), World Resources Institute, Conservation International, World Wildlife–US, The World Bank: Gland, Switzerland.

7. Cohen, J.I., and R. Bertram. 1989. *In Biotic Diversity and Germplasm Preservation, Global Imperatives*, L. Knutson and A.K. Stoner, eds., 459–76. The Netherlands: Kluwer.

8. Falk, D.A. 1990. *Annals of the Missouri Botanical Garden* 77:38.

9. Cohen, J.I., et al. 1991. Ex Situ Conservation of Plant Genetic Resources: Global Development and Environmental Concerns. *Science* 253:866.

Part 1

Themes in Species and Genetic Resource Conservation

1

Conservation: Tactics for a Constant Crisis*

Michael E. Soulé

Is wildlife conservation failing? In the United States, species diversity appears to be declining at an accelerating rate (1). Even the Endangered Species Act of 1973 (ESA) has not significantly slowed the deterioration of the nation's biological estate, although this is largely the result of lack of support from the federal administration. Currently there are over 4000 species and subspecies recognized as candidates for endangered species status, but the listing process administered by the U.S. Fish and Wildlife Service is bogged down because of lack of funding. There are no recovery plans for nearly half of the 600 or so species in the United States that have been officially listed as threatened or endangered, and the score or so of recovering species is balanced by an equal number that may be extinct (2).

The situation is generally much worse in other nations. Biologists with extensive experience in developing countries are saying that by almost any quantitative standard conservation is failing, and that current approaches to conservation, such as traditional parks and reserves, are unlikely to succeed (3, 4). Worldwide, only about 3% of the land is set aside in 5000 nature reserves or protected areas (5), but many of these reserves are deteriorating (6). Because the moist tropics are far richer in species diversity than other biogeographic regions, and because deforestation will probably eliminate almost all of the tropical forests outside of protected areas by 2100 (7), biogeographers estimate that from 25 to 50% or more of tropical species will vanish in the next century or sooner (Fig. 1) (8). Even if humanity were to depart the earth, recovery of biotic diversity by evolutionary mechanisms would require millions of years, depending on how deep, taxonomically, the extinction crisis cuts (9).

Such dire predictions are now leading to a reappraisal of conservation's goals and tactics. In this article, I conclude that this reappraisal would be more fruitful

---

* This chapter was originally published in *Science*, vol. 253 (1991), pp. 744–50. It is reprinted here with permission.

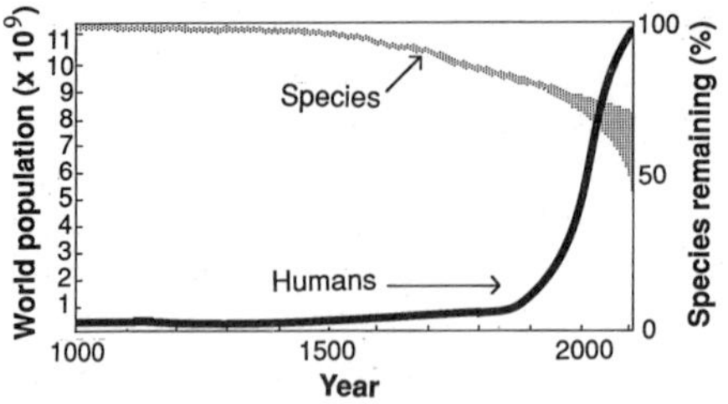

Figure 1. The expected inverse correlation between human population size and the survival of species worldwide. Extinction rates depend on the size of the habitat fragment and occur at a decreasing rate as habitat fragments age. Anthropogenic extinctions before A.D. 1000 are ignored. The shape and width of the extinction curve reflect the uncertainty of the predictions; the curve is based in part on the assumption that most of the extinct species will be small organisms with geographically limited distributions.

if there were a deeper appreciation of the biological and social contexts of conservation actions, particularly how both biogeography and political geography dictate different conservation tactics in different situations. I also argue for an actuarial approach to the viability of protected areas—one that considers the social factors determining the half-life of nature reserves.

THE BIOSPATIAL HIERARCHY

Effective conservation is impossible without some knowledge of biotic (biological) diversity (biodiversity). For most scientific purposes, "life" is classified taxonomically, based on similarity and presumed evolutionary relationship. For purposes of protection, however, the living components of nature are usually classified in a "biospatial" hierarchy of nested sets. In practice, there are about five levels to this hierarchy: (i) whole systems at the landscape or eco-systems levels, (ii) assemblages (associations and communities), (iii) species, (iv) populations, and (v) genes (10). Place, not evolutionary relationship, is the basis for the biospatial hierarchy, because most conservation strategies are geographically anchored (11, 12).

The targets at the top of the biospatial hierarchy are ecosystems (or landscapes and seascapes making up interacting ecosystems), including such topographic features as entire drainages. A frequently cited example is the Yellowstone National Park region, including the adjacent Grand Teton National Park and other federally managed lands. Ideally, ecosystem conservation protects the contained biotic communities: habitats, species, populations, and genes, not to mention all ecological interactions, processes, and some of the traditional, human cultural practices that have been historically associated with the ecosystem.

At the second level, an arbitrary number of biotic assemblages can be defined within ecosystems, although the species themselves show little correlation in their distributions when climate changes (13). Nevertheless, state, federal, and international conservation programs often base their conservation strategies on the completion of the network of biotic community types—the so-called coarse-filter approach. The discovery of "gaps" in the network of assemblages is most often based on systems of biogeographic classification (12, 14).

The third biospatial level, species, is defined as groups of populations that routinely exchange genes or are phenotypically similar (15). The selection of protected areas is frequently based on the presence of one or more endangered species, often large-bodied or attractive ones. In addition, regions with high species diversity, such as tropical forests, coral reefs, or regions with large proportions of local endemic species, such as isolated mountain ranges or oceanic islands, are frequently identified as targets of conservation. Another reason for focusing on species is that the management of protected areas is often facilitated by attending to a relatively small number of so-called keystone or indicator species; these species may not be endangered themselves, but they are used to monitor the status of a much larger assemblage of species (16–18).

Next is populations. Populations, whether mobile or sedentary, are dynamic assemblages of individuals which maintain genetic and sometimes social information in lineages that may ramify and merge as individuals are born, reproduce, and die. Endangered populations, and those of species that mediate important ecological processes, are often targets of conservation, so that their viability is a major concern (18, 19). Theoretical treatments of population viability are influencing public policy, such as the debate over the spotted owl in the Pacific Northwest (20).

At the small end of the biospatial hierarchy of conservation targets are genes. Genes are sometimes conserved ex situ (21, 22) as seed collections, in tissue culture or germplasm collections, or as cryopreserved semen, ova, embryos, and tissues. The extraction of genes from nature annually produces multibillion dollar benefits for agriculture, biotechnology, and public health (23). In nature, genetic variation maintains the fitness and evolutionary flexibility of natural populations (16). Reserves in seminatural areas have been set aside to preserve the wild relatives of commercially important plants, especially to protect genes and gene combinations providing resistance to pests, drought, and other climatic factors (24).

THE SIX CLASSES OF INTERFERENCE AND THE NORTH-SOUTH DISTINCTION

The five levels of the biospatial hierarchy are being undermined by six major classes of human interference (25), as shown in Fig. 2. These six factors are (i) the loss of habitat; (ii) the fragmentation of habitat-producing deleterious area, edge, demographic, and genetic effects; (iii) overexploitation; (iv) the spread of exotic (introduced and alien) species and diseases; (v) air, soil, and water pollution; and (vi) climate change. These factors have all been discussed in great detail (16, 19, 22, 26, 27). The intensities of shading in the two parts of Fig. 2 are subjective, but suggest that the present and future hazards posed by the six factors are not equal in strength or concordant in rank across the range of conservation targets, or from economically poorer to economically richer nations.

Clearly the impact of a given factor depends on the time, the place, and the circumstances. As indicated in Fig. 2, economics, culture, as well as the temperate-tropical disparity in species diversity and other biogeographic patterns, explain the differences in biotic vulnerability between tropical, poor countries, and temperate, wealthier ones. The vastly greater number of species in the tropical nations, the much smaller geographic ranges of tropical species on average (28), in addition to the high rates of habitat destruction in most of these countries, means that species in the tropics are particularly vulnerable to habitat loss and fragmentation. Similarly, not all parts of the planet will be equally susceptible to the impacts of acid rain, ozone thinning, or greenhouse warming; for example, the effects of greenhouse warming will be much greater at high than low latitudes, except, perhaps, for marine systems (29). Other aspects of biogeography are relevant to geographic heterogeneity in biotic vulnerability; on oceanic islands, for example, introduced predators are typically more damaging than on continents (16, 25, 30), and introduced animals (goats, pigs, rats, mongooses, snakes, and predatory snails for instance) and plants may have catastrophic effects (31).

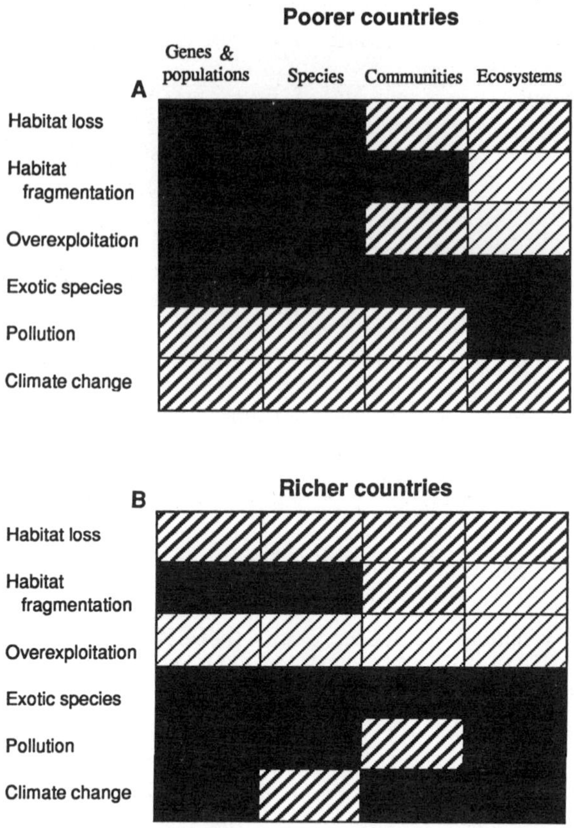

Figure 2. Relative impacts of factors affecting terrestrial biotic diversity in (**A**) poor and (**B**) rich countries. Shading indicates intensity of impact: solid = highest; thick lines = intermediate; thin lines = lowest. Ecosystems refers to landscape level formations including, for example, mangrove habitats, coral reefs, riverine/riparian systems, forests, and savannas. The distribution of impacts on aquatic and marine systems differs somewhat from those shown here.

Although it is difficult to generalize, one can point to some rough principles about the global vulnerability of terrestrial biodiversity (32). Habitat loss, fragmentation, and the direct and indirect effects of exotic species are problems everywhere (Fig. 2A), but overharvesting of economically important species is now of greater concern in poorer countries. Pollution and climate pose major threats in the temperate zone nations (Fig. 2B). As discussed below, north-south differences in socioeconomic variables and biogeography mean that conservation tactics must be tailored to the location.

THE SEVEN SOURCES OF BIOTIC DEGRADATION

The six classes of interference may constitute the most obvious proximal causes of biotic attrition, but the more fundamental causes are rooted in the contemporary human condition, especially as they are amplified by the explosive growth in human numbers in the last three centuries (Fig. 1). These more fundamental causes are listed in Table 1. The following brief descriptions of these factors are neither systematic nor exhaustive, but even this superficial treatment demonstrates why simple approaches (such as a network of protected areas alone) will fail.

Population growth. The continuous increase in human numbers exacerbates nearly every other environmental problem (33, 34). The population reached 1 billion about 1800, and appears to be headed toward 10 billion by 2046 and 12 billion by 2100, according to recent World Bank and United Nations projections. Ecologists argue that such numbers are incompatible with many ecological and evolutionary processes, including the persistence of large predators, the continuation of annual migrations of birds (35), speciation in large organisms (36), and the

Table 1. Categories of fundamental human factors that contribute to the erosion of biological diversity.

Factor	Example of impact on conservation
Population growth	Population pressures
Poverty	Hunger, deforestation, trade in rare and endangered species, failure of grass roots support
Misperception	Desire for quick results and denial of long-term failures
Antropocentrism	Lack of support for nonutilitarian causes
Cultural transitions	Unsustainable resource management during colonization and rapid social change
Economics	Failure of planning because of internationalization of markets and erratic pricing of commodities
Policy implementation	Civil disruption, wars, corruption, failure of law enforcement

protection and maintenance of native biotas in the face of increasing pressure from human beings and introduced species. For nonhuman species, this "demographic winter" will last until human beings decide to reduce their numbers to levels compatible with the restoration of pre-explosion biotic processes (37). Human populations are already declining in many industrialized countries.

Poverty. The problem is not merely the sheer magnitude of human numbers, however; it is compounded by poverty, the aspirations of people the world over for a better quality of life, and by social and political forces that impede the smooth transition to minimum (let alone "optimal") levels of prosperity, health, and justice (38). Disparities in income produce disparities of impacts. The per capita contribution to atmospheric pollution (and, hence, global climate change) is often orders of magnitude higher for citizens of the industrialized countries than for those in poorer nations (34), and economic pressures from the former contribute to unsustainable land use practices in the latter. Habitat destruction and extinction, however, will occur most rapidly in the tropics (Fig. 2A), where lack of economic opportunity, demographic momentum, and restrictions on reproductive choice are the engines that power the destruction of life.

It is probable that the price of raising human economic welfare to a standard similar to that in the wealthier countries will be biotic devastation in the tropics on a scale inconsistent with the persistence of wildlands except, perhaps, in remote, nonarable regions (39). Ehrlich and Wilson (40) point out that the magnitude of human aspirations, including demands on natural resources, if multiplied by the expected increases in human numbers, would require the human co-option of most remaining wildlands for grazing, farming, energy production, mining, transportation, and other uses. Therefore, the loss of most tropical wildlands in the next 50 years or so, an epochal catastrophe for earthly life, appears a virtual inevitability.

Misperception and time scale. Gradual environmental degradation goes almost unnoticed (41), whereas governments often overreact to sudden events of lesser overall impact. This short-term mentality is also reflected in current social mores and public policies favoring quick profits and results. The problem is that the benefits of conservation projects can only be measured on a scale of centuries. This difference in time scales between economic development projects and some conservation projects leads to conflicts because the business of conservation is keeping options open, whereas business as usual (economic development) often forecloses them.

Anthropocentrism. Many conservationists argue that current cultural values are antithetical to effective conservation policies, and that a new ethic or a revolutionary change in human consciousness is necessary before significant progress is possible (42). There are many calls for less human-centered, more biocentric economic policies. The anthropocentric orientation of most societies (43) however, augurs poorly for behavioral revolutions. If charitable donations reflect how Americans rank society's needs, it is evident that humanitarian concerns are dominant; money flows primarily to religious organizations and to medical, cultural, and social welfare causes. Figure 3 shows that only 1.5% of donated monies go to support environmental (nonhuman) groups and causes. This percentage is likely to increase, though, as donors learn about the environmental foundations of physical and social welfare.

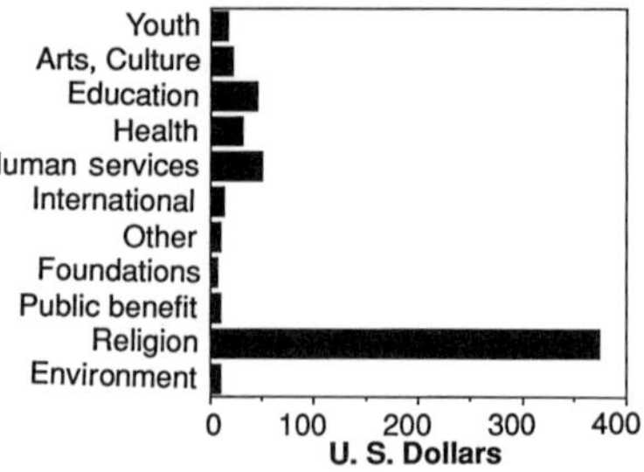

Figure 3. Average charitable contributions per household in the United States. [Adapted from (67)]

Mindful of biases favoring our own species, nearly every book, report, or "strategy" written to promote or guide the conservation of biodiversity presents a list of utilitarian justifications, including the free services and amenities provided by nature (for example, water purification and storage, habitat for fish and livestock, vistas), and the promise of life-extending pharmaceuticals and agro-industrial products that are yet undiscovered in the tissues of organisms (23). Unfortunately, the political effectiveness of narrowly utilitarian arguments for large protected areas in the tropics and elsewhere is weak, in part because the promise of long-term economic and health benefits to society as a whole appears abstract to individuals and corporations more concerned with survival and short-term economic gains.

Cultural transitions. The most destructive cultures, environmentally, appear to be those that are colonizing uninhabited territory and those that are in a stage of rapid cultural (often technological) transition (44). The cultural groups that appear to be the least destructive to natural systems are those that have been occupying the same place for centuries or more (45). Overharvesting of wild animals, of aquatic and marine organisms, and of forests, is predictable, therefore, when human groups (i) have little or no experience in their current geographic setting or (ii) are undergoing integration into the world economy. Wealthy, well-educated, industrialized cultures may have the potential for minimizing environmental damage, but show little promise of this at present. Because most of the world's people are not only poor, but in a transitional phase between traditional agrarian self-sufficiency and a modern, high-input agricultural or industrial-urban society, relatively little value is placed on the protection of nature, and even where nature is highly valued, such valuation is often left out of economic calculus.

Economics. Environmental destruction and the erosion of biological diversity in the tropics and elsewhere is exacerbated by systems of commerce that create demands from the industrialized north for products, the production of which causes massive habitat destruction (46). The "cool chain" industry, for example, produces fresh produce such as fruit, vegetables, cut flowers, and mariculture produce (such as shrimp) in the poorer countries and ships them in refrigerated carriers to the richer countries (47). This new industry contributes to the destruction of many habitats including lowland forests, mangrove, estuarine, and reef habitats. Better known are the coffee, sugar cane, banana, cacao, forest products, and cattle industries that account for the loss of a large proportion of tropical forests in developing countries (23, 48). In addition, a major contributor to forest and woodland destruction is the cutting of trees for the production of fuel wood and charcoal for domestic cooking and heating uses. Before the international price-fixing agreements among petroleum producers, most people in developing countries could afford to cook with kerosene. Now they must rely on wood, charcoal, and dung, contributing to the deterioration of forests and soils (49).

Notwithstanding the grave moral, social, and geopolitical implications of current economic disparities, the redress of such imbalances is unlikely to occur in time to save most seminatural biological systems from massive attrition. Few would question the goals of economic and social justice or their fashionable surrogate, sustainable development, but the premise that a new economic order would, alone, solve the biodiversity crisis (50) is suspect. The North American, let alone the Costa Rican experience (4), suggests that social justice and other progressive changes cannot protect biological diversity in the face of rapidly changing economic conditions including the internationalization of markets, increasing human numbers, the loss of cultural and ecological traditions, not to mention ethnic and religious conflicts. Even wealthy countries such as the United States and Canada justify the removal of the last remnants of ancient forests on the grounds of economic necessity; attempts to save that remaining 15% of original forests in the Pacific Northwest have yet to prove successful (18). In addition, corruption and bureaucratic inefficiency appear to be virtually indelible.

Policy implementation. There are many reasons for the inability of modern states to enforce laws and implement conservation policies, especially policies that require short-term sacrifices for the sake of long-term benefits. For example, the setting aside and long-term protection of land from the national estate is improbable in societies with many poor or landless people, powerful oligarchies, or corruptible judges and bureaucrats.

In countries where adequate resources are lacking for the protection and management of protected areas, even relatively secure reserves are subject to the removal of trees and to the poaching of game. Most poor nations simply lack the resources to preserve biotic diversity *in situ* (51). Such attrition is frequent during "normal" times (52), but during periods of social unrest, the loss of biodiversity can be catastrophic (53).

Many conservation and development projects are destined to fail in a statistical sense, given their unstable social or political contexts. Wars and the breakdown of civil administration can undermine decades of successful policy implementation. In Africa, recent wars in Ethiopia, Sudan, Liberia, Libya, Morocco, Somalia, South Africa, Zimbabwe, Uganda, Chad, Angola, Mozambique, Rwanda, Burundi, and other countries have led to the partial or complete collapse of nature reserves, the destruction of habitat, and the local extinction of endangered species (53). The frequency of events such as wars should be built into the planning processes of responsible agencies and organizations. This is not to say that we should abandon reserves in regions where civil chaos is frequent. Rather, expectations and policies must be tuned to appropriate distributional parameters—for example, to the mean and variance of persistence times of protected areas in similar situations and to the kinds of damage that protected areas are likely to suffer, including the killing of most large animals. The lower the mean and the higher the variance, the greater the emphasis there must be on redundancy, on alternative approaches, and on backup, *ex situ* projects. It would be prudent, in other words, to think of nature reserves as ephemeral islands, and to plan accordingly.

The human condition is dynamic and unpredictable and will remain so for at least a century, if for no other reasons than the momentum of the population explosion and the unsatisfactory economic and social status for billions of people during the 21st century. The "biotic condition," therefore, will also be tenuous during this interval. Fortunately, conservationists have an increasing number of tools with which to deal with the crisis.

TACTICS AND CONFLICTS

The eight paths to biotic survival. What tools are available to protect living nature from humanity? Table 2 presents a brief survey of eight conservation tactics or systems (5). The tactics are defined roughly in order of least to most artificial or intrusive.

1) *In situ* refers to those conservation systems based on bounded wild areas with relatively little human disturbance; it includes most protected areas, from wilderness parks to the core areas of biosphere reserves (54). Persistence may depend to some extent on the economic benefits, as generated, for example, by tourism, but protected areas tend to degrade, even in the best of circumstances, and few, if any are large enough to maintain viable populations of large predators and omnivores without *ex situ* supplementation (16, 19, 26, 55).

2) *Inter situ* refers to conservation systems or activities in regions where native species still persist, but which are outside the boundaries of established protected areas. Most of the lands belonging to this category are nonarable; typically, they are relatively infertile, cold, steep, rocky, or arid. In the United States, most such regions are administered by the Bureau of Land Management and the U.S. Forest Service.

3) *Extractive reserves* permit certain kinds of resource harvesting on a (theoretically) sustainable basis. Examples include rubber tapping, the collection of edible fruits and nuts, thatch grasses, and, perhaps, even limited logging and hunting. Sustainability of such practices, however, depends on a low population density, a stable economy, and careful management (56). In practice there may be little difference between extractive reserves and *inter situ* projects, except that the latter are more circumscribed.

Table 2. The relative potential significance of eight different conservation systems for the protection and maintenance of natural biological diversity. The "0" indicates little or no role; "X, XX, and XXX" indicate low, moderate, and high significance, respectively. The order of the systems does not imply a ranking of value.

Targets of conservation	Conservation system							
	In situ	Inter situ	Extractive reserves	Restoration projects	Zooparks	Agroecosystems & agroforestry	Living ex situ	Suspended ex situ
Entire systems (ecosystems)								
Processes or functions	XXX	XX	XX	XX	XX	X	0	0
Biosocial (traditional human uses)	X	XX	XXX	XX	X	X	0	0
Biogeographic assemblages	XXX	XX	XX	X	XX	0	0	0
Indigenous and endemic species	XXX	XX	XX	X	XX	X	XX	X
Local populations of species	XXX	XX	XX	X	XX	X	X	X
Genetic variation within species								
Wild relatives of domesticates	XXX	XX	XX	X	X	X	XXX	XX
Traditional domesticated varieties	X	X	X	0	X	XX	X	XXX
Noneconomic genetic variation	XXX	XX	XX	X	X	0	X	X
Ownership	Public & private	Private & public	Public & private	Private & public	Private	Private	Private & public	Private & public

4) *Ecological restoration projects* refers to intensive management activities intended to increase species richness or productivity in degraded habitats. Among the necessary conditions for such activities are political and institutional stability.

5) *Zooparks* refers to facilities in secure locations where a mix of local and exogenous species can be maintained under seminatural conditions—in other words, sanctuaries for sensitive species of diverse provenance (57). The assumptions underlying the establishment of such reserves are that protected areas, in many places, are not viable for social or political reasons and the inevitability of highly recombined biotic communities in the future given current rates of species introductions (58). This category differs from *in situ* reserves because of the conscious introductions of target species.

6) *Agroecosystems and agroforestry projects* are highly managed, production-oriented systems with a wide range of dependence on artificial chemical and energy inputs (59). The number of native species that can survive in such systems is highly variable, depending mostly on the proximity of garden, farm, and plantation to wildlands, the use of artificial chemical inputs, and the tolerance of farmers to wildlife (60).

In addition to zooparks, there are two kinds of *ex situ* tactics or backup systems (14). These are essential where particular reserves are likely to fail or lose significant numbers of their species.

7) *Living ex situ* programs refers to botanical gardens, zoos, aquaria, and similar institutions that maintain and propagate living organisms for noncommercial (education, research, conservation) purposes in a highly controlled, usually urban, context.

8) *Suspended ex situ* programs are completely artificial; living material is metabolically slowed or arrested. Among these projects are germplasm storage facilities such as seed banks, tissue culture collections, and cryopreserved collections of gametes, zygotes, and embryos.

As shown in Table 2, this typology of tactics manifests a current trend—the privatization of conservation. For many reasons, nonprofit groups and individuals increasingly are complementing if not supplanting government agencies in protecting biodiversity. Private zoos, botanical gardens, and others are taking responsibility for the captive propagation of endangered species. Responsibility for the restoration of degraded forest, pastures, and farmlands on both public and private lands is being assumed by private groups. Organizations like The Nature Conservancy and Conservation International are acquiring new sites for protected areas (61), though governments are usually the ultimate owners.

SOCIAL CONTEXT AND THE DEBATE OVER TACTICS

Current discussions have tended to oversimplify the diversity in conservation approaches by exaggerating the differences between the so-called species approaches and ecosystem approaches. The former emphasizes the protection, both *in situ* and *ex situ*, of endangered, often charismatic vertebrates, whereas the objective of the latter is to set aside and manage natural areas based on systems of landscape classification that will capture as much species and ecological diversity as possible (62). Critics of species-level approaches have emphasized the shortcomings of the Endangered Species Act and point out that most of the federal dollars are directed at a few birds and mammals (62). Some of these critics argue that success in captive breeding and cryopreservation will lead to complacency about the need for more and better protected areas. Supporters of endangered species might counter that if it were not for the charismatic species, the public appeal of conservation would be much less, that endangered species justify many of the larger protected areas in the United States and elsewhere, and that endangered species legislation is providing the economic leverage to bring developers and government agencies into negotiations about the preservation of large areas of habitat for general biodiversity conservation in the United States (63).

Such adversarial discussions, however, often ignore social context. As shown in Fig. 4, conservation tactics can be ranked according to the degree each is integrated into the local human community and the degree that each is dependent on artificial (technological) means and invasive management practices. Implicit is the idea that different tactics require different degrees of social and technical sophistication.

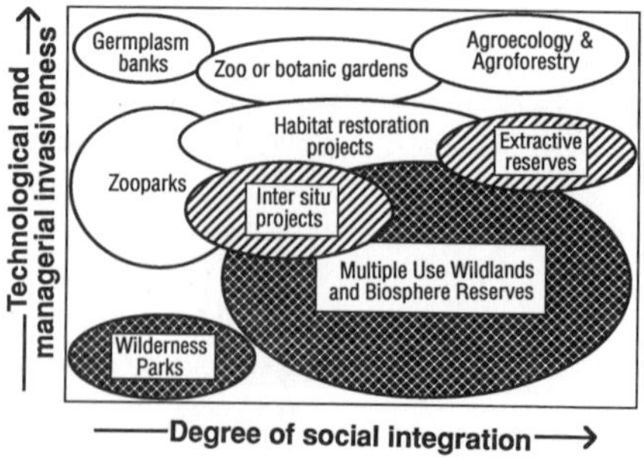

Figure 4. Descriptive distribution of conservation tactics according to the degree of social integration at the local level, and the degree of technological input or management intensity. Shading indicates relative degree of human interference with natural processes; darker shades indicate less interference. The positions shown for each tactic are meant to suggest the center of the probable zone of action for the tactic. The term "Biosphere Reserves" refers to multiple use, production-oriented projects, with a relatively sacrosanct core protected area.

A more prescriptive classification is shown in Fig. 5. It distributes the tactics in a plane of human population pressure and political stability. It is based on the untested assertion that the persistence of conservation projects, particularly protected areas, is related to the frequency and degree of political unrest and the rate of population growth. The combination of the two figures suggests that the choice of tactics should be influenced by the probable impact of demographic, economic, and social conditions as discussed above. For example, *ex situ* tactics are prescribed where political instability is frequent and where population pressure is building.

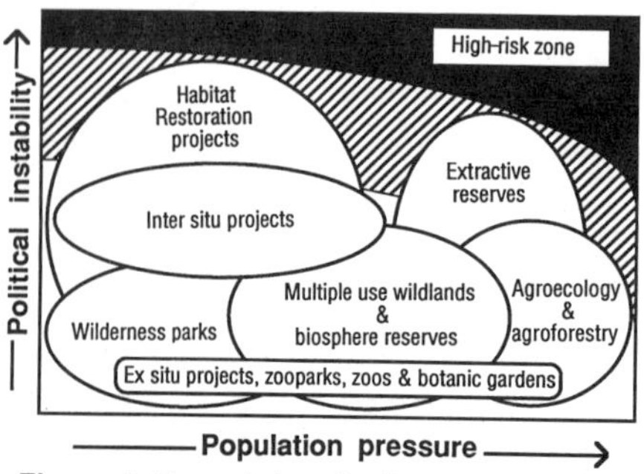

Figure 5. Prescriptive distribution of conservation tactics based on the probability of increasing population pressure and the likelihood of political instability or violent conflicts. Backup, *ex situ* facilities are placed in relatively secure, politically stable locations.

Much of the debate in the United States over approach and tactics stems from uncertainty and bias about landscape and geography, the importance of socioeconomic conditions and the stability of political structures, confidence in new legislative and legal remedies, and the identity of target organisms. For example, conservationists with experience in the species-rich tropics—where infrastructure is fragile at best, episodes of social chaos inevitable, human populations are doubling every few decades, laws are ignored, and hunting of rare animals and deforestation are a way of subsisting—should support a pluralistic approach that includes *ex situ* backup for protected areas. On the other hand, those with experience in wealthy, stable, temperate zone regions—where most species have wide geographic ranges and where there exists extensive areas of low productivity, government-owned lands—are more likely to promote systems of protected areas linked by corridors in multiple use zones that can be managed for conservation and sustainable forms of exploitation (64). They will also have more faith in legislative remedies and law enforcement. Figure 5 illustrates this tactical pluralism.

CONCLUSIONS

Today, the conservation of biodiversity is virtually equivalent to the *ex situ* protection of wildlands. In the future, however, such reserves will come to be seen actuarially, their life times dependent on many biogeographic, social, and political factors. Unless a much denser and more secure network of protected areas is established soon, the importance of less appealing alternatives will be greater than conservationists would wish.

This awareness has led some observers to call for a greater emphasis on adjunctive approaches, including *inter situ* projects—the management of wildlife in nonarable lands outside of traditional reserves (65). Though appropriate in certain places, these lands are not immune to overexploitation, desertification, and to other forms of abuse, as the recent history of Tibet, the Sahel of Africa, and the American Southwest have shown. The *inter situ* tactic is an important backup, however, especially in socially and demographically stable nations and regions. The point is that every tactic has its limitations; sole reliance, for instance, on ecological restoration or on cryopreservation technologies would be premature, if not immoral, because these technologies could protect only a tiny fraction of species diversity for the foreseeable future, especially in tropical seas and forests.

Progress in conservation is hampered by the lack of a clearly articulated public policy on biodiversity. The United States and many other countries lack a coherent conservation strategy. In part, this may stem from confusion about tactics, as discussed above. The United States should join the nations that have developed a national conservation or biodiversity strategy. There is also a need for new institutions such as a National Institutes of the Environment (similar to the National Institutes of Health) to provide the intellectual leadership and sustainable funding for planning and research in biodiversity. In addition, a high level review of federal agencies is necessary so that either the authority for the protection of biodiversity is vested in a new agency with clear directives, or the organic acts (if any) of the agencies should be restructured, making conservation a prime directive of the U.S. Forest Service, the Bureau of Land Management, and the National Wildlife Reserve System.

Everywhere, nature reserves must be defended and bolstered by social experimentation in "sustainability." But there is too much at risk to gamble on any one social ideology, theory, or approach. All human institutions are transient expedients, and the conservation systems that are fashionable today will certainly undergo many changes in the next century. Opportunism and tolerance must be the watchwords of the science, the politics, and the art of nature protection (66). The issue, therefore, is not the "failure" of conservation; it is whether it can stay the course. During the construction of cathedrals in the Middle Ages, planners and artisans were not dismayed that "success" might require centuries. Like those workers, conservation scientists and practitioners must accommodate their objectives to the social complexity and temporal scale of their enterprise (67).

The author is professor of conservation biology, Board of Environmental Studies, University of California, Santa Cruz, Santa Cruz, CA 95064.

REFERENCES AND NOTES

1. K. A. Kohm, Ed., *Balancing on the Brink of Extinction* (Island Press, Washington, DC, 1991).
2. W. Reffalt, in (1), pp. 77–85.
3. J. A. McNeely, in (4), pp. 150–157; D. Western, in *ibid.*, pp. 158–165; D. Janzen, in *Conservation Biology: Science of Scarcity and Diversity*, M. E. Soulé, Ed. (Sinauer, Sunderland, MA, 1986), pp. 286–303.
4. A. F. Ugalde, in *Conservation for the Twenty-first Century*, D. Western and M. C. Pearl, Eds. (Oxford Univ. Press, New York, 1989), pp. 145–149.
5. *1989 United Nations List of National Parks and Reserves* (United Nations, New York, 1990).
6. *Threatened Parks Register* (International Union for the Conservation of Nature and Natural Resources Commisson on National Parks and Protected Areas, IUCN, Gland, Switzerland).
7. R. A. Houghton, *Environ. Sci. Technol.* 24, 414 (1990).
8. D. Simberloff, in *Dynamics of Extinction*, K. K. Elliot, Ed. (Wiley, New York, 1986); W. V. Reid and K. R. Miller, *Keeping Options Alive: The Scientific Basis for Conserving Biodiversity* (World Resources Institute, Washington, DC, 1989); E. O. Wilson, *Sci. Am.* 261, 108 (September 1989). Arthropods would probably account for more than 95% of the extinctions because insects and other arthropods constitute most of the world's species diversity and because many tropical arthropods may have quite localized distributions [see T. L. Erwin, *Col. Bull.* 36, 74 (1982)].
9. Many millions of years are required to replenish taxonomic diversity at the family level or higher (D. Jablonski, *Science* 231, 129 (1986); *ibid.* 253, 754 (1991); D. M. Raup, *ibid.* 231, 1528 (1986). Not only is the current rate of extinction many orders of magnitude higher that the historical average rate of speciation, but contrary to intuition, the process of speciation for large organisms is now severely compromised by habitat loss and fragmentation [see M. E. Soulé, in *Conservation Biology: An Evolutionary-Ecological Perspective*, M. E. Soulé and B. A. Wilcox, Eds. (Sinauer, Sunderland, MA, 1980)].
10. Individual organisms are rarely considered to be targets of conservation. Their conservation status, if any, usually derives from their potential genetic and generative contribution to the lineage or population, not because of their value or rights as individuals per se, a premise that distinguishes the conservation movement from the animals rights movement. For a discussion of the "rights" argument, see T. Regan, *The Case for Animal Rights* (Univ. of California Press, Berkeley, 1983). Regan considers the conservation argument to be fascistic (pp. 361–362) on the grounds that conservation emphasizes aggregative (population and community) considerations which, he says, cannot be reconciled with the animals rights view. Even so, a single individual can have instrumental value in conservation; habitat and species diversity is often maintained by natural disturbance [see S. T. A. Picket and P. S. White, Eds., *The Ecology of Natural Disturbance and Patch Dynamics* (Academic Press, Orlando, FL, 1985); L. D. Harris, *Environ. Manage.* 12, 675 (1988)].
11. M. I. Dyer and M. M. Holland, *BioScience* 41, 319 (1991).
12. J. M. Scott, B. Csuti, J. D. Jacobi, J. E. Estes, *ibid.* 37, 782 (1987); J. M. Scott, B. Csuti, K. Smith, J. E. Estes, S. Caicco, in (1), pp. 282–297.
13. For summaries of this literature see D. J. Futuyma, in *Speciation and Its Consequences*, D. Otte and J. A. Endler, Eds. (Sinauer, Sunderland, MA, 1989),

pp. 557–578; M. L. Hunter, Jr., in (1), pp. 266–281; M. E. Soulé, *Conserv. Biol.* 4, 233 (1990).

14. Office of Technology Assessment, *Technologies to Maintain Biological Diversity* (U.S. Congress, OTA-F-330, U.S. Government Printing Office, Washington, DC, 1987).
15. Information on endangered species issues is available in *The Endangered Species UPDATE* (School of Natural Resources, University of Michigan, Ann Arbor, MI 48109–1115) and *Species* (Species Survival Commission, c/o Chicago Zoological Society, Brookfield, IL 60513).
16. O. H. Frankel and M. E. Soulé, *Conservation and Evolution* (Cambridge Univ. Press, Cambridge, 1981).
17. H. Salwasser, in (1), pp. 247–265. The use "management indicator species," however, is controversial [see P. B. Landres, J. Verner, J. W. Thomas, *Conserv. Biol.* 2, 316 (1988)].
18. D. S. Wilcove, *Trends Ecol. Evol.* 4, 385 (1989).
19. M. E. Soulé, *Viable Populations for Conservation* (Cambridge Univ. Press, Cambridge, 1987).
20. J. W. Thomas *et al.*, *A Conservation Strategy for the Northern Spotted Owl* (U.S. Department of Agriculture, Forest Service, Portland, OR, 1990).
21. Conservation tactics that focus on life in natural place are called *in situ* and those that conserve it elsewhere are *ex situ*. For information on *ex situ* conservation, see W. G. Conway, in (4), pp. 199–209; A. H. D. Brown, O. H. Frankel, D. R. Marshall, J. T. Williams, Eds., *The Use of Plant Genetic Resources* (Cambridge Univ. Press, New York, 1989).
22. E. O. Wilson and F. M. Peter, Eds., *Biodiversity* (National Academy Press, Washington, DC, 1988).
23. N. Myers, *The Primary Source: Tropical Forests and Our Future* (Norton, New York, 1984).
24. H. H. Iltis, in (22), pp. 98–105; J. T. Williams, in *ibid.*, pp. 240–247.
25. J. Diamond, in (4), pp. 37–41, has used a more compact classification—"the evil quartet."
26. M. E. Soulé, *Conservation Biology: Science of Scarcity and Diversity* (Sinauer, Sunderland, MA, 1986).
27. D. Simberloff, *Annu. Rev. Ecol. Syst.* 19, 473 (1988); C. M. Shonewald-Cox, S. M. Chambers, B. MacBryde, W. L. Thomas, *Genetics and Conservation* (Benjamin-Cummings, Menlo Park, 1983); D. Western and M. C. Pearl, Eds., *Conservation for the Twenty-First Century* (Oxford Univ. Press, New York, 1989); M. E. Soulé and B. A. Wilcox, Eds., *Conservation Biology: An Evolutionary-Ecological Perspective* (Sinauer, Sunderland, MA, 1980).
28. J. Terborgh and B. Winter, in *Conservation Biology: An Evolutionary-Ecological Perspective*, M. E. Soulé and B. M. Wilcox, Eds. (Sinauer, Sunderland, MA, 1980), pp. 119–134; A. H. Gentry, in *Conservation Biology: Science of Scarcity and Diversity*, M. E. Soulé , Ed. (Sinauer, Sunderland, MA, 1986).
29. R. L. Peters and T. Lovejoy, Eds., *Consequences of Greenhouse Warming for Biological Diversity* (Yale Univ. Press, New Haven, in press).
30. I. Atkinson, in (4), pp. 54–75; J. A. Savage, *Ecology* 68, 660 (1987).
31. S. L. Pimm, *Trends Ecol. Evol.* 2, 106 (1987); P. M. Vitousek, in (*22*), pp. 181–189.
32. Similar analyses might be useful for aquatic and marine systems.
33. P. R. Ehrlich and A. H. Ehrlich, *Extinction: The Causes and Consequences of the Disappearance of Species* (Random House, New York, 1981).
34. P. R. Ehrlich and A. H. Ehrlich, *The Population Explosion* (Simon & Schuster, New York, 1990).

35. See, for example, J. Terborgh, *Where Have All the Songbirds Gone?* (Princeton Univ. Press, Princeton, NJ, 1989).

36. M. E. Soulé , in *Conservation Biology: An Evolutionary-Ecological Perspective*, M. E. Soulé and B. M. Wilcox, Eds. (Sinauer, Sunderland, MA, 1980), pp. 151–170.

37. M. E. Soulé , M. E. Gilpin, W. G. Conway, T. Foose, *Zoo Biol.* 5, 101 (1986). There is a false paradox about conservation programs and time scale. A critic might ask, "Why the haste if conservation projects must last centuries?" The problem is the that the current rate of biotic destruction demands immediate actions. This is not inconsistent with the objective that they persist a long time.

38. R. Repetto, Ed., *The Global Possible: Resources, Development, and the New Century* (Yale Univ. Press, New Haven, 1985); H. E. Daly and J. C. Cobb, Jr., *For the Common Good: Redirecting the Economy Toward Community, the Environment, and a Sustainable Future* (Beacon, Boston, 1989).

39. P. M. Vitousek, P. R. Ehrlich, A. H. Ehrlich, P. A. Matson, *BioScience* 36, 368 (1986).

40. P. Ehrlich and E. O. Wilson, *Science* 253, 758 (1991).

41. R. Ornstein and P. R. Ehrlich, *New World, New Mind* (Doubleday, New York, 1989).

42. B. Devall and G. Sessions, *Deep Ecology* (Gibbs M. Smith, Layton, UT 1985); A. Leopold, *A Sand County Almanac and Searches Here and There* (Oxford Univ. Press, Oxford, 1948); *World Conservation Strategy: Living Resources Conserved for Sustainable Development* (International Union for the Conservation of Nature, Nairobi, Kenya, 1980).

43. D. Ehrenfeld, *The Arrogance of Humanism* (Oxford Univ. Press, New York, 1981); C. Birch and J. C. Cobb, Jr., *The Liberation of Life: From Cell to the Community* (Cambridge Univ. Press, Cambridge, 1986).

44. P. S. Martin and R. G. Klein, Eds., *Quarternary Extinctions: A Prehistoric Revolution* (Univ. of Arizona Press, Tucson, 1984); S. T. Olson, in *Conservation for the Twenty-first Century*, D. Western and M. C. Pearl, Eds. (Oxford Univ. Press, New York, 1989), pp. 50–53.

45. P. M. Chandler, *Agriculture and Human Values* 8, 59 (1991); T. H. McGovern, G. Bigelow, T. Amorosi, D. Russell, *Human Ecology* 16, 225 (1988).

46. R. B. Norgaard, in *Biodiversity*, E. O. Wilson and F. M. Peter, Eds. (National Academy Press, Washington, DC, 1988), pp. 206–211.

47. D. E. Goodman and W. H. Friedland, personal communication.

48. P. H. Raven, in (22), pp. 119–122; M. J. Plotkin, in *ibid.*, pp. 106–118.

49. S. Postel in *State of the World* (Norton, New York, 1989), pp. 21–40.

50. For example, S. Hecht and A. Cockburn, *The Fate of the Forest: Developers, Destroyers and Defenders of the Amazon* (Harper Perennial, New York, 1990); to this extent, the working principle of some international conservation organizations that economic development is a necessary, let alone sufficient, precondition for conservation is untested.

51. H. H. Iltis, *Environment* 25, 55 (1983).

52. G. E. Machlis and D. L. Tichnell [*The State of the World's Park* (Westview, Boulder, CO, 1985)] report that 95% of tropical reserves report poaching of wildlife.

53. For example, F. Kayanja and I. Douglas-Hamilton, in *National Parks, Conservation and Development: The Role of Protected Areas in Sustaining Societies*, J. A. McNeely and K. R. Miller, Eds. (Smithsonian Institution Press, Washington, DC, 1984), pp. 80–86.

54. The Biosphere Reserve concept of Unesco's Man and the Biosphere Program is, in part, an attempt to integrate economic development and conservation by sharing the management and benefits of protected areas with local peoples. It attempts to avoid the extremes of banishing people to save nature and banishing nature for the sake of people. It has been difficult to apply in practice [see D. Hales, in (4),

pp. 139–144]; M. Batisse, *Nat. Resour.* 22, 1 (1986); S. R. Kellert, *Environ. Conserv.* 13, 101 (1986).

55. U. S. Seal, in *Endangered Birds: Management Techniques for Preservation of Threatened Species*, S. A. Temple, Ed. (Univ. of Wisconsin Press, Madison, WI, 1978), pp. 303–314; W. D. Newmark, *Nature* 325, 430 (1987); P. F. Brussard, *Ecol. Appl.* 1, 6 (1991); R. L. Peters and J. D. Darling, *BioScience* 35, 707 (1985).
56. L. Silberling, *BioScience* 41, 284 (1991); J. R. Browder, *ibid.* 40, 626 (1991); J. R. Browder, *ibid.* 41, 286 (1991).
57. An example is North American ranches harboring African ungulates. Entrepreneurs might consider the purchase of strategically located islands and other real estate where secure facilities could be located. See also M. E. Soulé, in *Conservation of Threatened Natural Habitats*, A. V. Hall, Ed. (South African National Science Programmes Report 92, CSIR Foundation for Research Development. P.O. Box 395, Pretoria, South Africa), pp. 46–65; M. E. Soulé , in (4), pp. 297–303.
58. M. E. Soulé , *Conserv. Biol.* 4, 233 (1990).
59. M. A. Altieri and L. C. Merrick, in (22), pp. 361–369.
60. Although the vast majority of native species currently are unable to survive in intensively managed agricultural zones, especially in the tropics, sound agroecological practices create a healthy environment and contribute to self-sufficiency and the maintenance of crop genetic resources, especially if practiced in an economically and political stable environment. In addition, they may effectively reduce wood-collecting, hunting, and other pressures on nearby wildlands [see M. A. Altieri and D. K. Letourneau, *Crop Protect.* 1, 405 (1982); S. R. Gliessman, E. R. Garcia, A. M. Amador, *Agro-Ecosystems* 7, 173 (1982)].
61. Among the most recent and impressive acquisitions by The Nature Conservancy is the purchase in 1990 of the 130,000-ha Gray Ranch in southwestern New Mexico, which includes an entire mountain range (Las Animas).
62. J. M. Scott, B. Csuti, K. Smith, J. E. Estes, S. Caicco, in (1), pp. 282–297. See also Part IV of (1).
63. See L. A. Greenwalt, in (1), pp. 31–36; M. J. Bean, in *ibid.*, pp. 37–42; D. D. Murphy, in *ibid.*, pp. 181–198.
64. R. Reed, *Nat. Areas J.* 7, 2 (1987); L. D. Harris and J. Eisenberg, in (4), pp. 166–181.
65. D. Western, in (4), pp. 158–165.
66. Others have called for greater respect for pluralism [D. Western, in (4), pp. vi–xv; R. F. Noss, in (1), pp. 227–246].
67. *Giving and Volunteering in the United States: Summary of Findings* (Independent Sector, 1828 L Street, NW, Washington, DC, 1988).
68. The manuscript was much improved thanks to the comments of D. B. Botkin, P. R. Ehrlich, D. Goodman, W. P. Gregg, Jr., R. E. Grumbine, D. F. Hales, J. A. McNeely, P. Romans, J. M. Scott, D. Wilcove, and G. Zegers.

2

An Evolutionary Basis for Conservation Strategies*

Terry L. Erwin

Conservation strategies have been remarkably anthropocentric from their inception in the Middle Ages to the present (1, 2). During dynastic and feudal times, parts of kingdoms were set aside as hunting grounds for the aristocracy, thus preserving everything that dwelled therein. This, plus severe natural and cultural control of human populations resulted in environmental protection for centuries. Today, with a burgeoning and expanding human population of 5.3 billion, no more than 4500 areas are protected globally (1); that is equivalent to a mere 3.2% of our planet's landmass. National parks, wildlife refuges, biosphere reserves, military reserves, Indian reservations, and other forms of legally protected areas have been established for aesthetic, political, or practical purposes in the last 150 years. Many reserves in less-developed nations are paper parks only; many in the more developed are lamentably endangered by touristic herds, and certain wilderness parks are threatened by short-sighted national energy policies.

Today, conservation strategy is based on a perceived impending loss of biodiversity due to tropical deforestation or disappearing habitats where populations of "interesting" species, subspecies, or even varieties (especially in temperate areas) reside. Campaigns usually focus on loss of potentially useful resources, such as plants with pharmaceutical properties or large animals that capture human interest. In practice, this results in saving fauna and flora in a few "available" acres where a well-known target taxon lives. Science has been too slow in providing inventory data to do much more; thus, what should be a major collective effort between conservation and science is often nonexistent, or in some cases, discord.

In the past 3 billion years, more species and their natural assemblies with their particular interactions have come and gone than are now present on Earth (3). One fact of evolution is that species go extinct, and others come into existence.

---

* This chapter was originally published in *Science*, vol. 253 (1991), pp. 750–52. It is reprinted here with permission.

Today, because of unprecedented human impact, species are increasingly going extinct and the speciation process, which creates future biodiversity, is being severely pressured through the removal of contiguous related biotic habitats. The pattern of continental habitats, often vast biomes, is being reduced to one of scattered island-like habitats and, just as on real islands, major extinctions are destined to occur. If this disruption of natural systems continues into the 21st century, we can expect the evolutionary process as we know it to become degraded and retarded.

There is no unified scientific method behind conservation strategy that addresses the nature and quantity of biodiversity, nor what it means environmentally either to save it or lose it outside direct human interest. In fact, there is little altruism or science in the fight to save the rain forest, the spotted owl, or the Antioch blue butterfly. Rather, politics and economics weigh heavily on most decisions. It seems that degradation and conversion of the environment is proceeding so rapidly that getting something preserved—anything at all—is acceptable regardless of the yardstick. Worst of all is that legitimate arguments within the scientific and conservation communities allow decision-makers an out in politically difficult choices. In order to supplement positive conservation practices and provide an alternative to negative ones, an effort to establish a sound scientific underpinning must be made. Scientific rationale may transcend cultural changes through time, whereas economic and political grounds certainly will not.

What is biodiversity? Is it important, and if so, important to what? Is it possible to separate contemporary human needs from what is really necessary for the long-term environmental health of the planet? How can we hope to manage 30 or more million species? Given the myriad of societal demands and an ever-increasing population, what can realistically be achieved even if a global effort is sustained in environmental management? Should conservation strategy be scientifically or culturally based? These and others are the tough questions with which political systems must deal. For scientists, the question is what can we provide from our science that will help generate a long-term, transcultural foundation on which conservation strategy can be based?

Biodiversity can be equated with species richness, that is the number of species, plus the richness of activity each species undergoes during its existence through events in the life of its members, plus the nonphenotypic expression of its genome. Biodiversity evolves through numerous processes that vary from locality to locality, habitat to habitat. Species richness at a site is a readily observable index of the number of interactions among and between species and how these species are grouped as a living unit at that site. A species richness index then is a reasonable and knowable tool that can be used in setting policy and making decisions about biotic conservation and management. To understand the significance of a biodiversity index across geography, one needs context. Relationships between species and a knowledge of lineages to which they belong provide that context.

Radiation of lineages of organisms occurring on both continents and islands proceeds stepwise and requires contiguous habitats of various kinds through which sequences of phylogenetically related species pass as the lineage to which they belong rises to dominance (within the context of the occupied habitat) and ebbs to extinction (4, 5). Centers of endemism, or relict occurrences of organisms, are the last remaining footholds of past radiations. Elsewhere these endemic organisms have been replaced by better adapted lineages to an ever-changing contemporary environment. This model taken to its extreme, given current trends, indicates that within a few hundred years this planet will have little more than lineages of domestic weeds, flies, cockroaches, and starlings, evolving to fill a converted and mostly desertified environment left in the wake of nonenvironmentally adaptive human cultural evolution.

What should we know to aid in countering the planet's impending biotic destruction? Assuming that it is the species radiation part of the evolutionary process, the generator of biodiversity, which is endangered, and that is what we (altruistically) decide to protect through scientifically based choices rather than cultural ones, we need to know where lineages (not individual species) originate innovations in their evolution and how these become distributed over some part of the planet. The disciplines involved to achieve this are phylogenetics and biogeography, together referred to as systematics. We need to use this science to tell us where the critical areas are that need sound environmental management—that is, where we need to protect the active processes of contemporary evolution. The most powerful tool to emerge during the past 20 years as a robust and comparative science, with both practitioners and theoreticians, is phylogenetics (6, 7) and its methods and applications (8, 9). Phylogenetics is well suited to provide predictions of as yet unobserved qualities that are directly applicable to conservation decisionmaking (10) and because its tools are now computer-based it can be applied in a short time to many groups for detecting congruence in patterns of occurrence of radiating lineages (9). Site congruence, which can be mapped easily, of many evolving lineages can then become the target of conservation activities.

A cladogram illustrating the hypothetical phylogenetic relationships among seven known species that make up a monophyletic lineage of organisms is shown in Fig. 1. According to such an analysis, species A and B have not demonstrated radiation—that is, the ability to evolve into a more broadly adaptive and widespread lineage through time. Both are found to be geographically restricted endemic forms (relics) occupying small areas. Current conservation strategy places highest priority for protection on such areas as 1 and 2 (11). Endemic forms such as A and B are often unusual or rare, and even interesting to many scientists (12), but they are predictably on their way to extinction. These forms carry information about past evolutionary flourishes; they are important to protect, but they are only half the picture. The relatively more recent sublineage in Fig. 1 (stem C + D + E + F) is where phylogenetic theory predicts radiation and dynamic changes in taxa are occurring today and will occur in the future. Species

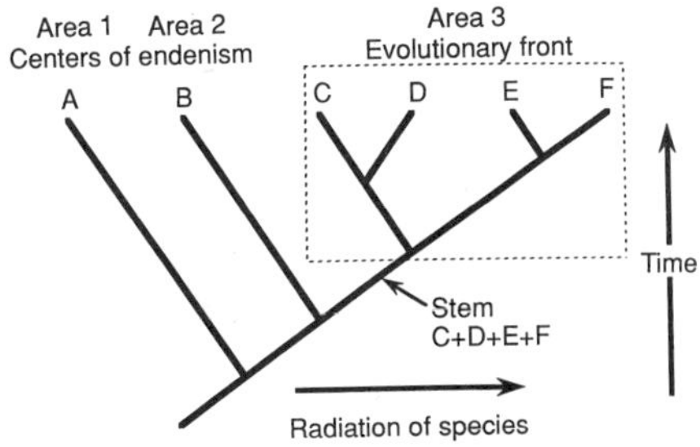

Figure 1. Simple cladogram of seven species in a monophyletic lineage. More complex lineages may have more than one evolutionary front.

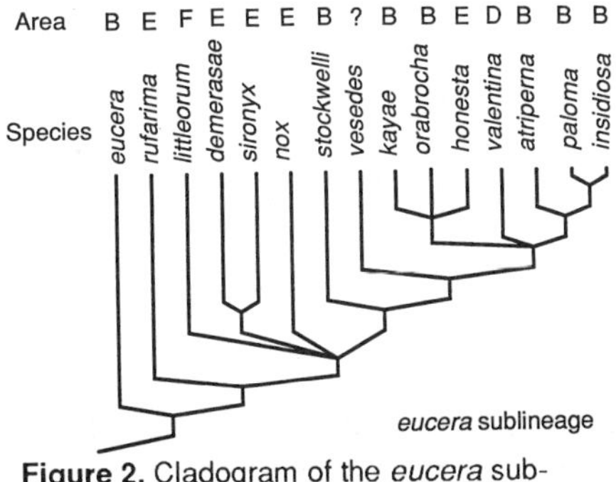

Figure 2. Cladogram of the *eucera* sublineage of the carabid beetle genus *Agra* (9).

such as C, D, E, and F are sometimes widespread and may even be regarded as "weedy" species, but does that make them less important? Their stem has become the multi-species sublineage that holds the most promise for continued evolution of this line of biodiversity under natural conditions. For example, in the *eucera* sublineage of the carabid beetle genus *Agra* (9) (Fig. 2), current interest would focus on areas D and F, each of which contains a relatively primitive and rare species. The cladogram (Fig. 2) shows that areas B and E contain both recent radiation and older species of the sublineage. If the *eucera* sublineage were something of general conservation interest, then the investment for protection would be better put into areas B and E to maximize salvaging this kind of beetle diversity now and in the future. Vane-Wright *et al.* (13) provide a novel index for cladogram analysis that needs careful testing in its application to making choices in conservation of taxic diversity. Congruence across many groups with their method may be the best way to find centers of radiation for conservation purposes.

Conservation strategy should incorporate methods to detect such contemporary evolution for the good of future maximum biodiversity. Conservation of only an accumulation of mostly nonradiating endemic taxa, the current conservation strategy (11), is like saving living fossils, something of human interest, but perhaps not beneficial to the protection of evolutionary processes and environmental systems that will generate future biodiversity.

Through analyses of diverse groups and detection of congruent patterns among radiating lineages (8), evolutionary fronts (centers of radiation) can be detected and targeted for long-term protection. Site protection and future ecological reconstruction of natural corridors (Fig. 3) between important centers will be essential to allow continued species radiation because climatic shifts may displace species' ranges (in isolated parks great extinction will occur); evolution proceeds from centers of radiation outward through sequences of contiguous habitats latitudinally and altitudinally and there become disrupted from time to time allowing speciation.

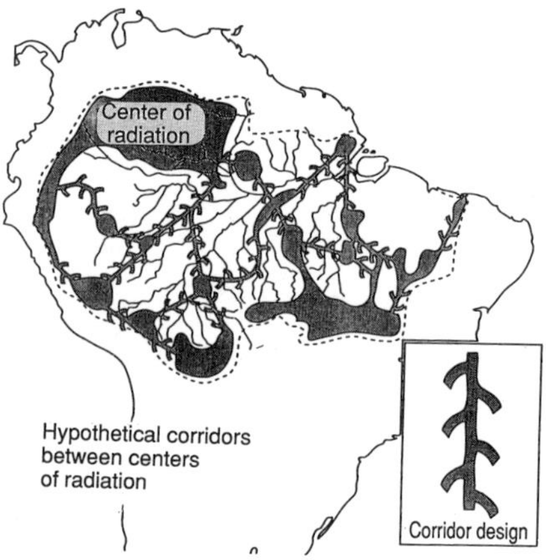

Figure 3. Future corridors of ecological reconstruction between hypothetical centers of radiation—for example, in the Amazon Basin, to allow species movements and radiation. Inset is a design for a corridor that maximizes soil and habitat types in small areas.

Evolutionarily dynamic lineages today create future biodiversity. Such lineages are the cornerstone of natural environmental health. Science has the philosophy and tools to detect these lineages through phylogenetic systematics. Conservation strategy can use the patterns detected in cladistic studies to defend contemporary centers of radiation from destruction on the premise that today's maximum biodiversity, as well as tomorrow's, are in and stem from such centers. Acceptance of a nonhuman yardstick to measure environmental health—that is, evolutionary processes—and implementation of a scientific approach in conservation policies will provide a strategy to achieve a lasting stability for global environmental health because the basis for conservation will not be tied to the whims of human culture. The goal of conservation strategy should be the protection of future maximum biodiversity as well as preservation of contemporary species of human interest.

The author is a curator of entornology (Coleoptera) at the National Museum of Natural History, Smithsonian Institution, Washington, DC 20560.

REFERENCES AND NOTES

1. J. A. McNeely *et al.*, *Conserving the World's Biological Diversity* (International Union for the Conservation of Nature, Gland, Switzerland, 1990).
2. W. V. Reid and K. R. Miller, *Keeping Options Alive: The Scientific Basis for Conserving Biodiversity* (World Resources Institute, Washington, DC, 1989).
3. D. M. Raup, in *Biodiversity*, E. O. Wilson, Ed. (National Academy Press, Washington, DC, 1988), pp. 51–57.
4. T. L. Erwin, in *Taxonomy, Phylogeny and Zoogeography of Beetles and Ants: A Volume Dedicated to the Memory of Philip Jackson Darlington, Jr., 1904–1983,* G. E. Ball, Ed. (Junk, The Hague, 1985), pp. 437–472.
5. E. O. Wilson, *Am. Nat.* 95, 169 (1961).
6. W. Hennig, *Phylogenetic Systematics* (Univ. of Illinois Press, Urbana, 1966).
7. G. Nelson and N. I. Platnick, *Systematics and Biogeography: Cladistics and Vicariance* (Columbia Univ. Press, New York, 1981).
8. M. G. Pogue and M. F. Mickevich, *Cladistics* 6, 319 (1990).
9. T. L. Erwin and M. G. Pogue, *Proc. Brasilian Acad. Sci.* (1988), pp. 161–188.
10. T. L. Erwin, *Am. Entomol.*, in press.
11. R. A. Mittermeier, in *Biodiversity*, E. O. Wilson, Ed. (National Academy Press, Washington, DC, 1988), pp. 145–54.
12. R. E. May, *Nature* 347, 129 (1990).
13. R. I. Vane-Wright, C. J. Humphries, P. H. Williams, *Biol. Conserv.* 55, 235 (1991).
14. I thank B. D. Farrell, W. N. Mathis, R. J. McGinley, M. G. Pogue, and F. C. Thompson for helpful comments, G. L. Venable for the illustrations, and Smithsonian Institution Programs BIOLAT, Neotropical Lowlands Research Project, and Scholarly Studies for funding that led to the concepts developed in this 18th contribution to the BIOLAT Series.

3

Balancing Species Preservation and Economic Considerations*

Harold J. Morowitz

How much is a species worth? We generally take an anthropocentric view of that question. The species *Homo sapiens*, as judged by the lives and well-being of individuals, is infinitely precious in our public ethic. A tiny arachnid, found only in the sands of Suvarov Islands, isolated in the mid-Pacific, is likely to get a much lower rating. The question becomes, "What is the value of a given species to human society?" Once the term "value" is introduced, the question moves to economics and ethics, both of which use that construct, but in very different senses. From a narrow economics point of view, we need a monetary metric of a species value to balance benefits against costs of preservations (1). Viewed from environmental ethics no such direct measure is possible (2). These considerations apply to ecosystems as well as to individual taxa. We are often left trying to balance the "good" of ethics with the "goods" of economics.

Some conservationists have argued for the virtue of the preservation of almost all species (3). There are techno-optimists who downplay the species problems (4). Extremist advocates of artificial intelligence envision a silicon chip–based "life" to succeed carbon-based humans (5). Some traditional economists might argue that the amount we are collectively willing to expend to preserve a species is an appropriate utility measure. But traditional theory does not deal effectively with goods not exchanged in organized markets. Free air and water pollution are examples of this approach. One senses that there has been far too little dialogue between environmental biology and economics.

The National Academy Forum on Biodiversity (3) devotes 30 of its 500 pages to economic issues, and the newly formed Society for Ecological Economics has begun to approach value problems. But one senses that there is not a full engagement of either of the contributing disciplines. Economics students are not

---

* This chapter was originally published in *Science*, vol. 253 (1991), pp. 752–54. It is reprinted here with permission.

required to study biology, and the curriculum of ecologists does not usually include economics. As noted in a recent business publication, "Environmental economics has been relegated unfairly to the margins of the economic community" (6). Difficult concepts such as the irreversibility accompanying species extinctions, the preferences of future generations, the problem of present benefits and future costs, and the distinction between commodity value and moral value render the economic mensurability of species preservation extremely difficult. Some conservationists argue that such an approach is wrong from a philosophical perspective (2). We are often left contrasting what is economically beneficial to individuals versus what is beneficial to society as a whole.

A few ideas are basic to our considerations. Species have almost certainly existed for over 3.7 billion years. It is in the nature of these taxa to arise, flourish for a time, and perish. For many biological groupings, the majority of species are no longer extant. Even the notion of what constitutes a species is not universally agreed on.

At some time in the last several million years, along the evolutionary line of hominids, reflective thought arose, coming to fruition in *Homo sapiens*. The appearance of a species with this novel noetic property is a discontinuity, a biological event of catastrophic impact and, of necessity, it profoundly changed the development of local and global ecosystems. In our reductionist fervor we sometimes have been remiss in underestimating the effect of reflective thought on global ecology. Environmentalists have not focused on how strange and different organisms human beings really are.

The agrarian revolution during the past 10,000 or more years converted vast land areas from forests, savannahs, and plains to fields and pastures. The large accompanying species loss was an inevitable sequela of the emergence of human civilization as we know it. This epoch, which was devastating in terms of biological diversity, is nevertheless referred to as "the humanization of the earth" by Dubos (7). Human society in the present context is inextricably intertwined with agricultural monoculture, the worst of all worlds from a diversity perspective. As the Irish potato famine and the desertification of the Sahal show, agricultural practices may also have devastating social and economic consequences. Humans, because of their engineering abilities, create another geosphere, the technosphere, to add to the traditional lithosphere, hydrosphere, atmosphere, and biosphere.

The continuing growth of human population requires ever more land for housing and agriculture, resulting in progressively increased habitat destruction. The conclusion is clear: the amount of unexploited lands and waters will decrease with increasing human population. Although the exact function is uncertain, the sign of the first derivative is ensured. Similarly, the total number of planetary species will be some monotonic function of unused habitats. Ergo, biodiversity decreases with increasing human population.

No discussion of managing global habitats and preserving species can avoid the population imperative. Ignoring this problem, whether for political, ideological, or theological reasons, which is the present stance of many world political leaders, is a policy that inevitably leads to habitat loss and species depletion. There has been an unwillingness to face the conclusion that environmental planning and concern for conservation in the absence of population control are tasks worthy of the metaphor of Sisyphus.

Following agriculture, a second great change has occurred: the industrial revolution and burning of fossil fuels. In chemical terms, for several hundred million years there has been a dehydration and reduction of buried biogenic CH_2O

converting it to CH_4, CH_2, and C. The sudden reversal of this process to yield CO_2 is a major global consequence of industrialization. The acquisition and use of fossil fuels has resulted in habitat changes and loss of species.

From a direct economic perspective, the most severe changes in biota would be those that alter the global climate or geochemical distribution of essential materials or impact severely on cosmopolitan taxa. For example, loss of the genus *Rhizobium* (a most unlikely happening) would change the entire global nitrogen balance to a degree that would require industrial nitrogen fixation or the farming of reduced nitrogen, which are clearly energetically expensive operations. If the loss of the Amazon rain forest were to lead to widespread deleterious climatological effects or major global mineral nutrient imbalance, then this would similarly be of direct economic importance.

Let us examine some of the more difficult to evaluate public goods. Species preservation has been argued from the viewpoint of potential future sources of natural products. Here the costs might be compared with those of the chemical research and biotechnology required to synthesize and test new compounds and produce them industrially. This is a case where conservationists are making direct economic assertions whose strength is subject to analysis by economic methods. Even this is complicated by the unknown future value of such commodities as tropical plant-based medicinals, which are now a rapidly expanding growth industry (8).

Much more difficult to deal with as a public good is the value of knowledge. A lost species may potentially possess some novel biological features that will forever be inaccessible to study. For example, the sperm and blue whales and the elephants, as the largest marine and terrestrial mammals, certainly merit extreme efforts of preservation on grounds of physiological uniqueness. It is difficult to put knowledge arguments in economic terms, but such knowledge is certainly part of my utility function and ultimately may influence agriculture, medicine, and industrial processes.

On the grounds just discussed, I would also argue for the careful conservation of primates, prosimians, and their immediate predecessors. The transition along the primate line to hominids is an ill-understood, vastly profound, and clearly globally overwhelming phenomenon. The anatomy, physiology, and behavioral biology of the primates are clearly major avenues to understanding aspects of ourselves and our culture.

The search for physiological uniqueness centers on larger organisms, whereas the search for novel biochemical features extends over all sizes. In the center of the size range is a vast array of insect species dominated by the Coleoptera and Diptera. In any case, if an argument is to be made to undertake the study of economic costs of preserving an individual species, the uniqueness or lack of uniqueness of that taxon should certainly be a component of the reasoning. Biodiversity per se may be of great intellectual importance to biologists, but to a broader society the argument hinges on something of more public value than our desire as biologists to understand all life, as laudable as that desire might be. The politics of species preservation should perhaps be shifted to the politics of habitat preservation.

Another group of taxa have a special but difficult to assign value because of cultural reasons or because they inspire a sense of awe and wonder. This has been referred to as amenity value. The mammals of Africa, the redwood trees of the western United States, and the Antarctic penguins are in this category. It is hard to develop a metric of awe and wonderment, yet it clearly is a factor in human response. If a habitat draws paying tourists to observe the flora and fauna, then we have some measure of human response. An example of a culturally significant species is the American bald eagle, which has acquired a symbolic importance far beyond the purely biological.

In dealing with issues of knowledge or cultural association, we clearly deal with public goods or collective goods as distinguished from the private goods of market economics. Here the market analogy becomes less useful and the invisible hand has to be replaced by aspects of the social compact. Regulation may be introduced for other than economic reasons, but this requires some kind of consensus. Bossons (9) argues,

> Difficulties created by the public nature of ecological resources and their externalities suggest that the approach needed to rationalize protection of these resources must be reversed with respect to the approach used by competitive markets to satisfy consumers' demands.

Many environmental battles are being fought on these grounds.

There is a school of economic analysis (10, 11) that maintains that environmentally unsound practices are often economically unsound and involve governments fostering habitat destruction to protect politically influential industries. This leads to (11) "the use of limited natural resources at practically no cost." A number of examples are given (10) from the logging industry in the United States. The author maintains that in many cases the government is in fact subsidizing the clear-cutting of forests to produce a product that would be noncompetitive in the market without the subsidy. This is the reverse of the role a government should play in dealing with public goods.

What becomes clear is that it is not true that a species is a species is a species. The debate about preservation and management versus letting nature take its course must be argued for each taxon and habitat in some detail based on an understood and agreed upon way of assigning values. If preserving a species is to be used as a cover statement for preserving a habitat, it would be better to get the actual reasons up front so they can be debated on merits. Except in those very few cases where cost and benefit have calculable monetary values, conversion factors will have to be developed in terms of more abstract benefits. As has been pointed out by Baden (12), "not all values can be denominated on a spreadsheet."

It is necessary to stress that none of the trade-offs necessary to establish the relations between different value systems can be accomplished until biologists, economists, and technologists are willing and able to carry out discussions. A rational approach to problems demands this kind of communication. One would envision that the recently proposed National Institute for the Environment would be a locus for this activity, which at present lacks a home.

At the beginning of this century, humankind inherited a great diversity of biota. The industrial revolution inevitably compromised habitats and led to large-scale extinctions. We have reached a stage where there is general agreement that ecosystems, including the global ecosystem, must be managed (13). This requires, at the very least, more effort devoted toward an improved understanding of ecological theory. It also urgently requires some national and international consensus as to the goals of that management. Public goods are clearly the province of governments.

We would be remiss not to repeat the assertion that as human population goes up, biological species diversity goes down. We might be able to moderate the rate of decline, but we cannot fend off the inevitable. As species number goes down, we might, of course, change our valuation system and subsequent responses; they are, after all, cultural, not metaphysical. The answer to "How much is a species worth?" is "What kind of world do you want to live in?"

The author is Robinson Professor of Biology and Natural Philosophy at George Mason University, Fairfax, VA 22030–4444.

REFERENCES AND NOTES

1. A. Randall, in *Biodiversity*, E. O. Wilson, Ed. (National Academy Press, Washington, DC, 1988), chap. 25.
2. W. Ehrenfeld, in *ibid.*, chap. 24.
3. E. O. Wilson, Ed., *ibid.* (National Academy Press, Washington, DC, 1988).
4. J. L. Simon and A. Wildavsky, in *The Resourceful Earth, A Response to Global 2000*, J. L. Simon and H. Kahn, Eds. (Blackwell, Oxford, 1984), pp. 171–183.
5. H. Moravec, *Mind Children; The Future of Robot and Human Intelligence* (Harvard Univ. Press, Cambridge, 1990).
6. *Business Soc. Rev.* 71, 60 (fall 1989).
7. R. Dubos, *The Wooing of the Earth* (Scribners, New York, 1980).
8. R. Eisner, *Scientist* (10 June 1991), p. 5.
9. B. Bossons, *Int. J. Soc. Econ.* 17, 3 (1990).
10. R. E. Rice, *Ecologist* 20, 141 (July/August 1990).
11. C. Liepert and U. E. Simonis, *Int. J. Soc. Econ.* 15, 37 (1988).
12. J. A. Baden, *Financier* 13, 37 (1989).
13. D. Botkin, *Discordant Harmonies* (Oxford Univ. Press, Oxford, 1990).
14. I thank R. Rice, D. Wake, R. Morck, and W. Rosen for helpful discussions.

4

Genetic Resource Conservation in Natural Habitats: Biological and Socioeconomic Dimensions

J.T. Williams

Since all domesticated species of plants derive from wild species, the genetic resources used by breeders represent a range from the progenitor wild species through primitive cultivated forms to the modern cultivars and plants containing gene and other sophisticated manipulations. There are thus three components to the continuum of diversity—time, ecology, and population genetics—and these all have implications for the types of genetic resources to be conserved, as well as the methods of conservation. Additionally, it is necessary to be clear on which user communities will use the resources, since it is impossible to conserve more than a range of samples representing the total evolutionary continuum.

For decades, it was plant breeders who discussed the methods of conserving the spectrum of diversity of plant genepools. Most of the breeders were concerned with major staple food crops, some forages, and a limited number of miscellaneous tree fruit or industrial crops. As a result, when the International Board for Plant Genetic Resources (IBPGR) was created in 1974, action programs were inevitably geared to these concerns. In more recent times, a great deal of interest in conservation of plant genetic resources has emerged from major concerns on threatened or endangered species, cultural heritage of indigenous peoples, and major acceleration of loss of ecosystems and reservoirs of biological diversity, particularly in the tropics. In parallel, the concept of sustainable use of the environment has become a priority for implementation, with the need for positive linkages between development and conservation. Inequalities in expertise to use genetic diversity between developing and industrialized countries raised ethical and economic issues, and a great deal of rhetoric was generated on the central role of plant genetic resources. Unfortunately, all these debates clouded the definitions, the needs for, and the limitations of *in situ* conservation of plant genetic resources. This chapter attempts to clarify some of the muddled thinking, point out what is being done, and outline the scientific and other constraints that need to be addressed.

THE BASIC RESOURCE

For species of actual socioeconomic value, such as crops, and for many species of potential value, such as wild medicinal plants, underutilized crops, and many forages and forest trees, the primary target for conservation is a range of genetic variation within the species. Within-species (infraspecific) variation is often not the only target, because many species interact through geneflow with closely related species. Depending on the amount of geneflow, targets may thus be a group of related species, the group being limited in some cases or numerous in others.

In the case of annual crops, domestication led to a wide range of adaptations in cultivation. Ecologically, plants became modified so that they could no longer exist in natural habitats, and genetically, adaptations accumulated according to the selection pressures of the cultivation methods and to some degree to the needs of the cultivators. Races developed that are recognizable morphologically and represent particular populations that are genetically diverse but balanced in the sense that there is an equilibrium in time and space with the maintenance of heterozygosity. These populations are called landraces: populations that arose in response to primitive agriculture with low inputs. The landraces were the populations out of which early plant breeders first selected cultivars and capitalized on the genetic potentials within the populations to produce higher yields. Because of deliberate plant breeding, landraces become obsolete due to newer advanced cultivars inevitably replacing the primitive landraces of traditional agriculture.

Since landraces and all derived forms, whether advanced cultivars or genetic stocks, cannot grow in wild habitats, their maintenance as sources of genetic materials for plant improvement and their conservation rely constantly on human intervention. The conservation techniques used are known as *ex situ* methods, since seed stocks are usually conserved in research institutions.

Empirically genetic resources, conserved *ex situ*, are maintained as populations so that individual genotypes can be selected or used as sources of specific genes or adaptations for transfer to other genotypes or cultivars in purposive breeding.

In building up *ex situ* germplasm collections, ideally the indigenous variation of specific geographical regions is included. Originally, when crops were domesticated in certain regions, variation was spread unevenly with concentration in the area of origin and in a limited number of secondary centers. Rarely do the large *ex situ* collections of major crop landraces represent the past traditional patterns of diversity. This is because the products of modern plant breeding had already made major inroads into the centers of diversity before most modern genetic resource conservation programs were established.

Additionally, in a number of more minor crops, traditional diversity often represents local selections or genetic forms more akin to botanical varieties than to landraces in the sense of major crops. This is because the species' domestication often related, not to settlement patterns, but to nomadic ways of life and shifting cultivation. In the case of many minor tree fruits of the tropics, there is evidence of forms, hardly removed genetically from the wild state—and probably owing their cultivation to an inherent weedy or colonizing ability—reverting over time to the wild and being re-selected or "protected" by man. Primitive forms of such species have been less researched than primitive forms of more major crops, and the populations needed for genetic conservation differ in type and number.

Species of potential enhanced use, which are relatively unselected and which have never evolved under cultivation or protection, are wild species; but the resource base for conservation contrasts with those wild species that are currently used by man. The latter may represent wild relatives of domesticated plants of interest to plant breeders: wild species gathered by man such as bamboo and many nonwood forest products for food fiber, raw materials for industrial processes, wild forage species exploited through grazing, and others.

There is a tendency to lump all wild species together in terms of conservation requirements but with little consideration of the resource base, and this ignores the scientific literature (vide 9, 10). This tends to result from the expansion of interests from agriculture to nature conservation without adequate critical use of the extensive scientific literature, and from the highlighting of the role of primitive farmers in the evolutionary continuum. As a result, *in situ* conservation methods applicable to wild species have been extended beyond the wild habitats to the artificial areas of primitive agriculture. As succeeding sections of this chapter will point out, it is extremely timely to consider the resource base needed and the strategy for conservation based on scientific knowledge for specific plant groups, rather than the continual development of frameworks and priorities according to vested interests of organizations.

To the present, the greatest portion of international funding is expended on ecosystem conservation, with only a small percentage (well under 5%) being targeted to sustaining and enhancing *ex situ* conservation. This, despite the latter having laid the scientific groundwork that can be applied with some modifications to *in situ* work and to a system that will sustain the future food security of the world (4). Little relevant scientific work has gone into the broader actions on nature conservation, and hence it is surprising that there has appeared to be a conflict between the proponents of *in situ* and the proponents of *ex situ* conservation when, *for particular purposes,* they are complementary (32). Hawkes (15) summarized this succinctly: "It is quite clear that much more thought and research initiatives need to be applied to the problem of *in situ* conservation." Furthermore, a statement made in 1984 still holds: "New developments in *in situ* conservation must not detract from the overwhelming need for more effective *ex situ* conservation programmes for crop species" (16).

TYPES OF *IN SITU* CONSERVATION

Historical Background

The crops of most immediate interest for *in situ* conservation are the perennials that are vegetatively propagated (13) and those with seed that cannot survive cold storage, i.e., those with recalcitrant seeds (19). Even in these cases *in situ* conservation will be an adjunct to rapidly developing *in vitro* conservation and utilization systems. For forest trees, similarly, *in situ* conservation provides the major opportunity to conserve the diversity, and *ex situ* seed collections or managed plantings are important when access to specific, well-studied, or threatened populations is desired (24) and for those where there is active breeding. However, active breeding is done far less on forest trees than agricultural crops, so the emphasis for conservation is more on *in situ* methods until much more is known about patterns of genetic variation and the potential of many species. In most cases, forest trees are wild species that have not been domesticated or are widely planted but genetically in a virtual wild state. Genetic erosion in forest species is therefore due to shrinking of the natural resources rather than replacement of primitively cultivated forms, as is the case with crop plants. Whereas for crop genepools landraces constitute a major genetic resource, the parallel for forest trees is the resource of the natural forests.

The wild species related to most forages may parallel, in terms of genetic resources, the resource base of forest trees. Or, in the case of many crop relatives, it may occur across a range of habitats in natural, semi-natural, and weedy environments.

With regard to types of *in situ* conservation, this has generally been assumed to fall within the framework of designated reserve areas: those categorized by the International Union for the Conservation of Nature and Natural Resources (IUCN) as scientific reserves and strict nature reserves, national and provincial parks, protected landscapes, resource reserves, multiple-use management areas, biosphere reserves, and wild heritage sites. It is assumed that when enough of these types of reserves are available, a framework of protected areas and habitats will be in place to conserve *in situ* a pool of the world's biological diversity. The UN Educational, Scientific, and Cultural Organization (UNESCO) represents an international focal point for the establishment of preserved natural areas, and its Man and Biosphere Program encompasses representation of all biotic provinces of the world.

However, this approach has suffered greatly over the past 25 years because of a lack of adequate inventories, design of the reserves for representativeness of particular plant communities within the ecosystem, and not enough attention to the notional time scale related to anticipated economic purposes (8, 10).

Additionally, over the past 15 years concepts for *in situ* conservation of useful plants have been evolving, although in an uncoordinated way. Hawkes (15) summarized the formulation of *in situ* conservation as being in the early stages and being particularly worrisome, especially in relation to size of areas and confidence that protected areas indeed include adequate genetic diversity. In fairness, the nature conservation community has moved in the direction of species conservation as well as ecosystem conservation, but the parameters relevant to economic plant genepools must depend on an understanding of the patterns of distribution and diversity of each target species, including breeding systems and viable population size.

While ecosystem conservation concepts have been developing, thinking has also evolved toward *in situ* conservation planning involving a diverse range of activities, including the role of people. Finally, consensus has emerged on two courses of action. The first is to locate reserve areas in places with highest species diversity—most important for biodiversity conservation, and of some interest where several of the species are related to crops. The second is to consider reserve areas in places with lower species diversity, but with substantial genotypic diversity (or simply with a useful species or population as a special case).

There are also two keys to successful practical action. First, there must be the continuous curation and monitoring of genetic reserves in relation to the maintenance and use of the germplasm they contain. This need has been stressed by this author for the past few years, although its implementation has been hampered by the lack of adequate training. Second, the axioms derived from biogeography (16) must be built into the planning. These are summarized as follows:

1. None of the major food crops and their progenitors are associated with climax vegetation. However, such a link is found in some other crops, such as rubber, tropical fruits, and some roots and tubers. A genepool for conservation may include some species dependent on very narrow ranges of disturbance, e.g., in ecozones, and some on stable environments.

2. The natural distribution of a particular target species can vary from extreme localization to dispersion across continents, and it may be distributed in a disjunct fashion because of topographic features.

3. Geographic range alone as the primary basis for delineation of diversity is misleading since allelic variability can vary within and between populations.

4. Densities of wild species can vary from dense stands to widely separated individuals.

5. Current concepts on areas of diversity of wild relatives have been derived from studies on the origin, diversity, and spread of domesticates. There are widespread misconceptions that *in situ* conservation in a few areas of the Vavilovian centers (14) will adequately conserve enough wild genotypes for future crop enhancement when wild relatives in other areas will also be valuable.

Types of Reserves Proposed

A synthesis of developments over the past decade leads to diverse types of reserves for genetic conservation of useful diversity (Table 1). To a degree these represent diverse opinions and aims.

Table 1. Types of reserve areas.

Natural Communities	Level of Management Intervention	Wild Species	Domesticates	Security of Target of Conservation	Remarks
1. Biosphere reserves, national parks, strict nature reserves	low	yes	no	low	Many existing. Inventory and analysis of useful diversity needed.
1a. Gene sanctuary	medium	yes	no	low	Special reserve areas designated for high species diversity, e.g. citrus in NE India.
2. Extractive reserves	medium	yes	no	high	Few; specific crop/product harvested by local people.
Modified Natural Communities					
3. Enriched genetic/extractive reserves (planned evolution garden reserves/managed nature reserves)	high	yes	yes	high	Proposed to ensure sustainability of (2) above
Multiple-use management areas					
4. Buffer zones (manipulation zones)	high	yes	yes	high (in reserve areas)	Managing diversity around protected areas to relieve pressure on the latter.
5. Sites of special scientific interest (Hawkes, 1991)	high	yes	yes	varies	Areas where much research and analysis merit special consideration.
6. Landraces in their areas of cultivation/genetic reserves (Dinoor, 1976)	high	no	yes	low	A number of areas established for landraces.

In numbers 2 and 3 in Table 1, preservation of indigenous human cultures is the major goal, and in numbers 4 and 6 recognition of the active role of peasants in genetic resource conservation is the major aim, with clear emphasis placed on folk science and traditional uses of the cultures that have nursed them (22).

Over the past decade, there has been a huge amount of interest in genetic conservation with the scientific endeavors added to by proponents of the rights of indigenous people, the rights of farmers who were associated over time with crop selection, and of the needs for sustainability. Many debates occurred, but one specific one has led to a degree of confusion. On the one hand, clear international priorities for scientific work on the more major crop and forage species (17) lead to a set of practical actions and programs that rely heavily on *ex situ* conservation. On the other, conservation and continued use of indigenous crops, usually unimproved genetically, in self-sustained agroecosystems as part of development (see reference 1) rely on conservation *in situ*, albeit in artificial ecosystems. There are many problems associated with organizations advocating native crops in relation to whether there is indeed justification. In the case of quinoa in the Andes, its justification has been analyzed by Tripp (28). Similar analyses are needed so that the underutilized crops that fully justify further improvement and wider production can be identified and action taken. This approach is being taken by the International Centre for Underutilised Crops (12) and will amplify international priorities for genetic conservation and focus on a limited number of crop genepools rather than trying to cover all useful indigenous crops. Amplification of priorities in this way, as a response to needs for crop diversification or sustainability, will help the more political aspects of some of the debates (11) where conservation of landraces by farmers (34) and conservation of minor unimproved crops are often confused. In such instances, various ownership and intellectual property rights systems have been hotly debated (27).

Despite such issues, the amount of scientific work on the population structure of wild species is so small that *in situ* conservation is unlikely to be greatly enhanced in the best scientific way in the immediate future, nor as rapidly as justified. A summary of the current disadvantages of *in situ* conservation is provided by Hawkes (15) as follows:

1. Right reserve size unknown.
2. Supervision and monitoring not yet clarified.
3. Vulnerable to outside influences.
4. Materials not available to plant breeders.
5. Only a limited amount of diversity can be stored in established reserves.
6. No generally accepted methods of storing landraces.

INTERNATIONAL REQUIREMENTS

Leaving aside the value of local *in situ* conservation schemes (such as examples 2, 4, and 6 in Table 1), which may or may not be secure in terms of responsibility accepted by local authorities, and funding and future wishes of local people, there still remains the need for some better planning and implementation for species of international conservation interest. To illustrate this, the known *in situ* conservation of wild relatives of rice in Asian countries can be used (Table 2). (Of course more reserves, not yet inventoried, will contain these species.) A number of general points emerge as follows:

1. Nations, on their own, are unlikely to designate areas to encompass the widest range of variation of the species since the reserves are for multipurpose conservation.

2. A responsible regional or international organization needs to work actively with the nations to see that, in total, the species are conserved *in situ* and cover the ecogeographical ranges and that the reserve areas are subject to appropriate management.

3. Analyses of the patterns of genetic variation across the ranges are needed.

4. Vaughan (29) points out that in some cases species are below minimum population size, so mention of a species on a list of sites or in an inventory does not mean effective genetic conservation.

This example illustrates the strong need for international interests and expertise to be involved in planning and monitoring reserves in partnership with local organizations. The international interests must include technical expertise on specific species groups, often unavailable at the level of the local communities, nongovernmental organizations, and national institutions. Hence, international frameworks linking organizations require the backing of urgent scientific activities (11). In planning, a large degree of genetic redundancy within and among conservation sites is needed, at least until more is known about safe levels of loss and how they are balanced (26). In a sense, this would parallel past action on *ex situ* conservation where as much as possible was salvaged for conservation, and then collections were sorted out to reduce unnecessary duplication.

Significantly, despite all the organizations involved, there exists no global agenda for species inclusiveness, no guidelines for clear scientific management or curation, and no guidelines for local people use. It is not, therefore, funding that is

Table 2. *In situ* conservation of rice relatives in Asia [modified from Vaughan (29)].

Species	Country	Reserve Area
Oryza rufipogon	India	Bhoothathankettu Forest Reserve
O. officinalis	India	Parambikulam Game Reserve Karulai Range Teak Forest Reserve
	Indonesia	Ujung Kulon National Park
O. nivara	India	Karulai Range Teak Forest Reserve
	Sri Lanka	Yala National Park Ruhuna Wildlife Sanctuary Polonnaruwa ruins
O. granulata	India	Karulai Range Teak Forest Reserve
	Thailand	Pukai/Mae Sai Valley
O. eichingeri	Sri Lanka	Ritigala Strict Nature Reserve Wilapattu National Park Yala National Park Ruhuna Wildlife Sanctuary
O. ridleyi	Thailand	Forest Reserve near Chiang Mai Tong Nya Chank Park

the key issue now, but the need for science, so that available funding can be well targeted and ultimately cost effective. The need for science-based priorities is also true for the removal of constraints that currently hinder programs in *ex situ* collections (31). New skills and scientific expertise are needed, not just funding.

SPECIFIC SCIENTIFIC REQUIREMENTS

Scientific inputs are needed on a broad front. They include:

- Enhanced inventory of protected areas and mapping of ecological heterogeneity; Enhanced ecogeographic survey of target species for use in reserve design and designation (including data on densities);
- Enhanced analyses of patterns of variation within species and closely related species;
- Targeted knowledge on key interactions in complex communities necessary to support management decisions and avoid major changes in population structure and species composition of reserves designated;
- Better linkages between databases on *ex situ* collections, economic plant databases, threatened plant databases, and other conservation databases to allow relational querying.

Local intensive scientific work, e.g., in Israel at Ammiad on wild tetraploid wheat, *Triticum dicoccoides* (25), and wild barley, *Hordeum spontaneum*; in the UK, on wild white clover; on buffer zones in Ecuador and other areas for a range of plant materials; has provided enough evidence for pragmatic requirements for establishing *in situ* reserves to maintain diversity of target plant species. Frankel (1988) discusses the genetic conservation of four categories of wild species: those of great significance for the stability of an ecosystem; those that have not been domesticated yet but are extensively planted and used though genetically virtually wild; those that may be found useful in the future; and wild relatives of domesticated crops. There are genetic principles for *in situ* conservation appropriate to each with associated implications on reserve structure and design. Furthermore, if *in situ* conservation of these target species is to be effective, overall ecosystem stability is essential. Organizations and international bodies concerned with ecosystem conservation must build into their priorities and actions the need for long-term adaptation for ecosystem stabilizers and for species that might be useful in the future and ensure short-term fitness for all four categories, largely by extrapolation of concepts of minimum population size to avoid adverse changes in breeding systems such as the deleterious effects of inbreeding in small populations.

The maintenance, in conserved areas, of variance in target species requires, just as does consideration of fitness, consideration of numbers in populations. Despite the major applications of population genetics to *ex situ* conservation practices, the role of population genetics in rare plant conservation or ecosystem stabilizers and species of unknown but potential use in the future has rarely been appreciated (7). Most management practices have been directed toward habitat preservation and have focused on ecological rather than genetic considerations. Conservation of genetic diversity, *in situ*, is a challenging task, particularly in tropical ecosystems.

One area of science that justifies expansion and careful planning is that related to understanding patterns of diversity. Although there are several ways of measuring genetic variation, biochemical and molecular techniques allow rapid indications of differences between populations and degree of variation within populations. Whereas much of the research involves collecting materials and subsequent analysis of seedlings or other progeny in laboratories, the methods have not been applied adequately to assessing variation across wide areas in order to enable targeted collecting for *ex situ* conservation, nor have they been applied widely to reaching decisions on areas for *in situ* conservation. As expertise develops, it is not beyond the bounds of development assistance to strengthen a limited number of institutions in the regions to perform rapid screening. New *in vitro* collecting methods (33) or long-distance transport of cuttings in the tropics (2) can be combined with such screening. Proposals to the Program Committee of IBPGR to develop Restriction Fragment Length Polymorphisms (RFLP) technology in combination with field surveys did not lead to the IBPGR's research program developing new techniques and strategies to address targeted collecting or to identifying hot spots of diversity. However, sufficient work has been done elsewhere to justify the initiation of programs for regional screening centers, and such work could greatly accelerate designation of reserve areas to cover targeted plant genetic conservation *in situ*.

In order for new locally based screening to be effective, it will require scientific advisory panels to advise on interpretation of rapid screening. For instance, local, naturally occurring populations may not be optimally reproductively fit nor ideal for selection for enhancement (6, 23). Natural occurrence is not the sole criterion for conservation *in situ* or for use in germplasm enhancement (24).

SOCIOECONOMIC DIMENSIONS

Recognizing that current levels of funding for genetic conservation of crops, forages, and trees are inadequate, large new influxes of funding proposed by international bodies must be matched by addressing a number of constraints:

1. The capacity of national institutions to manage reserve areas is limited. When reserves are to be established for special purposes other than ecosystem conservation, specialist training is required. The number of scientists available to address very specific needs of *in situ* target plant genetic resources conservation and curation is low. New international training courses are urgently needed to bridge the current gap between specialist genetic resources and wider conservation biology. In particular, the application of the teaching to basic, strategic, and applied research needs attention during the training so that scientists can put these into practise for maintenance of ecosystems and for development issues. For instance, wide-ranging interdisciplinary training and research would enable a scientist to identify and design *in situ* reserve areas while appreciating both traditional and advanced technologies for resource management and to develop a strategy for sustainable new systems that integrate nature conservation and agriculture (26).

2. Reserve areas are under the sovereign jurisdiction of nations. Sustained political support must be built. There are far too many

expectations of secure reserves that in practice see a great deal of attrition, as in the case of perennial teosinte (maize relatives) in Central America (30). Individual governments' adherence to international conventions, undertakings, or commissions, or short-term project funding, do little in practice to rectify such problems. Concerted effort by organizations within the countries in partnership with international organizations will help greatly in this area.

3. Current international interests appear compartmentalized, and concerted action will only be possible when the groundswell of scientific rather than political opinion urges program reorientations. For this to be effective, socioeconomic aspects must be included. Few scientific programs adequately address the latter, and an illustration of where consideration of such aspects helps in the research strategies can be seen in the case of an exercise to plan a 10-year research strategy (including conservation) for two grossly overexploited tropical commodities: bamboo and rattan (18).

4. Socioeconomic aspects are extremely important in recent thinking concerning sustainable development, including the rights of indigenous peoples and cultural survival (3). In genetic resources science there is a lot to be gained from collating and assessing indigenous knowledge. Nonetheless, at the local level it is important to separate conservation and utilization of plants of international priority from the availability of minor species (and their conservation) to sustain rural development, including local plants of low international priority. Grassroots experience has often blurred these differences, even though both topics require due consideration in a socioeconomic context. The well being of the rural poor is a major criterion for development assistance, not the continuation of traditional ways for their own sake, and certain agroecosystems of marginal lands should not be replaced hastily. There are cases of promoting minor crops in this respect that have not been justified. Action in this whole area needs vision so that, for instance, specific priorities and funding for underutilized crops stand out clearly. To the present, priorities for a range of scientific activities have not been clarified in the socioeconomic context of rural development, especially when there are competing interests: feeding the population, increasing rural income, and creating exportable agricultural production.

5. The value of biodiversity has received a lot of attention recently. Wild species and their genetic resources contribute to agriculture, forestry, industry, and medicine (21), but biological resources make a much greater contribution to local economics than they do to national and international economies. Economic incentives and disincentives can play an important role in ensuring community benefits (20). There is a range of mechanisms currently being explored: debt-for-development swaps,

international funds, developing markets for local products, and others. There are major opportunities in genetic resources work for international organizations to work with nongovernmental organizations.

CONCLUSION

Carefully planned, monitored *in situ* conservation of target plant genetic resources is slow to emerge, even within the wider framework of ecosystem conservation.

A number of scientific constraints have been identified that need urgent attention to translate international rhetoric into practical action. There is also the need for appropriate training. Nevertheless, in many countries the scope for improving *in situ* conservation with existing technologies is still large. The experiences of the past two decades probably suggest a new system for the delivery of effective programs, and the design of such a system requires not only policy dialogue among international development agencies and countries but a clearer focus for scientific R&D so essential to any scheme. At the same time, the unwarranted gap between *ex situ* and *in situ* conservation, which has grown in recent years, must be bridged.

REFERENCES

1. Altieri, M.A., and L.C. Merrick. 1987. *In situ* conservation of crop genetic resources through maintenance of traditional farming systems. *Econ. Bot.* 41:86–96.
2. Aronson, J., G. Wickens, and E. Birnbaum. 1990. An experimental technique for long-distance transport of evergreen or deciduous cuttings under tropical conditions. IBPGR/FAO *Plant Genetic Resources Newsletter* 81/82:47–48.
3. Clay, J. 1991. The knowledge of nations. *Ceres* 127:29–33.
4. Cohen, J.I., J.T. Williams, D.L. Plucknett, and H. Shands. 1991. *Ex situ* conservation of plant genetic resources: global development and environmental concerns. *Science* 253:866–72.
5. Dinoor, A. 1976. Germplasm and the phytopathologist. FAO *Plant Genetic Resources* 32:36–38.
6. Eriksson, G., I. Ekhert, and A. Jonsson. 1972. Meiotic and pollen investigations as a guide for localization of forest tree seed orchards in Sweden. In *Proc. Forest Tree Breeding Symposium*, 1–28. Government Forestry Exp. Sta., Tokyo.
7. Falk, D.A., and K.E. Holsinger, eds. 1991. *Genetics and Conservation of Rare Plants*. New York: Oxford University Press.
8. Frankel, O.H. 1983. Genetic principles of *in situ* conservation of plant resources. In *Conservation of tropical plant resources*. S.K. Jain and K.L. Mehra, eds., 55–65. Bot. Survey India, Howrah, India.
9. Frankel, O.H., and E. Bennett, eds. 1970. *Genetic Resources in Plants—Their Exploration and Conservation*. Oxford: Blackwell.
10. Frankel, O.H., and M.E. Soulé. 1981. *Conservation and Evolution*. Cambridge: Cambridge University Press.
11. Food and Agriculture Organization (FAO). 1991. Report of the Fourth Session of the Commission on Plant Genetic Resources. Rome.
12. Haq, N. 1991. Profile: the international centre for underutilised crops. *Diversity* 7(4):16–17.
13. Hawkes, J.G. 1975. Vegetatively propagated crops. In *Crop Genetic Resources for Today and Tomorrow*, O.H. Frankel and J.G. Hawkes, eds., 117–21. Cambridge: Cambridge University Press.

14. Hawkes, J.G. 1983. *The Diversity of Crop Plants.* Cambridge, MA: Harvard University Press.

15. Hawkes, J.G. 1991. Summary of final discussions and recommendations. International Workshop on Dynamic *In Situ* Conservation of Wild Relatives of Major Cultivated Plants. *Israel J. Bot.*, forthcoming.

16. Ingram, G.B., and J.T. Williams. 1984. *In situ* conservation of wild relatives of crops. In *Crop Genetic Resources: Conservation and Evaluation*, J.H.W. Holden and J.T. Williams, eds., 163–79. London: Allen and Unwin.

17. International Board for Plant Genetic Resources (IBPGR). 1985. Ecogeographic surveying and *in situ* conservation of crop relatives. Rome.

18. International Fund for Agricultural Research (IFAR). 1991. Tropical Tree Crops Program. Research needs for bamboo and rattan to the year 2000. Arlington, VA.

19. King, M.W., and E.H. Roberts. 1979. The storage of recalcitrant seeds—Achievements and possible approaches. Rome: IBPGR.

20. McNeely, J.A. 1988. Economics and biological diversity: Developing and using economic incentives to conserve biological diversity. Gland, Switzerland: IUCN.

21. McNeely, J.A., K.R. Miller, W.V. Reid, R.A. Mittermeier, and T.B. Werner. 1990. *Conserving the world's biological diversity*. Gland, Switzerland, and Washington, DC: World Bank, World Resources Institute, IUCN, CI, and WWF.

22. Nabhan, G.P. 1979. Cultivation and culture. *Ecologist* 9:259–63.

23. Namkoong, G. 1969. The non-optimality of local races. In Proc. Tenth S. Conf. Forest Tree Breeding. 149–53. Austin, TX: University of Texas Press.

24. National Research Council (NRC). 1991. Managing global genetic resources: Forest trees. Washington, DC: National Academy Press.

25. Noy-Meir, I., Y. Anikster, M. Waldman, and A. Ashiri. 1988. Population dynamics research for *in situ* conservation: Wild wheat in Israel. IBPGR/FAO *Plant Genetic Resources Newsletter* 75/76:9–11.

26. Okigbo, B. 1990. Sustainable agricultural systems in Africa. In *Sustainable Agricultural Systems*, C.A. Edwards, R. Lal, P. Madden, R. Miller, and C. House, eds., 323–52. Ankeny: Soil and Water Conservation Society.

27. Sedjo, R.A. 1992. Preserving biodiversity as a resource. *Resources* 106:26–29.

28. Tripp, R. 1990. Does nutrition have a place in agricultural research? *Food Policy* (Dec.):467–74.

29. Vaughan, D.A. 1989. Collection, conservation, and potential use of the wild relatives of rice in Asia and Australia. In *Review of Advances in Plant Biotechnology, 1985–88,* A. Mujeeb-Kazi and L.A. Sitch, eds., 179–90. Mexico, DF, and Manila, Philippines: CIMMYT and IRRI.

30. Wilkes, G. 1991. Personal communication.

31. Williams, J.T. 1989. Practical considerations relevant to effective evaluation. In *The Use of Plant Genetic Resources*, A.H.D. Brown, O.H. Frankel, D.R. Marshall, and J.T. Williams, eds., 235–44. Cambridge: Cambridge University Press.

32. Williams, J.T. 1991. The time has come to clarify and implement strategies for plant conservation. *Diversity* 7(4):37–39.

33. Withers, L.A. 1986. *In vitro* methods for germplasm collecting in the field. IBPGR/FAO *Plant Genetic Resources Newsletter* 69:2–6.

34. Worede, M. 1991. Crop genetic resource conservation and utilization: An Ethiopian perspective. In *Science in Africa: Achievements and Prospects*, 103–23. Washington, DC: American Association for the Advancement of Science.

PART 2

Case Studies of Biodiversity Conservation in Natural Habitats

5

Conserving Crop Genetic Resources in Latin America Through Farmers' Participation

Miguel A. Altieri and Camila Montecinos

Latin America is an ecologically diverse region. It is a major repository of biodiversity and genetic resources that provides ecological services to the biosphere and for the advancement of agriculture, forestry, industry, and medicine. It is also a culturally heterogeneous continent; more than 100 ethnic groups still manage local ecosystems with indigenous technologies. Some of these management systems are now considered sustainable and proper to guide modern resource management.

The 1980s were called "the lost decade" for the region. Latin American economies went through a major economic crisis with extraordinary social and environmental costs. Despite numerous internationally and state-sponsored development projects, poverty, food scarcity, malnutrition, health deterioration, and environmental degradation continue to be widespread. As Latin American countries are pulled into the existing international order and change policies in order to serve the unprecedented debt, governments increasingly embrace economic models that promote export-led growth. In some countries such as Chile, Mexico, and Brazil, positive macroeconomic indicators highlight the "success" of the model. On the other hand, deforestation, soil erosion, industrial pollution, pesticide contamination, and loss of biodiversity (including genetic erosion) proceed at alarming rates and are not reflected in the economic indicators. So far, there is no clear system to account for the environmental costs of such models.

The crisis has demonstrated that conventional development strategies are fundamentally limited in their ability to promote equitable and sustainable development. Such development failures legitimized the role of grassroots nongovernmental organizations (NGOs) as new actors in local development in the region. Most NGOs question "conventional wisdom" in technological design and diffusion by promoting new kinds of development and resource management based on local participation, skills, and resources.

Grassroots efforts to conserve local genetic resources are a case in point. Several NGOs and farmer organizations are engaged in numerous conservation, selection, storage, and distribution projects that, despite funding and technical difficulties, have relevance to small farmers' needs across rural communities in the region.

NGOs' efforts in genetic and environmental conservation have met with mixed success. In addition to the lack of proper technical capabilities and clear methodological guidelines, a key reason is that conservation at the grassroots level is difficult to implement where institutional arrangements, market forces, policies, and research efforts are biased against such types of work. A major challenge will be to enhance the technical capabilities of farmers and NGO staffers on *in situ* conservation along agroecological lines, and also to create a new policy framework that enhances local development and conservation efforts.

In this chapter, we provide case studies of grassroots efforts to conserve crop genetic resources in Latin America. We attempt to document genetic erosion and loss of species and to provide an overview analysis of the primary threats to traditional varieties in the region. We also highlight the values of those traditional varieties.

Based on several field visits and interviews with NGO technicians, we provide an assessment of the status and trends in conservation efforts, delineating the trends in community participation in seed banking and local breeding efforts. We make suggestions on what is needed to strengthen efforts to conserve genetic resources and meet human needs. Besides stressing the need for, among other things, more surveys, collecting work, and ethnobotanical work, we make a plea for more funds and policy changes, as well as new institutional arrangements and collaboration.

THREATS TO TRADITIONAL VARIETIES

The major threat to traditional varieties in Latin America is the process of agricultural modernization. Table 1 presents a list of endangered and/or threatened crop genetic resources in Latin America.

In 20 years, a handful of scientifically bred, modern varieties (HYVs) have replaced ancient crop varieties or "landraces" as part of international and national efforts to develop agriculture (1). This process of loss of genetic resources is called "genetic erosion" and is the result of a complex chain of events depicted in Figure 1.

As peasants directly link to the market economy, economic forces increasingly influence the mode of production characterized by genetically uniform crops and a mechanized and/or agrochemical package. Landraces and wild relatives are progressively abandoned, becoming relicts or extinct (2). In some areas, land scarcity (resulting mostly from uneven land distribution) has forced changes in land use and agricultural practices resulting in the disappearance of habitats that formerly maintained wild progenitors and weedy forms of crops (3).

Although genetic erosion is a major concern in Latin America, very few systematic studies have been conducted on the causes, rates, and implications of this genetic change. Most studies that relate to traditional varieties have concentrated on the rate pattern of adopting of HYVs, on whether HYVs could or would be adopted by small farmers, and on the possible negative impacts of such adoption (4, 5).

Preliminary assessments of genetic erosion in Mexico and Peru, however, indicate several trends that may be generalized to most crops in the region (1, 6):

1. Genetic erosion is occurring because farmers are changing their farming systems, pushed by social, economic, and technical forces.

Table 1. Endangered or threatened crop genetic resources in Latin America needing priority collections.

Crop	Regions Where Priority Collection is Needed*
Legume crops:	
Phaseolus beans	Mexico, Andes, Chile, Ecuador
Peanut/groundnut	Meso-America
Cowpeas (Asiatic *Vigna*)	Brazil
Pigeon pea (*Cajanus*)	Meso-America
Root and Tuber Crops:	
Potato	Mexico, Andes, Central America
Sweet Potato	Meso-America, Andes, Brazil, Southern South America
Cassava	Brazil, Meso-America
Yam	Brazil, Meso-America
Vegetable Crops:	
Tomato	Andes
Amaranthus	Meso-America, Andes
Peppers	Mexico, Peru
Cucurbits	Meso and South America
Edible Oil Crops:	
American Oil Palm	Brazil, Meso-America
Sunflower	Meso-America
Beverage Crops:	
Grape	Mexico
Cocoa	Meso-America, Brazil
Forage Grasses:	
Paspalum	South America
Carpet grass (*Axonopus*)	South and Meso-America
Tripsacum	South and Meso-America
Bromegrass (*Bromus*)	Temperate South America
Bluegrass (*Poa*)	Temperate South America
Forage Legumes:	
Lucernes (*Stylosanthes*)	Tropical South America, Meso-American
Desmodium	Meso-America, Tropical South America
Desmanthus	Meso-America
Centrosema	Meso-America, Tropical South America
Leucaena	Meso-America, Tropical South America

* Regions listed in each column in descending order of importance for collection.

2. The pattern and rate of adoption of HYVs is very uneven across geographic regions.

3. As adoption of HYVs occurs, farmers tend to subdivide their farming systems into commercial (mostly devoted to HYVs) and subsistence sectors where they still grow native varieties.

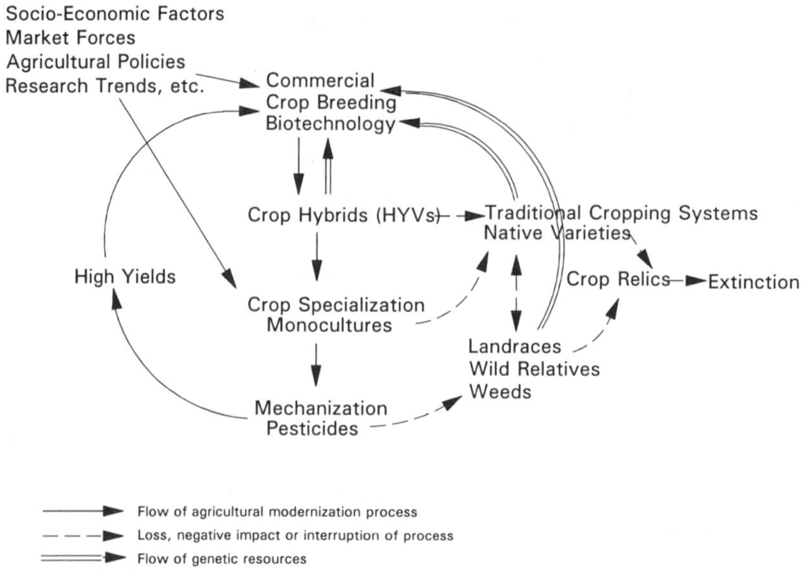

Figure 1. A simplified model of the process of agricultural modernization and crop genetic erosion [modified after Querol (2)].

4. Greatest loss of traditional varieties has occurred in lowland valleys close to urban centers and markets.
5. The least genetic erosion occurs in higher zones (especially mountain areas), more distant from urban centers and markets.

Although the data are still inadequate to predict future trends of crop genetic resources conservation and loss in Latin America, it is possible to identify several factors that both promote and constrain genetic change (1). In general, loss of crop genetic resources is promoted by the following factors:

- Changes in numbers and distribution of human populations.
- National and international crop improvement and extension programs.
- Changes in consumption patterns in urban and rural populations.
- Increases in demand from urban markets.
- Improvements in transportation and communication systems.
- General factors in the agricultural economy such as human capital, labor availability, credit, tenure, and the availability of technological components such as fertilizers and irrigation.

On the other hand, replacement by HYVs is slowed by factors associated with risk aversion, cost of new seeds, and marginality of certain areas because of high risk, land fragmentation, and isolation from markets. Culinary preferences and increased levels of impoverishment also promote conservation. For example, in the higher altitudinal zones of the Andes, resource-poor farmers maintain a diet predominantly reliant on native potatoes. These varieties have better culinary characteristics than improved varieties, and they can also be kept viable indefinitely. Improved potato varieties degenerate quickly and ultimately fail as seed (7).

LOSS OF INTRASPECIFIC DIVERSITY

Case Study 1

In studies conducted in three agroclimatic zones (high-, mid-, and low-altitude zones) of the Mantaro Valley in Peru, Franco *et al.* (8) found that 87% of all farms on the lower zones were planted with improved potato varieties. The adoption of HYVs dropped sharply as altitude increased. In the highest zone, only 2% of the farms were planted in HYVs. These authors concluded that improved varieties were grown for commercial purposes in both low- and mid-altitude zones. However, even in these lower zones native varieties were often retained and managed under traditional tillage practices. The total area devoted to these varieties was smaller and they were grown under greater environmental stress than previously (1).

Case Study 2

In Chiapas, Mexico, Brush *et al.* (6) found a similar agroecological pattern with traditional and improved maize varieties. Tall and local maize landraces predominated on hillside milpas (maize fields). These were managed with hand tools, had low chemical inputs, and were usually intercropped. On flat land milpas, short-statured maize hybrids (i.e., Tuxpeño) predominated, requiring more chemical inputs and frequent purchase of seeds. Other key differences between the two zones included:

1. The flat agricultural land was dominated by two varieties, even though the number of varieties found in both areas was similar. Each farmer planted an average of three varieties per farm. However, in the flat system only two varieties accounted for 79.3% of the total area, whereas in the slopes five varieties accounted for 82.5% of the area.

2. Tuxpeño seed had to be renewed every three or four years, whereas landraces were rarely renewed. Eighty-two per cent of the farmers changed Tuxpeño seed; 39% of the farmers never changed landraces.

3. Forty-six percent of the Tuxpeño seed was purchased; only 16% of the seed of local varieties was purchased.

4. Yields of Tuxpeño variety were higher (2.8 t/ha) than yields of landraces (2.15 t/ha), but exhibited higher production costs.

This particular case study suggests that peasants choose between risk-reducing technology and higher yields, and apparently there was no clear cultural distinction made between improved varieties and landraces. Farmers in the flat system adopted improved varieties because these were perceived as more productive and less risky than landraces. In the slopes, however, maintenance of native maize diversity was preferred as a strategy to cope with drought and the insecurity associated with credit and high-input technology.

Case Study 3

In Central Chile, small farmers have adopted hybrid maize varieties with better marketing perspectives, abandoning "criollo" maize varieties (i.e., maiz chadino) that clearly require less inputs and have lower production costs (i.e., they produce their own seed) but that have no commercial demand from consumers in Santiago. A similar situation occurs in the island of Chiloe in southern Chile, where farmers are progressively abandoning traditional potato varieties as commercial demand for the "Desiree" potato rapidly increases. Traditional or native crops and varieties are also abandoned because farmers adopt new and botanically unrelated crops.

Case Study 4

In the cooler portions of temperate valleys and basins of Bolivia, Perú, and Ecuador, the cultivation of tarwi (*Lupinus mutabilis*), once an abundant and nutritious crop, has declined because it cannot compete with legumes introduced from Europe (i.e., kidney beans, chickpeas, lima beans, and lentils). These new crops are dietary rather than agronomic competitors, and the implications of this replacement are many-fold (9):

1. Tarwi is hardy and adaptable. It tolerates drought, frost, a wide range of soils, and many insect pests and diseases. When intercropped with maize or broad beans, tarwi functions as an insurance crop in the face of environmental stress. The new legumes yield poorly under marginal conditions, thus making the systems more risky. In the event of a major plant disease epidemic affecting the new legumes, tarwi could become again an important crop.

2. Tarwi has a high protein and oil content and, therefore, provides double nutritional benefits for campesinos for whom meat is a luxury. The new legumes do not exhibit as a high protein content as tarwi. Tarwi can produce up to 1.75 t/ha of protein.

3. Because of its local adaptability and growth habit, tarwi fits well into rotational schemes with maize, potatoes, and broad beans. It is an excellent green manure producing about 50 t/ha of green matter and is able to fix as much as 400 kgs nitrogen per hectare. Although other legumes may be important in crop-rotation systems, their potential as green manures remains to be proven under Andean conditions.

Case Study 5

National agricultural policies in Mexico are stimulating genetic erosion through the process of modernization of agriculture that promotes hybrid sorghum, associated mechanization, and the use of agrochemical inputs. Basically, the government opened the rural economy to the influences of the global system of markets that transmits international prices to local communities. The impact has been that most farmers (peasants included) are unwilling to continue producing basic food crops for domestic consumption because of the new opportunity to profit from commercial crop production. Mexico finds itself using substantial amounts of foreign exchange to import maize and other agricultural produce that otherwise it could produce with idle human and natural resources. In the meantime, sorghum is grown as a cash crop for cattle feed at the expense of several varieties of maize and food self-sufficiency of the country (10).

In summary, there is insufficient data to correctly assess the geographical pattern and rates of genetic erosion in Latin America. What is clear is that it occurs unevenly and proceeds faster in lowland areas subject to modernization and areas closer to urban centers and markets. In marginal areas, farmers continue to rely on traditional varieties, since this is the only sensible strategy to face economic and environmental uncertainty. Recent reports from potato-collecting expeditions indicate that it is becoming increasingly more difficult to find wild species. About 21 species of wild tomatoes (including species such as *S. burkartii*, *S. garcia-barrigae*, *S. palustre*, *S. chavinese*, *S. pumilum*, etc.) have still not been found again or re-discovered in the living state. Explorers mention over-grazing by goats, deforestation, soil erosion, and expanding modern agriculture as the major threats to wild species (11). Similarly, in Chiloe, Negrón (12) was able to find only two potato varieties in 1990 in the area of Ten-Ten, Putemún, and Yutuy, whereas in 1975 he had collected 14 different varieties.

TRADITIONAL FARMING SYSTEMS AS REPOSITORIES OF CROP GENETIC DIVERSITY

Latin American farmers, without access to scientific information, external inputs, capital, credit, and developed markets, by using locally available resources, have often developed farming systems with sustained yields. A salient feature of traditional farming systems is their degree of plant diversity (Table 2) in time and space in the form of polycultures and/or agroforestry patterns (13,

Table 2. Useful plants originally from Latin America and extensively used by farmers throughout the region (51).

Plant Species	Family	Region
	GRAINS	
Amaranthus spp.	*Amaranthaceae*	Sudamerica
Bromus mango	*Gramineae*	Sudamerica Chile
Canavalia spp.	*Leguminosae*	Sudamerica
Chenopodium spp.	*Chenopodiaceae*	Region Andina
Dipteryx odorata	*Leguminosae*	Sudamerica
Erythrina spp.	*Leguminosae*	N. de Sudamerica
Lupinus spp.	*Leguminosae*	Region Andina
Omphalea megacarpa	*Euphorbiaceae*	Sudamerica
Oryza spp.	*Gramineae*	Sudamerica
Phaseolus spp.	*Leguminosae*	Sudamerica
Tripsacum spp.	*Gramineae*	Sudamerica
Zea mays	*Gramineae*	Sudamerica
	LEGUMES	
Arachis spp.	*Leguminosae*	Sudamerica
Elaeis melanococa	*Palmae*	Amazonia
Jatropha curcas	*Euphorbiaceae*	Sudamerica

Table 2. (Continued)

Plant Species	Family	Region
Lippia citriodora	*Verbenaceae*	Sudamerica
Madia sativa	*Compositae*	Sudamerica Chile
	ROOTS	
Arracacia xanthorrhiza	*Umbelliferae*	Zona Andina
Callathea allouia	*Marantaceae*	Amazonia
Canna edulis	*Cannaceae*	Zona Andina
Dioscorea spp.	*Dioscoraceae*	Sudamerica-Brasil
Lepidium meyenii	*Cruciferae*	Peru-Bolivia
Manihot spp.	*Euphorbiaceae*	Sudamerica
Maranta arundinacea	*Marantaceae*	N. de Sudamerica
Mirabilis expansa	*Nyctaginaceae*	Zona Andina
Oxalis tuberosa	*Oxalidaceae*	Zona Andina
Pachyrrhizus spp.	*Leguminosae*	Sudamerica
Polymnia sonchifolia	*Compositae*	Zona Andina
Solanum spp.	*Solanaceae*	Zudamerica
Tropaeolus spp.	*Tropaeolaceae*	Zona Andina
Ullucus tuberosum	*Basellaceae*	Zona Andina
Xanthosoma spp.	*Araceae*	Amazonia
	FIBERS	
Abutilon oxycarpum	*Malvaceae*	Sudamerica
Carludovica palmata	*Cyclanthaceae*	Sudamerica
Fourcroya cubensis	*Amaryllidaceae*	Sudamerica
Funifera brasiliensis	*Thymelaeaceae*	Brasil
Furcraea spp.	*Agavaceae*	Sudamerica
Gossypium spp.	*Malvaceae*	Sudamerica
Minthostachys setosa	*Labiatae*	Peru
Scirpus californicus	*Cyperaceae*	Sudamerica
Wissadula spp.	*Malvaceae*	Sudamerica
	VEGETABLES	
Amaranthus spinosus	*Amaranthaceae*	Sudamerica
Boussingaultia cordifolia	*Basellaceae*	Sudamerica
Crotolaria anagyroides	*Leguminosae*	Sudamerica
Curcubia maxima	*Cucurbitaceae*	Zona Andina
Cyclantera pedata	*Cucurbitaceae*	Sudamerica
Jatropha urens	*Euphorbiaceae*	Sudamerica
Justicia pectoralis	*Acanthaceae*	N. de Sudamerica
Lycopersicon spp.	*Solanaceae*	Zona Andina
Peperomia pellucida	*Peperomiaceae*	Sudamerica
Pereskia bleo	*Cactaceae*	N. de Sudamerica
Plukenetia volubilis	*Euphorbiaceae*	N. de Sudamerica
Spilanthes oleracea	*Compositae*	Brasil
	FRUITS	
Anacardium oscidentale	*Anacardiaceae*	N. de Sudamerica
Ananas spp.	*Bromeliaceae*	Amazonia
Annona spp.	*Annonaceae*	N. de Sudamerica

Table 2. (Continued)

Plant Species	Family	Region
Bertholletia excelsa	*Lecythidaceae*	Amazonia
Bunchosia armeniaca	*Malpighiaceae*	Zona Andina
Carica spp.	*Caricaceae*	Zona Andina N. de Sudamerica
Caryocar spp.	*Caryocaraceae*	Zona Andina
Cereus spp.	*Cactaceae*	N. de Sudamerica
Cyphomandra spp.	*Solanaceae*	Zona Andina
Eugenia spp.	*Myrtaceae*	Brasil Central
Fragaria chiloensis	*Rosaceae*	Sudamerica
Grias cauliflora	*Lecythidaceae*	N. de Sudamerica
Guilelma gasipaes	*Palmae*	Amazonia
Hancornia speciosa	*Apocynaceae*	Amazonia
Inga spp.	*Leguminosae*	N. de Sudamerica
Lucuma spp.	*Sapotaceae*	N. de Sudamerica Zona Andina
Malpighia glabra	*Malpighiaceae*	N. de Sudamerica
Mammea americana	*Guttogerae*	N. de Sudamerica
Manilkara bidentata	*Sapotaceae*	Sudamerica
Melicossus bijugatus	*Sapindaceae*	N. de Sudamerica
	STIMULANTS	
Banisteriopsis caapi	*Malpighiaceae*	Sudamerica
Datura spp.	*Solanaceae*	Sudamerica
Erythroxylum spp.	*Erythroxypaceae*	Zona Andina
Euterpe oleraea	*Palmae*	Sudamerica
Mirabilis ialapa	*Nyctaginaceae*	Sudamerica
Nicotiana spp.	*Solanaceae*	Zona Andina
Paullinia spp.	*Sterculiaceae*	Sudamerica
	CONDIMENTS	
Bixa orellana	*Bixaceae*	Sudamerica
Capsicum spp.	*Solanaceae*	Sudamerica
Chenopodium ambrosioides	*Chenopodiaceae*	Sudamerica
Nectandra cinnamonoides	*Lauraceae*	Zona Andina
Tagetes minuta	*Compositae*	Sudamerica
Tropaeolum majus	*Tropaeolaceae*	Sudamerica
	INSECTICIDES	
Lonchocarpus spp.	*Leguminosae*	Sudamerica Peru
Phyllanthus acuminatus	*Euphorbiaceae*	Centro America
Tephorosia singaporu	*Leguminosae*	Colombia
	VARIOUS	
Acacia spp.	*Leguminosae*	Sudamerica
Copernicia prunifera	*Palmae*	Amazonia N. de Sudamerica

14). These systems represent a strategy to promote diversity of diet and income source, stability of production, minimization of risk, and maximization of returns under low levels of technology (15). Traditional multiple cropping systems are estimated to still provide as much as 15 to 20% of the region's food supply (16). Traditional agroforestry systems throughout the American tropics commonly contain well over 100 plant species per field, species used for construction materials, firewood, tools, medicine, livestock feed, and human food (17). In Mexico, for example, Huastec Indians manage a number of agricultural and fallow fields, complex home gardens, and forest plots totalling about 300 species (18). Small areas around the houses commonly average 80 to 125 useful plant species (19). Similarly, traditional agroforestry systems managed by Amazonian ethnic groups such as the Bora and Kayapó commonly contain about 100 or more plant species. Of these plants, many provide for building materials and fuelwood, some are fruit trees and vegetables, and a great variety constitute ornamentals, medicinal plants, spices, and cash crops (17). Table 3 provides a list of Amazonian tree species important to indigenous people, including assessment of their conservation status.

Many traditional agroecosystems are located in the Latin American centers of diversity (i.e., Meso-America, Andes, southern Chile, Brazil-Paraguay), thus containing populations of variable and adapted landraces as well as wild and weedy relatives of crops (20, 21). In the Andes, farmers cultivate as many as 50 potato

Table 3. Conservation status of important tree species used by Amazonian indigenous groups (52).

Species	Common Name	Management	Conservation Status
Teobroma cacao	Cacao	Plantation	1
Swietenia macrophilla	Mahogany	Extraction	2
Amburana acreana	Brazilian cherry	Extraction	2
Couma rigida	Sorva latex	Extraction	3
Bertholletia excelsa	Brazil nut	Plantation Extraction	2
Paullinia cupana	Guarana	Plantation	1
Aniba rosaeodora	Rosewood oil	Extraction	3
Croton cajucara		Extraction	1
Leopoldina piassaba	Piassaba fiber	Extraction	2
Attalea funifera		Extraction	2
Dpteryx odorata	Tonka bean	Extraction	1
Copaifera langsdorffi	Copaiba balsum	Extraction	1
Copaifera multijuga	Copaiba balsum	Extraction	1
Manilkara kuberi	Massaranduba	Extraction	1
M. bidentata	Balata	Extraction	1
Hevea brasilensis	Rubber	Plantation Extraction	1
Orbignya phalerata	Babassu palm	Extraction	2
Mauritia flexuosa	Buriti palm	Extraction	2
Euterpe olearacea	Acai palm	Extraction	2
Grias spp.	Sachamangua	Extraction	2
Manihot esculenta	Cassava	Plantation	1
Myrciaria dublas	Camu camu	Extraction	2

1 abudant
2 declining
3 rare

varieties in their fields (22). The resulting genetic diversity confers at least partial resistance to diseases specific to particular strains of the crop and allows farmers to exploit different microclimates and derive multiple nutritional and other uses from within-species genetic variation (14, 20). Clawson (14) described a number of systems in which traditional tropical farmers plant multiple varieties of each crop, providing both intraspecific and interspecific diversity, which thus enhances harvest security.

A number of plants within and/or around traditional cropping systems are wild or weedy relatives of crop plants. The ecological amplitudes of wild relatives may exceed those of the crops derived from or otherwise related to them, a feature exploited by plant breeders to enhance the resistance or adaptive range of crops (20, 23, 24). For example, in the Centro International de la Papa's (CIP) germplasm collection there are more than 19 species of wild potato and about 1,450 classified accessions. Many of these have been collected in traditional farming systems (11). In these settings, landraces and wild and weedy relatives have co-existed and co-evolved over a long period of time with each other and with human cultivators. Natural hybridization and introgression have often occurred between crops and wild relatives, increasing the variability and the genetic diversity available to farmers (20). Through the practice of nonclean cultivation, farmers have inadvertently increased the gene flow between crops and their relatives. For example, farmers in Mexico allow teosinte to remain within or near maize fields so that when the wind pollinates the maize, some natural crosses occur (25). Although crosses such as these are not immediately evident, the following year when the new maize crop is planted from last year's seeds, the maize-teosinte seeds produce hybrid plants. Such hybrids and their descendants are phenotypically distinct and fertile and thus capable of passing on their genetic traits. The teosinte is said to increase corn yields (25). In northwestern Mexico, farmers recognize that exchange of traits occurs between sympatric cultivated and wild/weedy squashes (26) and suggest that a similar pattern occurs for other local crops as well. The encouragement of specific weeds by peasant farmers in their agroecosystems may represent progressive domestication, a process described by Davis and Bye (27) for *Jaltomata*, a herbaceous perennial used by the Tarahumara in Mexico. Farmers derive other benefits from the presence of tolerable levels of weeds in their fields. Certain weeds are directly used for medicinal and culinary purposes (28), and in many cases, weed communities are managed within crop fields, resulting in increased biological insect-pest control (29) and enhanced organic matter accumulation and soil conservation (30).

Through this continual association, relatively stable equilibria among crops, weeds, diseases, cultural practices, and human habits have developed (31). In fact, the great variety of primitive crop cultivars corresponds well with the heterogeneity of the social and ecological environment characteristic of Latin American rural communities (22, 32).

The stabilized equilibria in traditional agroecosystems are complex and very difficult to modify without upsetting the balance and risking loss of genetic resources, not to mention negative effects on the social organization (33). In fact, it is argued here that many landraces and wild and weedy relatives can be preserved only in agroecosystems under traditional management and, furthermore, only if this management is guided by the local intimate knowledge of the plants and their requirements (18). On this basis, we resist the implementation of topdown rural development approaches (i.e., those derived exclusively from decisions of policy makers and researchers external to the community of local farmers) that do not reflect indigenous social, ecological, and ethnobotanical considerations (34).

A Case Study on Grassroots *In situ* Conservation: The Centro de Educacion y Tecnologia (CET) in Chiloé

The Archipelago of Chiloé, a group of islands in southern Chile, is considered to be one of the centers of origin of the potato *Solanum tuberosum* L. (35). Collecting expeditions by several researchers throughout the years determined a great diversity of potato native varieties (36). In 1975, Contreras *et al.* collected 146 different samples of native varieties, the most prevalent ones being the "michunes coloradas y moradas" and the "clavelas." These native varieties are highly adapted to the range of ecological conditions found in the region and are of key importance for subsistence production (37).

Since the early 1940s the Chilean government made has made several introductions of European and North American varieties (some of which had been bred from Chilotan material). Farmers in areas close to urban and market centers (especially on the big island) have abandoned most native varieties and adopted these new introduced varieties such as Desiree, Industrie, Cóndor, Ginecke, and others, which now have greater commercial demand.

The introduction of new varieties, and their diseases, contributed to the extinction of native varieties. Around 1950, *Phytophthora infestans* devastated most potato fields, affecting mostly native varieties that had never been exposed to the exotic pathogen and therefore lacked the necessary genetic tolerance.

In an effort to slow genetic erosion and recover some of the native potato germplasm, the Centro de Educacion y Tecnologia (CET) initiated an *in situ* conservation program at its "peasant training center" in Notuco, near Chonchi, and in several neighboring rural communities.

In 1988, CET technicians surveyed several agricultural areas of Chiloè and collected hundreds of samples of native potatoes still grown by some small farmers throughout the big island. In 1989, CET established a "live" collection (garden-seed bank) of 96 native potato varieties at its Notuco center, each planted in rows of 5 to 10 plants in a half-hectare plot area. These varieties are grown year after year and are subjected to selection and seed enhancement.

In 1990, CET technicians initiated an *in situ* conservation program involving 21 farmers in five different rural communities (Dicham, Petanes, Huitauque, Notue, and Huicha).

Each farmer is given a sample of five different native varieties that they are responsible to grow within their potato fields. After harvest, farmers return part of the seed production to CET (for the garden-bank), exchange seeds with other farmers, or plant the seeds again for further reproduction of genetic material. Figure 2 illustrates the conservation and exchange dynamics of the 96 varieties maintained at CET's garden-seed bank and grown by the 21 collaborating farmers.

It is expected that more farmers will become involved in the project, and that CET will be able to select varieties based on farmers' needs and desirable characteristics. Selected varieties will be propagated and distributed among participating farmers. Excess seeds could also be sold to other farmers or exchanged for seeds of traditional varieties still not available in CET's collection. This strategy will allow a continuous supply of valuable seeds to go to resource-poor farmers for subsis-

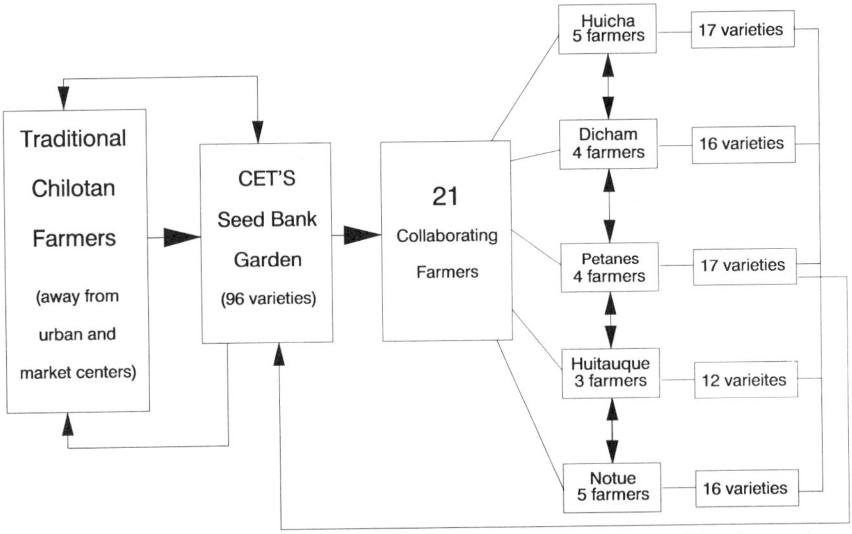

Figure 2. The strategy of grassroots *in situ* potato genetic conservation of the Centro de Educacion y Technologica (CET) in Chiloeto, southern Chile.

tence, and will also build a repository of vital genetic diversity for future regional crop improvement programs.

STATUS AND TRENDS IN GENETIC CONSERVATION

For centuries, peasant communities in Latin America have engaged in the selection, maintenance, and utilization of genetic diversity, while developing and improving their production systems. In this section, we analyze the conservation and utilization efforts of genetic resources conducted by NGOs, scientists, and farmers' groups at the community level. Most of these efforts possess a degree of artificiality and/or *ex situ* conservation, since they include recollection and storing of seed samples in banks outside the farming system.

General awareness of the importance and urgency of crop genetic conservation has grown widely among different sectors in Latin America. The international agricultural research centers located in the region have expanded their conservation efforts to previously underrepresented or almost ignored species, such as quinoa, yam, ulluco, oca, and others. Several national research institutes have renewed their interest in different crops and even wild species; years of budget cutbacks have been followed by fresh funds coming from northern governments or northern private enterprises. Universities have suffered serious financial problems and policy uncertainties; nevertheless, it is not uncommon to find committed faculty members who manage to keep conservation efforts moving with the help of farmers and students.

Transnational corporations have also shown greater interest in genetic conservation. For example, Ciba-Geigy has been collecting medicinal and aromatic plants among the Jivaro people in the Amazon (38), and Nestlè is setting up a center for quinoa germplasm conservation in Ecuador. Ciba-Geigy has been searching for promising pharmaceuticals, while Nestlè is interested in developing commercial quinoa products and by-products.

NGOs working with small farmers and farmer organizations are also part of this trend. "Genetic conservation," "genetic erosion," and "biodiversity" are concerns that very few grassroots organizations ignore today.

Unfortunately, concerns do not always result in appropriate actions. We can still see that not enough is being done in the way of genetic conservation, that what is being done is often wrong or inadequate, that resources are painfully scarce and overwhelmingly concentrated in just one kind of strategy (gene banks), and that many technical and social factors related to genetic conservation deserve more detailed analysis and discussion in order to be properly addressed.

Grassroots Genetic Conservation Efforts in Latin America

There are three main types of grassroots genetic conservation efforts in Latin America:

1. In their search for cheaper and more autonomous ways to do agriculture, small farmers all over Latin America are turning to varieties and landraces they can reproduce themselves or that are better adapted to local conditions. Other groups are exclusively focusing their efforts to find ways to grow their own seeds without making clear distinctions between local varieties with a wide genetic base, exotic seeds, or even those with very narrow genetic bases. These initiatives may be started by farmers themselves or may respond to some external proposal, but they are often linked to the work and technical assistance of external groups, such as NGOs.

2. A growing number of development NGOs have learned from their field experience that autonomous seed production and conservation of local crop varieties are critical factors in any development strategy. In collaboration with farmers, some of these NGOs work with crops or varieties considered important or interesting, with no consideration to their origin or their genetic base. In such cases, both conservation and seed reproduction work are done with materials varying from a well-adapted, rather recent, modern wheat variety to a native, local potato landrace. Other NGOs work with any commercial crop, native or exotic, but simultaneously focus on local landraces of native crops that can either be commercially important or belong to the "forgotten crops" that are (or used to be) critical for subsistence and nutrition. Some members of the latter group also include medicinal and wild plants.

3. Professors and scientists from universities and public research stations interested in *in situ* genetic conservation have also initiated some efforts focusing on local varieties of native species, ranging from potatoes, maize, and beans to many "forgotten" crops, medicinal plants, and wild species (37, 39, 40, 41). They conduct genetic conservation in a stricter sense and make a clear distinction between *in situ* conservation and what is only local seed production. Many of these projects that consider direct use

of the collected resources usually lack the most fundamental financial support. Therefore, it is not uncommon that these initiatives finally do more for *ex situ* than for *in situ* conservation, and most materials finally end up in national or university gene banks that also suffer from financial scarcity. Conservation is usually accomplished through links and agreements with farmers as individuals (and not as members of an organization), and therefore its potential coverage becomes limited; lack of financial resources makes wide coverage even more restricted.

Crop and Species Coverage

What is conserved is definitely related to who conserves, but three species are clearly the most important subjects of genetic conservation in the region: maize, beans, and potatoes. Maize and/or beans are present in most conservation projects all over Latin America. Major work on maize has been done in Mexico by the International Center for the Improvement of Corn and Wheat (CIMMYT), public, and citizens organizations. However, as far as the authors know, Brazilian NGOs are leading efforts to link conservation and breeding of maize at the farmer level (42).

Potato conservation efforts are prevalent in Peru, Bolivia, Chile, and Ecuador. Major work has been done by CIP and the national research institutes of these countries, but NGOs and farmers' organizations, especially in Peru and Bolivia, are increasingly taking part in this effort, either because the "official" work is not bringing clear benefits to small farmers or because it is becoming directly relevant as "improved varieties" reach small farmers and genetic erosion takes its toll (43).

Conservation efforts in tomatoes, *Capsicum*, and Cucurbitae are far behind the latter crops, especially among NGOs. One reason for this seems to be a rather generalized conception that these are not basic or especially nutritious food crops. Very early and massive introduction of "improved" varieties, and concurrent genetic erosion of landraces, especially in tomatoes and cucurbitae, may be other factors leading to less conservation efforts of these crops.

Quinoa, amaranth, and *Lupinus* are three crops that have received growing attention during the last decade. High nutritional value, hardiness, high pest resistance, and good yields under marginal conditions are all factors that have triggered this interest. However, rather complex processing requirements, as well as changing nutritional habits among the rural and urban population, seem to limit use and general interest. Although farmers' organizations and NGOs have not abandoned these crops, quinoa and amaranth are mostly the subject of collection work being done by groups from outside Latin America or by international centers and regional subsidiaries of transnational corporations. Massive quinoa collections have been gathered by Nestlè in Latin America and by the Sierra Blanca Associates based in the United States; perhaps one of the most important amaranth collections is in the hands of the Rodale Research Center, also in the United States (44).

Finally, crops such as yam, oca, ulluco, and other Andean crops are rather new targets of collection efforts (39, 45, 46), but it is not clear how much has been done or who has done the most.

Tree and shrub species seem to be the forgotten organisms of genetic conservation, at any level. A great deal of discussion focuses on the conservation of the Amazon, but very little is being done for the conservation of other forests in the region, as in Central America or in the Southern Cone of South America. Although numerous national parks, national reserves, and protected areas exist in the region, different kinds of pressures, including cattle ranching, mining, and the paper industry, all fueled by economic policies, are deforestating the last repositories of species such as alerce, *Araucaria*, and other *Nothofagus* species. The urgent need for further conservation does not stop there. There are many species in arid or semi-arid regions that are seldom even mentioned or considered in any type of conservation work. These ecosystems contain species or genera that could be of importance not only for ecological reasons, but also for their special capacity to thrive and produce under the marginal conditions many small farmers face. The genus *Prosopis* is just one example: it provides firewood, fodder, and food for the rural family, improves the soil, and may survive the driest years.

Strategies Employed

Four main strategies have been and still are being employed for genetic conservation at the grassroots level:

1. Collection, reproduction, and local distribution of any material considered interesting or important for commercial, nutritional, subsistence, cultural, or agroecological reasons. The *introduction* of material such as exotic and native species, as well as old landraces and new varieties, may be considered. Some of the new varieties may have been obtained from locally collected material.

   Materials reproduced are mainly those demanded by farmers or considered useful, important, or promising by technical staffers. Reproduction is assisted mostly by farmers, usually experienced ones. In order to ensure a good distribution, this strategy works best among organized farmers (42, 47).

   This strategy is common to almost all farmers' organizations working with genetic resources and to a good number of NGOs. Farmers' organizations tend to focus on commercially interesting crops/varieties, whereas NGOs tend to cover a much wider range.

2. A variant to the first strategy focuses exclusively on native species and is featured by many NGOs as exemplified by the work of CET in southern Chile and COINCIDE in Cusco, Perú; in both cases, the work is on potatoes.

3. Another variant, with an emphasis on breeding, usually focuses on the three "biggest" crops: maize, potatoes, and beans. Breeding is mainly done through mass selection, although some groups are also using crossing. This strategy is characteristic of NGOs [such as Projeto Tecnologias Alternatives (PTA)] that often work in cooperation with scientists from universities or governmental agencies.

4. A strategy of periodic, local "Farmers' Fairs" are usually organized by NGOs and/or professionals from universities or public institutes. They do not reward the best samples or specimens, but reward the farmer who keeps the widest diversity and knows best

the characteristics of what has been saved. The fairs focus on native crops (39).

This strategy may or may not promote reproduction or distribution of exhibited materials. Its basic aim is to encourage farmers to keep diversity in their fields, hoping that incentives (such as public recognition, diplomas, or tools for the farm) will make other farmers adopt or recover local variaties. This strategy requires little financial resources. It also encourages farmers' initiatives and leaves decisions to the farmers themselves. Its potential to promote living diversity is highest; its effectiveness, however, may vary significantly according to the incentives provided (39). These fairs have been successfully conducted in Perú (39) and Bolivia (48). A growing number of NGOs, university professors, and people working with governmental agencies in these countries are actively promoting this strategy.

CONCLUSION: ECONOMIC AND SOCIAL QUESTIONS OF GRASSROOTS CONSERVATION EFFORTS

Technical needs are not the only needs conservation efforts faced at the grassroots level. As mentioned before, lack of resources is endemic. If field work conducted so far did not show so clearly to so many that genetic resources are critical, and if personal commitments to farmers' needs were not so strong, almost no conservation projects at the community level would be alive in Latin America. Over 50% of the initiatives known by the authors do not have their own staff or budget, but must borrow from other projects, and they strongly rely on help from farmers and other local sources. Among those technicians who have received financial support, again over 50% have been working with less than U.S. $5000 per year. Over 60% of the programs have done collection work but do not have money to set up simple and adequate storing facilities or, even more seriously, to do a systematic monitoring of multiplication of material and performance tests in the field. This is an important reason for diversity loss and for the absence of breeding initiatives at the farm level.

Economic problems do not stop there. Some projects started with good financial support, but that support was later reduced or suspended. In some instances, the original support included funds to set up rather sophisticated storing facilities, which later became underutilized or even useless as technical problems arose and funding decreased or disappeared.

As plant breeding and biotechnologies are rapidly evolving and consuming large amounts of money, and as different kinds of intellectual property systems are being imposed, people working with genetic resources at the community level are increasingly asking relevant social and economic questions. Who is going to benefit from these conservation efforts? How can we guarantee that what we are conserving today does not return as an economic burden imposed by patent laws? As international trade discussions have evolved, many of these questions have turned into clear political concerns, and it is becoming increasingly common that many groups and organizations doing conservation work at the farm level will not continue sharing their information or collected samples unless a clear system of safeguards for farmers' rights and interests is established.

Political and social concerns have been further deepened by the fact that international agricultural research centers, which were recipients of resources that supposedly belonged to all humankind, are not addressing small farmers' needs and will most probably be or already are under constraints imposed by patent laws in industrialized countries (49, 50).

In summary, it should be stated that Latin America is a major repository of crop genetic resources and that small farmers have played and still play a major role as stewards of this legacy. These farmers have never been compensated for their service to humankind and no international mechanism is in place to secure such compensation. Several NGOs have initiated grassroots crop genetic conservation efforts, but their impact is still insignificant because of technical constraints as well as lack of funds and resources. Funds for conservation could be derived from payments made by those directly using biodiversity. Such funds should be allocated to support *in situ* conservation programs focusing on native species and varieties, and especially those linking local crop conservation with agroecological approaches.

REFERENCES

1. Brush, S.B. 1986. Genetic diversity and conservation in traditional farming systems. *Enthobiology* 6:151–67.
2. Querol, A. 1988. Recursos genéticos, nuestro tesoro olvidado. Lima, Peru: Industrial Gráfica.
3. Wilkes, H.G. 1983. Current status of crop plant germplasm. *CRC Crit. Rev. Pl. Sci.* 1:133–81.
4. Perrin, R., and D. Winkelman. 1976. Impediment to technical progress on small vs. large farms. *Amer. J. of Agric. Economics* 58:888–94.
5. Feder, G., R.E. Just, and D.Z. Zilberman. 1982. Adoption innovations in developing countries: A survey. World Bank Staff Working Paper No. 542. Washington, DC: World Bank.
6. Brush, S.B., M. Bellon, and E. Schmidt. 1988. Agricultural development and maize diversity in Mexico. *Human Ecology* 16:95–106.
7. Brush, S.B., H. Carney, and Z. Huaman. 1981. Dynamics of Andean potato agriculture. *Econ. Bot.* 35:70–88.
8. Franco, E., D. Horton, and F. Tardien. 1979. Producción y utilización de la papa en el valle del Mantaro, Peru. Social Science Unit Work Document No. 1979-1. Lima, Peru: Centro Internacional de la Papa (CIP).
9. Gade, D.W. 1969. Vanishing crops of traditional agriculture: The case of tarwi (*Lupinus mutabilis*) in the Andes. *Proc. of Assoc. of Amer. Geographers* 1:47–51.
10. Dewalt, B.R., and D. Barkin 1987. Seeds of change: The effects of hybrid sorghum and agricultural modernization in Mexico. In *Technology and Social Change.* 2d ed. H.R. Bernard and P. Pelto, eds., pp. 138–65. Prospect Heights, IL: Waveland Press.
11. Centro Internacional de la Papa (CIP). 1987. Strategies for the conservation of potato. *Genetic Resources IV.* Lima, Peru: CIP.
12. Negrón, J. personal communication.
13. Chang, J.H. 1977. Tropical agriculture: Crop diversity and crop yields. *Econ. Geogr.* 53:241–54.
14. Clawson, D.L. 1985. Harvest security and intraspecific diversity in traditional tropical agriculture. *Econ. Bot.* 39:56–67.
15. Harwood, R.R. 1975. *Small Farm Development — Understanding and Improving Farming Systems in the Humid Tropics.* Boulder, CO: Westview Press.
16. Francis, C.A. 1986. *Multiple Cropping Systems.* New York: MacMillan Pub. Co.
17. Altieri, M.A., and S.B. Hecht. 1990. *Agroecology and Small Farm Development.* Boca Raton, FL: CRC Press.

18. Alcorn, J.B. 1984. Development policy, forest and peasant farms: Reflections on Huastec-managed forests' contributions to commercial production and resource conservation. *Econ. Bot.* 38:389–406.

19. Alcorn, J.B. 1981. Huastec noncrop resource management. *Human Ecol.* 9:395–417.

20. Harlan, J.R. 1975. Our vanishing genetic resources. *Science* 188:618–22.

21. Browning, J.A., and K.J. Frey. 1969. Multiline cultivars as a means of disease control. *Ann. Rev. Phytopath.* 7:355–82.

22. Brush, S.B. 1982. The natural and human environment of the central Andes. *Mountain Res. Development* 2:14–38.

23. Frankel, O.H., and J.G. Hawkes, eds. 1975. *Crop Genetic Resources for Today and Tomorrow*. Cambridge: Cambridge University Press.

24. Prescott-Allen, R., and C. Prescott-Allen. 1983. *Genes From the Wild*. London: Earthscan Paperback.

25. Wilkes, H.G. 1977. Hybridization of maize and teosinte in Mexico and Guatemala and the improvement of maize. *Econ. Bot.* 31:254–93.

26. Merrick, L.C., and G.P. Nabhan. 1984. Natural hybridization of wild *Cucurbita sororia* group and domesticated *C. mixta* in southern Sonora, Mexico. *Cucurbit Genet. Coop. Newsletter* 7:73–75.

27. Davis, T., IV. and R. A. Bye, Jr. 1982. Ethnobotany and progressive domestication of Jaltomata (Solanaceae) in Mexico and Central America. *Econ. Bot.* 36:225–41.

28. Datta, S.C., and A.K. Banerjee. 1978. Useful weeds of West Begal rice fields. *Econ. Bot.* 32:297–310.

29. Altieri, M.A., A. van Schoonhoven, and J.D. Doll. 1977. The ecological role of weeds in insect pest management systems: A review illustrated with bean (*Phaseolus vulgaris* L.) cropping systems. *Pest Articles News and Summ.* (PANS) 23:195–205.

30. Chacon, J.C., and S.R. Gliessman. 1982. Use of the "non weed" concept in traditional tropical agroecosystem of south-eastern Mexico. *Agro-Ecosystems* 8:1–11.

31. Bartlett, P.F. 1980. Adaptation strategies in peasant agricultural production. *Ann. Rev. Anthropol.* 9:545–73.

32. Altieri, M.A. 1987. *Agroecology: The Scientific Basis of Alternative Agriculture*. Boulder, CO: Westview Press.

33. Grossman, L.S. 1984. *Peasants, Subsistence Ecology and Development in the Highlands of Papua, New Guinea*. Princeton, NJ: Princeton University Press.

34. Altieri, M.A., and L.C. Merrick. 1987. *In-situ* conservation of crop genetic resources through maintenance of traditional farming systems. *Econ. Bot.* 41:86–96.

35. Bukasov, S.M. 1933. The potatoes of South America and their breeding possibilities. Suppl. 58. *Bull. Appl. Bot. Gen. Pl. Breed.* (Leningrad).

36. Montaldo, A., and C. Sanz. 1962. Las especies de papas silvestres y cultivadas de Chile. *Agric. Técnica* 22:66–152.

37. Contreras, A.M. 1987. Germoplasma chileno de papas (*Solanum* spp.). *Anales Simposio Recursos Fitogeneticos*. Valdivia: UACH-IBPGR.

38. Fowler, M.C., *et al.* 1988. The laws of life. *Development Dialogue* 1–2.

39. Tapia, M., S. Franco, A. Rosas, and A. Florindez. 1990, November. Conservación in situ de los recursos fitogenéticos andinos para la producción de semillas con agricultores campesinos. Paper presented to the workshop, Preservacón de recursos fitogenéticos de cultivos andinos y producción de semillas para la agricultura campesina. Cusco, Peru.

40. Ramos, J. 1987. Reserva de la biósfera Reserva de Manatlán. Paper presented to the Latin American Workshop on Plant Genetic Resources. September 1988. Santiago, Chile.

41. Saragoussi, M. 1988, September. Extrativismo: Una forma de conservacao dos recursos geneticos. Paper presented to the Latin American Workshop on Plant Genetic Resources. Santiago, Chile.

42. Projeto Tecnologias Alternatives (PTA). 1987. Banco de sementes. Dossié. Mimeo.

43. Duenas, A., R. Mendivil, G. Lovatón, and A. Loaiza. 1990, November. Variabilidad y Erosión Genética en el Cultivo de Papa en Comunidades Campesinas del Cusco. Paper presented to the workshop Preservacón de recursos fitogenéticos de cultivos andinos y producción de semillas para la agricultura campesina. Cusco, Peru.

44. Rodale Press. *Amaranth Today* 1(1):5. Emmaus, PA: Rodale Press. Undated.

45. International Board for Plant Genetic Resources (IBPGR). 1982. Reunión sobre recursos fitogenéticos de interés agricola en la región andina. Lima, Peru: IICA-JUNAC, CIRF.

46. National Academy of Sciences (NAS). 1972. *Genetic vulnerability of major crops*. Washington, DC: NAS.

47. CIPCA. 1990. Los recursos fitogenéticos en la alternativa alimentaria de la región Grau. Paper presented to Recursos Fitogenéticos: Regionalización y Alternativas Alimentarias. Cusco, Peru.

48. Programa de Asistencia Agrobioenegetia al Campesino (PAAC), personal communication.

49. Komen, J. 1990a. CIMMYT: Biotechnology for wheat and maize breeding. *Biotechnology and Development Monitor*, No. 2 (March).

50. Komen, J. 1990b. IBPGR and genetic diversity. *Biotechnology and Development Monitor*, No. 4 (September).

51. Holle, M. 1987. Recursos genéticos vegetales de America del Sur: Pasado, presente y futuro. *Anales Simposio Recursos Fitogenéticos*. Valdivia: UACH-IBRG.

52. Prance, G. T. 1989. Economic prospects from tropical rainforest ethnobotany. In *Fragile Lands of Latin America: Strategies for Sustainable Development*, J. Browder, ed. Boulder, CO: Westview Press.

6

Managing Diversity in Traditional Agroecosystems of Tropical Mexico

Stephen R. Gliessman

Very little undisturbed forest is left within the humid tropical regions of Mexico, except for the land within the immediate boundaries of parks and reserves or in the more inaccessible areas of the Chiapas uplands. It has been estimated that 90% of the forests in the lowland humid tropics of Mexico were gone by the 1970s (1). Deforestation rates during the last decade continue to accelerate despite the growing concern of conservationists (2). Mexico is listed as one of the 10 countries that have lost the most forest during the last decade (2). Rates of forest clearing, estimated between 4.5 and 5.0% annually during the past 10 years, have led to the virtual elimination of tropical forests in some parts of tropical Mexico (e.g., the states of Tabasco, Veracruz, and Oaxaca) and their severe reduction in the rest (3). The actual loss of plant species in the humid Mexican tropics is debatable since no reliable inventories or national biological surveys exist, but Toledo (4) suggests that during the past decade at least 17 plant species have gone extinct and that another 477 are currently endangered. This represents 17% of the known floristic richness of Mexico. Even less is known about the probable impacts of deforestation on animal diversity.

No single or simple cause can be blamed for deforestation and consequent loss of biodiversity. It is the unfortunate consequence of many factors and conditions. The major components of this process include cattle ranching, colonization projects and agricultural development, forest fires, problems of land and tree tenure, timber and oil exploitation, and road building into previously undeveloped areas for an array of purposes. With the rapid expansion of all-weather highways into most tropical regions of Mexico, there has been a growing incentive for deforestation. Resource exploitation (mining, timber, oil, agriculture, etc.) accelerates, accidental or intentional forest fires become more common, and environmental degradation increases accordingly. New colonizers move onto tropical forested lands under the perception that these areas are a new agricultural frontier (5). Many deforested lands were at one time owned by the federal government and were used to alleviate the pressure for land from landless people (6).

In recent years, cattle production has taken the lead as the major form of development in such areas as the lowlands of Tabasco and the uplands of Chiapas (3, 7, 8). Large expanses of forested land, or land that had been in some other form of production, have been converted to pasture through clearing and planting of grasses. For example, in 1986, Tabasco had 1,649,620 hectares of grazing land out of a total area of 2,526,700 hectares (3). The rapid expansion of oil exploitation and a shift in agriculture toward export crops have also had severe effects on natural ecosystems and local food production systems (9). As the areas undergo deforestation and colonization, later immigrants find themselves confined either to progressively smaller parcels of land or to land of lower agricultural productivity, and the situation becomes worse.

AGRICULTURE IN THE TROPICS

The difficulties faced by agriculture in the humid tropics are well known (10, 11, 12). Many are related directly to problems that stem from farming under conditions of year-round high rainfall, heavy leaching of soil nutrients, poor soil fertility, rapid growth of pest problems, and increasing dependence on costly and imported inputs of fossil fuel–based chemicals. Therefore, the land that has already been cleared for crops is very difficult to keep in production, putting pressure on the increasingly limited forested land for new cropping areas.

The pressure to find new agricultural land has increased greatly, and since there is no new land available without moving onto forested land, it is essential that agricultural management strategies that restore productivity to already deforested land be given a high priority. To ensure that these new strategies are adopted and used, local inhabitants must be involved (13). Indigenous traditional farming and land-use knowledge that has a history of local adaptation and selection is an important starting point for the search for alternative systems (14–18).

Throughout much of the Mexican tropics, local land-use knowledge continues to form an important basis for much of the basic regional food production (Fig. 1). This knowledge reflectsexperience gained from past generations, which continues to develop with the ecological and cultural environment of local people involved. Studies of these traditional agroecosystems can contribute greatly to the development of ecologically sound management practices (14, 19–21). Such systems are able to provide products for local use and contribute to regional economies while maintaining a diverse vegetative cover over the soil (22). Complex crop mixtures, rotations, and practices developed by local farmers can provide environmental protection under tropical conditions as well as an array of harvestable products (21).

Many traditional farming systems make use of locally available resources rather than purchase chemical inputs. Often these are sufficient to satisfy local needs while contributing to production demands at the regional or national level. Production focuses more on the long-term sustainability of the system rather than placing overemphasis on maximizing yield and profit. The ability of these systems to maintain productivity on a sustained basis can lessen the need for the clearing of new lands. An agroecological approach to understanding how such systems function provides information that can reduce the pressure on rapidly diminishing forest reserves (23). Small-scale farmers, who have been displaced by the overall failure of national and international development strategies and are being blamed for more than 60% of the current deforestation in the tropics (2), are obviously in

Figure 1. A typical agricultural landscape in the lowlands of tropical Mexico, near Teapa, Tabasco. A corn/bean/squash intercrop is planted in the foreground, rice in a wet depression in the middle, corn on slightly higher ground to the right, casava (*yuca*) to the left, and a diverse forest garden with cacao in the interior behind.

the greatest need. A thorough knowledge of the interplay between physical, biological, and cultural factors of historically successful, traditional agroecosystems can foster an understanding of sustainable land-use systems. It seems apparent that such an approach must go hand in hand with efforts to preserve tropical rain forests. Several examples of such systems are presented below.

CASE STUDIES OF TRADITIONAL AGROECOSYSTEMS

Multiple-Crop Agroecosystems

The use of crop mixtures or multiple cropping can contribute greatly to increasing the sustainability of a cropping system on land formerly covered by tropical forest (24, 25). In such systems, more than one crop occupies the same piece of land either simultaneously or in some rotational sequence. Multiple cropping can increase yields, resources can be used more efficiently, and the land can be productive more continuously. Of course, there may be tradeoffs such as higher labor requirements or more complex management needs, but the importance of multiple cropping is well recognized (26), and the need for intensive agroecological studies of the effects of such mixed cropping has become more evident.

A traditional tropical multiple cropping system common in Mexico that has been studied in some detail is the corn, bean, and squash polyculture. Intercropping of corn (*Zea mays*) and beans (*Phaseolus vulgaris*) has been practiced in Central America since prehispanic times and continues to be important in food production today (27). Studies in Tabasco, Mexico (24), showed that corn yields could be

Table 1. Yields of polyculture corn/beans/squash as compared to monocultures planted at four different densities, Cardenas, Tabasco, Mexico (22).

	Monoculture Densities				Polycultures
Densities of corn*	33,000	40,000	66,000	100,000	50,000
Yield (Kg/ha)**	990	1,150	1,230	1,170	1,720
Densities of beans	56,800	64,000	100,000	133,200	40,000
Yield (Kg/ha)	425	740	610	695	110
Densities of squash	1,200	1,875	7,500	30,000	3,330
Yield (Kg/ha)	15	250	430	225	80

*Densities expressed as number of plants/ha.
**Yields for corn and beans expressed as dried grain, squash as fresh fruits.

increased as much as 50% above yields in monocultures when planted with beans and squash (Table 1).

There was some yield reduction for each of the two associated crop species, but the total yields for the two crops together were higher than in a monoculture. Studies of the ecological mechanisms of such yield increases are important before recommending widespread use of the cropping system. In a polyculture with corn, beans nodulate more and are potentially more active in biological fixation of nitrogen, which could be made directly available to the corn (28, 29, 30). This contributes to the future reduction of external inputs of fertilizer, as well as a sounder ecological basis for managing resources within the system.

At the same time, studies of the management practices employed in this polyculture have demonstrated the ecological basis upon which the practices can function. For example, despite lower squash yields in mixed plantings, farmers insist that the crop system benefits from the presence of squash through the control of weeds (31). Thick, broad, horizontal leaves cast a dense shade that blocks sunlight, and leachates in rains washing the leaves contain allelopathic compounds that can inhibit weed growth. Herbivorous insects are at a disadvantage in the intercrop system because food sources are less concentrated and more difficult to find in the mixture (32). Also, the numbers of beneficial insects often increase because of the availability of more attractive microclimatic conditions or the presence of more diverse pollen and nectar sources (33). In other systems, leaving weeds in the intercropped system can also be advantageous (34). *Chenopodium ambrosioides*, for example, has the potential for inhibiting plant pathogenic nematodes through the release of toxic root exudates (35). *Lagascea mollis* can produce chemicals that help control the invasion of weeds detrimental to the crops if it is allowed to form a dense cover after the critical establishment stage in crop development (the first three to four weeks), thus avoiding inhibition of the crop (31).

Diverse Home Gardens

Another agroecosystem that has been studied in some detail in tropical regions is the tropical home garden system. With trees forming the predominant structural element, home garden agroecosystems are ecologically well adapted to the tropics (36) and have even been proposed to be a managed "mimic" of tropical forest ecosystems (37). The landscape in many parts of the tropical lowlands of Mexico takes on the aspect of continuous, forest-like cover due to extensive use to

Figure 2. A forested landscape of merged home garden systems as viewed from the top of the Maya ruins at Comalcalco, Tabasco, Mexico.

the home garden system, especially when a shade-loving, economically valuable crop like cacao (*Theobroma cacao*) forms an important component (Fig. 2).

Many of the benefits of such systems for tropical farmers have been documented (38, 39). They are structurally very diverse, with an overstory of trees and an understory of herbs, shrubs, small trees, and vines. Animals are often an integral part of the gardens as well. Such diversity permits year-round harvesting of food products, as well as a wide range of other products used by local people such as firewood, medicinal plants, spices, and ornamentals.

In a home garden in the lowlands of Tabasco, a total of 89 useful plant species were found, providing food, firewood, medicine, spices, and general adornment to the household (Table 2)(38). Some plant species served more than one function. The Shannon diversity index, which took into account both the number of different species and how they were distributed in the garden, was over 3.5, a relatively high value for an agricultural system. To a certain extent, plants were also distributed in the garden depending on their uses. Trees were concentrated as shade over cacao, or in a more diverse, mixed orchard that combined useful and ornamental species. Annual crops (such as corn, chiles, squash, and beans) or herbaceous perennials (such as casava, papaya, and sugar cane) were concentrated in openings in the tree canopy and often protected with palm frond fences from free-ranging animals such as chickens, ducks, and turkeys, whereas other animals were kept in small pens (e.g., pigs). A large number of ornamental and medicinal species were clustered in an open area close to the family's thatched-roof house, which was made from local materials. Cacao and bananas were the most important understory perennial species, with a wide range of overstory trees including such species as coconut, citrus, breadfruit, avocado, mango, tamarind, sapote, chirimoya, cinnamon, all-

Table 2. The number of plant species in a home garden agroecosystem near Cupilco, Tabasco, Mexico, listed according to common uses. Total species numbers are less than the sum of the columns due to the multiple function of some species (adapted from 38).

Garden or Use Area	Size (ha.)	#Useful Spp.	#Native Spp.	#Tree Spp.	#Ornamental & medicinal
cacao area	0.3	21	15	15	1
mixed trees	0.5	35	22	24	5
ornamental/ medicinal area	0.1	74	48	31	24
annual areas	0.013	5	5	1	2
Totals	0.913	89	58	42	24

spice, and others. Food crops formed a mix of native and introduced species.

Interestingly, many of the emergent trees in the canopy were native forest species that had value for construction, firewood, nitrogen fixation, or other uses (e.g., *Acacia angustissima*, *Brosimum alicastrum*, *Bursera simaruba*, *Byrsonima crassifolia*, *Cecropia obtusifolia*, *Cedrella odorata*, *Colubrina arborescens*, *Crescentia cujete*, *Crotolaria maypurensis*, *Diphysa robinoides*, *Erythrina glabra*, *Glyricidia sepium*, *Heliocarpus donnell-smithii*, *Inga jiniquil*, *Muntingia calabura*, *Pogostemon cablin*, *Roystonea regia*, *Sapindus saponaria*, *Scheelia liebmanii*, *Spondias mombin*, *Swietenia macrophylla*, *Tabebuia rosea*). The high diversity of native forest species in this garden, and in others that have been examined in the Maya region of Mexico, add credence to the benefits of these systems in the preservation of biodiversity.

More research is needed in order to understand further the structure and diversity of home gardens. They are extremely variable in size and design. Introduced plant and animal species are integrated with native species. They respond to local variations in soil type, drainage patterns, cultural preferences, economic standing of the family, and family size and age patterns, reflecting a multiplicity of both ecological and cultural components. At the same time they are flexible, dynamic, and changing, depending on the needs of the family (40).

Traditional Wetland Agroecosystems

Physical or biological factors that are limiting to agricultural production require special attention if agriculture is to be sustainable and help reduce the need to clear new land for production. Many agricultural development projects in the tropics have attempted to eliminate or alter such limitations. This usually involves large levels of external inputs of energy or materials. There are many and well-known examples of massive irrigation, drainage, or desalination projects that attempted to alter existing ecological conditions, achieving only limited success regarding crop productivity and economic viability, and had little applicability for small-scale farmers (41). In many areas of tropical Mexico, where heavy rainfall and low-lying topography combine to generate conditions of water-logging, local farmers have found ways to accommodate excess soil moisture into the design and management of sustainable traditional agroecosystems.

An interesting and productive use of wetlands has been observed in Tabasco for the production of corn and other crops that integrates native wetland plants and animals into the system (42). Since this practice is observed most often in communities of indigenous background, the practice was probably handed down from generation to generation. In this system, corn is planted on higher ground around flood-prone areas during the wet-season months of June to December. As water levels drop during the dry-season months of February and March, farmers follow the receding water line with another corn planting, known locally as the *marceño* or March planting (Fig. 3). Similar systems have been observed as far north in Mexico as Veracruz (43).

Figure 3. A corn/squash intercrop in a low-lying area subject to flooding for the greater part of the year; planted during the short dry season between March and May, near Cardenas, Tabasco, Mexico.

During much of the year, the low areas are inundated to a depth of a few centimeters to as much as a meter. Marsh vegetation covers the area densely during the wet season and is cut quickly with machetes as the water level recedes. A very dense mat of organic matter is produced by this process (10 to 20 cm), into which seed is planted with a pointed stick. About a week following planting, fire is used to burn part of this organic mat, as well as kill back any weed seedlings or sprouts of the marsh plants. The timing of the burning is important to ensure that the fire burns only the dry leaves on top of the mat and does not reach the soil. Corn or other seed, planted 10 to 15 cm below the surface of the soil, is not harmed by the fire. Local short-cycle varieties of corn (2 to 3 months from planting to harvest) are most frequently used. The practice of using seed from the previous harvest for the subsequent planting favors the use of local varieties. The use of a 21/2-month variety known as "mejen," a Maya word meaning precocious or early maturing, again shows the link to the past that many of the present-day systems may have.

Corn grows very quickly in this system, and when fire is not used excessively and flooding is allowed to occur every year, weeding is usually not necessary. At approximately two and a half months the mature corn stalks are "doubled-over" just below the corn ear, facilitating final drying of the grain for another two to four weeks before harvest. Yields between 4–5 t/ha of dry grain are common, and in one year in five, yields can reach 10 t/ha. This is many times the average yield for mechanized production on lands that have been cleared and drained in the same region, but at a fraction of the input labor and cost (24).

Following harvest, all crop and noncrop residues are left on the soil surface. Soil profiles demonstrate the presence of a thick organic soil to a depth of 30 to 40 cm below the surface. During inundation, organic matter produced by the marsh plants, or left by the previous cropping cycle, are incorporated and conserved below water. Nutrient minerals that enter the system with surface drainage are captured by the highly productive aquatic sector of the ecosystem, leading to the formation of a soil that has organic matter levels over 30%, total nitrogen as high as 3%, and high levels of other important plant nutrients, as shown by analysis. The key element, then, in the management of this system is to take advantage of the benefits gained from a period of inundation and the recolonization by native species. When the system is drained artificially in an attempt to extend the cropping season, the organic layer in the soil can be reduced to 5 cm in less than two years, and yields drop dramatically.

FUTURE PERSPECTIVES

An agroecological focus for agricultural development in the humid tropics extends beyond crop yields, assessing the complex set of factors that contribute to agroecosystem sustainability. Local, traditional agroecosystems that have developed a degree of permanence despite the diverse and often limiting conditions facing farmers in the tropics are adapted to these factors. These systems have evolved through time with little dependence on purchased chemical inputs, a greater reliance on renewable resources, and ecologically based management strategies. Future research in agriculture for the humid tropics must take advantage of this knowledge and experience. It represents the blending of ecological principles with the knowledge of farmers and agronomists as to how best to manage the complexities of food producing agroecosystems. This can provide the sustainability in the production base so critical for giving small farmers the stability and viability they need to provide for their own needs, contribute to meeting the needs of the greater society, and ultimately reduce the pressure on tropical forests.

REFERENCES

1. Rzedowski, J. 1978. *Vegetacion de Mexico*. Mexico City: Editorial Lemusa.
2. Myers, N. 1991. Tropical deforestation: The latest situation. *BioScience* 41:282.
3. Gomez-Pompa, A., A. Kaus, J. Jimenez-Osornio, D. Bainbridge, and V. Rorive. 1991. Deforestation and sustainable agriculture in the humid tropics: A case study of Mexico. Report to the National Research Council, Committee on Deforestation and Sustainable Agriculture in the Humid Tropics, Washington, DC: NCR.
4. Toledo, V.M. 1988. La diversidad biologica de Mexico. *Ciencia y Desarrollo* 81:17–30.

5. Parsons, J.J. 1976. Forest to pasture: Development or destruction? *Rev. Biol. Trop.* (Suppl. 1) 24:121–38.
6. Gomez-Pompa, A. 1987. Tropical deforestation and Maya silviculture: An ecological paradox. *Tulane Studies in Zoology and Bot.* 26(1):1–17.
7. Nigh, R.B., and J.D. Nations. 1980. Tropical rainforests. *Bull. of the Atomic Scientists* 36(3):12–19.
8. Myers, N. 1979. *The sinking ark*. Oxford: Pergamon Press.
9. Scherr, S.J. 1985. *The oil syndrome and agricultural development: Lessons from Tabasco, Mexico*. New York: Praeger.
10. Janzen, D.H. 1973. Tropical agroecosystems. *Science* 182:1212–19.
11. Ruthenberg, H. 1980. *Farming systems in the Tropics*. Rev. ed. Oxford: Clarendon.
12. Goodland, R.J.A., C. Watson, and G. Ledec. 1984. *Environmental management in tropical agriculture*. Boulder, CO: Westview Press.
13. Chambers, R.A., A. Pacey, and L.A. Thrupp. 1989. *Farmer first: Farmer innovation and agricultural research*. London: Intermediate Technology Publications.
14. Gliessman, S.R., R. Espinosa, and M. F. Amador. 1981. The ecological basis for the application of traditional agricultural technology in the management of tropical agroecosystems. *Agro-Ecosystems* 7:173–85.
15. Gliessman, S.R. 1988. Local resource use systems in the tropics: Taking pressure off the forests. In *Tropical Rainforests: Diversity and Conservation*. F. Almeda and C.M. Pringle, eds., 53–70. San Francisco: Cal. Acad. Sci.
16. Gliessman, S.R. 1988. The home garden agroecosystem: A model for developing more sustainable tropical agricultural systems. In *Global perspectives in agroecology and sustainable agriculture*. P.A. Allen and D. Van Dusen, eds., 445–53. Santa Cruz: University of California Press.
17. Altieri, M.A., and M.K. Anderson. 1986. An ecological basis for the development of alternative agricultural systems for small farmers in the Third World. *Am. J. Alt. Agriculture* 1:30–38.
18. Altieri, M.A., and S.B. Hecht. 1990. *Agroecology and small farm development*. Boca Raton, FL: CRC Press.
19. Klee, G. 1980. *World systems of traditional resource Management*, ch. 1. New York: Halstead.
20. Altieri, M.A. 1987. *Agroecology: The scientific basis of alternative agriculture*. Boulder, CO: Westview Press.
21. Wilken, G.C. 1988. *Good farmers: Traditional agricultural resource management in Mexico and Central America*. Berkeley, CA: University of California Press.
22. Gliessman, S.R. 1989. The ecology and management of traditional farming systems. In *Agroecology and small farm development*. M.A. Altieri and S.B. Hecht, eds., 13–18. Boca Raton, FL: CRC Press.
23. Gliessman, S.R. 1992. Agroecology in the tropics: Achieving a balance between land use and preservation. In *Environmental Management* (in press).
24. Amador, M.F., and S. R. Gliessman. 1990. An ecological approach to reducing external inputs through the use of intercropping. In *Agroecology: Researching the ecological basis for sustainable agriculture*, S.R. Gliessman, ed., 146–59. New York: Springer-Verlag Ecological Studies Series.
25. Gliessman, S.R. 1985. Multiple cropping systems: A basis for developing an alternative agriculture. In *Innovative biological technologies for lesser developed countries—workshop proceedings*. Office of Technology Assessment, ed., 69–83. Washington, DC: OTA.
26. Francis, C.A. 1986. *Multiple cropping systems*. New York: MacMillan.

27. Pinchinat, A.M., J. Soria, and R. Bazan. 1976. Multiple cropping in tropical America. In *Multiple cropping*, R.I. Papendick, P.A. Sanchez, and G.B. Triplett, eds., 51–64. Amer. Soc. Agron. Spec. Publ. 27.

28. Boucher, D.H. 1979. La nodulacion del frijol en policultivo: el efecto de la distancia entre las plantas de frijol y maiz. *Agricultura Tropical (CSAT)* 1:276–83.

29. Bethlenfalvay, G.J., M.G. Reyes-Solis, S.B. Camel, and R. Ferrera-Cerrato. 1991. Nutrient transfer between the root zones of soybean and maize plants connected by a common mycorrhizal inoculum. *Physiol. Plant.* 82:423–32.

30. Gliessman, S.R. 1982. Nitrogen distribution in several traditional agro-ecosystems in the humid tropical lowlands of south-eastern Mexico. *Plant and Soil.* 67:105–17.

31. Gliessman, S.R. 1983. Allelopathic interactions in crop-weed mixtures: applications for weed management. *J. of Chem. Ecol.* 9:991–99.

32. Risch, S. 1980. The population dynamics of several herbivorous beetles in a tropical agroecosystem: The effect of intercropping corn, beans and squash in Costa Rica. *J. Appl. Ecol.* 17:593–612.

33. Letourneau, D.K. 1986. Associational resistance in squash monocultures and polycultures in tropical Mexico. *Environmental Entomology* 15:285–92.

34. Chacon, J.C., and S.R. Gliessman. 1982. Use of the "non-weed" concept in traditional tropical agroecosystems of southeastern Mexico. *Agro-Ecosystems* 8:1–11.

35. Garcia, R. 1980. *Chenopodium ambrosioides* L., planta con uso potencial en el combate de nematodos fitoparasitos. *Agricultura Tropical (CSAT).* 2:92–98.

36. Gliessman, S.R. 1990. Integrating trees into agriculture: The home garden agroecosystem as an example of agroforestry in the tropics. In *Agroecology: Researching the ecological basis for sustainable agriculture*. S.R. Gliessman, ed., 160–68. New York: Springer-Verlag, Ecological Studies Series 78.

37. Ewel, J., S.R. Gliessman, *et al.* 1982. Leaf area, light transmission, roots and leaf damage in nine tropical plant communities. *Agro-Ecosystems* 7:173–85.

38. Allison, J. 1983. An ecological analysis of home gardens (huertos familiares) in two Mexican villages. M.A. Thesis, Biology. Santa Cruz: University of California.

39. Alcorn, J.B. 1989. Indigenous agroforestry systems in the Latin American tropics. In *Agroecology and small farm development*. M.A. Altieri and S.B. Hecht, eds., 203–20. Boca Raton, FL: CRC Press.

40. Gonzalez Jacome, A. 1985. Home gardens in central Mexico. In *Prehistoric intensive agriculture in the tropics*, I.S. Farrington, ed. Oxford: BAR International Series 232.

41. Barkin, D. 1978. *Desarrollo Regional y Reorganizacion Campesina: La Chontalpa como Reflejo del Problema Agropecuario Mexicano*. Mexico City: Editorial Nueva Imagen.

42. Orozco-Segovia, A., and S.R. Gliessman. The marceño in flood-prone regions of Tabasco, Mexico. In *Pasado y Presente en los Sistemas Agricolas en Mexico*. A. Gonzalez J., and S.R. Gliessman, eds. Mexico City: SEP (in press).

43. Siemens, A.H. 1990. Reducing the risk: Some indications regarding prehispanic wetland agricultural intensification from contemporary use of a wetland/terra firma boundary zone in Central Veracruz. In *Agroecology: researching the basis for sustainable agriculture*. S.R. Gliessman, ed., 233–50. New York: Springer-Verlag, Ecological Studies Series 78.

7

Conservation of Maize Crop Relatives in Guatemala

Garrison Wilkes

As plant populations diminish to the point of extinction or their habitats are lost as ecosystems degrade, there is often no signal that they are failing. If humans are aware of these changes, they can sound the warning and advocate for the disappearing taxa or habitat loss. Because of my 30-year association with the wild relatives of maize, I suddenly find myself advocating for the *in situ* conservation of teosinte populations in both Guatemala and Mexico.

TEOSINTE

Teosinte is the closest wild relative of maize (1). Both maize and teosinte hybridize readily, the hybrid and subsequent generations being fully fertile. This co-evolutionary relationship can also be seen in the archeological record going back nearly 5,000 years. Yet, the two are separate and distinct taxa. In appearance they are very similar in growth pattern, a central stalk and terminal tassel, but maize has numerous kernels borne in a husk-enclosed massive cob, while teosinte possesses small spikes of 7 to 12 seeds each enclosed by a rachis segment, which disarticulates at a brittle zone of abscission between the segments. This ability to sow its seed separates teosinte as a wild plant from maize, a cultivated crop. Teosinte is the prototype wild relative of a major food crop with both weedy forms and large naturally occurring populations covering thousands of acres (2). The distribution of teosinte in Mexico and Guatemala exactly parallels the cultural zone of Meso-America where the origin of maize is thought to have taken place (3).

The wild taxa of *Zea*, collectively called teosinte, have both annual and perennial taxa. The annuals, all of them diploid (2n = 20), are found in geographically isolated population clusters, six in Mexico, two in Guatemala (with extensions to Honduras and Nicaragua). These occur within the subtropical zone and receive summer rains. Some of the populations are small, the smallest occupying less than 1 km^2, while the largest covers thousands of square kilometers. There are two perennial teosintes, both of which produce rhizomes, occurring only in restricted habitats in the state of Jalisco, Mexico. One, *Zea diploperennis*, Iltis, Doebley and Guzman, is diploid ($2n = 20$) and the other, *Z. perennis* (Hitch) Reeves and Mangelsdorf, is tetraploid ($2n = 40$) (4).

Throughout this paper the name "teosinte" refers to the annual teosintes. The perennial teosintes are the rarest of all the taxa and the discovery of a new taxon in the late 1970s resulted in the first and most spectacular story for *in situ* conservation of any wild relative of a crop plant: the preservation of an entire ecosystem of 140,000 ha with magnificent subtropical forests, the Sierra de Manatlán Biosphere Reserve in Jalisco, Mexico (5). This story is best told by one of the major participants, Hugh Iltis (6).

As described in Table 1, this chapter will deal with the annual taxa for which a conscious effort has been made to avoid the use of taxonomic categories (races or species and subspecies) for the various taxa of annual teosinte *Zea luxurians* (Durieu and Ascherson) Bird (7), *Zea mexicana* (Schrader) Kuntze (1), or *Zea mays* subsp. *parviglumis* Iltis and Doebley (8), and subsp. *mexicana* (Schrader) Iltis and *Zea mays* subsp. huehuetenangensis (Iltis & Doebley) Doebley. Instead, the geographically distinct teosinte localities will be referred to as populations. It is appropriate to do so since the unit of conservation of genetic resources is the population and the genes held therein, not the named taxa.

Habitats of Teosinte

The distribution of teosinte populations along the western and southern escarpment of Mexico and Guatemala is limited to a seasonally dry, subtropical zone

Table 1. Classification systems for Zea.

Wilkes (1967)	Iltis & Doebley (1984) Doebley (1990)
Sect. *Euchlaena*	Sect. *Luxuriantes*
	Z. diploperennis Iltis, Doebly & Guzman
Z. perennis (Hitchc.) Reeves & Mangelsdorf	*Z. perennis* (Hitch) Reeves & Mangelsdorf
Z. mexicana (Schrader) Kuntze	*Z. luxurans* (Duria & Ascherson) Bird
race Guatemala	
	Sect. *Zea*
race Balsas	*Z. mays*
	subsp. *parviglumis* Iltis & Doebley
race Huehuetenango	subsp. *huehuetenangensis* (Iltis & Doebley) Doebley
	subsp. *mexicana* (Schrader) Iltis
race Chalco	race Chalco
race Central Plateau	race Central Plateau
race Nobogame	race Nobogame
Sect. *Zea*	
Z. mays L.	*Z. mays* supsp. *mays* (L) Iltis

that is 800 to 1,800 meters above sea level and receives summer rains. Slightly higher on the Central Plateau of Mexico at 1,650 to 2,000 meters and in the Valley of Mexico at 2,250 meters, teosinte is usually associated with cultivated land. In the states of Jalisco, Guanajuato, and Michoacan, the plant grows mostly along stone fences bordering maize fields, not as a weed invading the fields, but as a rare survivor making its last stand in this narrow strip of untilled soil. In a few places such as Chalco in the Valley of Mexico, teosinte has successfully invaded maize fields as a mimic of the cultivated crop. The two, teosinte and maize, are so similar in appearance that farmers cannot readily distinguish them and thus do not weed teosinte from their fields during the early part of the agricultural cycle. Teosinte's growing season parallels that of maize, June through October (August through January in Huehuetenango, which has late rains). Germination begins with summer rains and by August or September (November and December in Huehuetenango) teosinte has reached the mid-flowering stage. Mature seed falls to the ground late October through December, where it remains dormant until the next summer rains (9).

Figure 1. Francisco Vazques observing margin of maize field where teosinte is growing against a stone wall. The site is just outside Agua Blanca on the road to Santa Catarina Mita, Department of Jutiapa, Guatemala, at an elevation of 1,000 m. The plants are dispersing seed.

Teosinte is not uniformly distributed on the western escarpment of Mexico and Guatemala, each taxon existing in semi-isolated populations, more or less in geographic isolation. Most of the geographic populations are spatially isolated by the broken topography of high mountains and deep valleys. The environments of these geographic populations are not identical, and teosinte has developed several physiologically and taxonomically differentiated populations, each of which has acquired a limited morphological, ecological, chromosomal, and genetic distinctness (10). The most distinct morphologically, chromosomally, and genetically are the teosintes from southeastern Guatemala in the departments of Jutiapa, Jalapa, and Chiquimula (Fig. 1).

The teosinte population in Huehuetenango, Guatemala, is allied to the Mexican teosintes. The "Florida" or "Guatemala" teosinte from southeastern Guatemala appears to be more primitive and ancestral, while the Mexican teosintes are more maize-like. Indeed the reciprocal teosinte introgression is recognized in more than half of the approximately 50 landraces of maize in Mexico (11).

Experiences With Teosinte

In 1960, the botanical literature on teosinte was old, nineteenth century, and fragmentary. At the time my field work began in Mexico in 1962, I was not sure teosinte could be located except at the well-known locality around the town of Chalco outside of Mexico City (12). What I soon learned, primarily from the wealth of personal knowledge of Professor E. Hernandez-Xolocotzi at the Mexican National School of Agriculture at Chapingo, was that teosinte could be located throughout Mexico (13). The most complete distribution maps for Mexican teosinte have been subsequently published by T. Angel Kato-Y of the Escuela de Postgraduates at Montecellas (14); J. Sanchez and L. Ordaz (15) of Mexico's Institute of Forestry, Agriculture, and Livestock Research (INIFAP); and one by H.H. Iltis, Duane Kolterman, and Bruce Benz (16).

Recently, my work with teosinte has taken me into genetic conservation, and during a sabbatical year 1985–86, participation in the reorganization and rejuvenation of the World Maize Genebank at the International Center for the Improvement of Corn and Wheat (CIMMYT). I have visited nearly all of the teosinte populations once each decade and some localities I have seen, on average, every third year. In the last decade I find myself becoming an advocate for the survival of teosinte's shrinking populations. Throughout these studies, I have been supported by the National School of Agriculture at Chapingo, the Mexican Maize Gene Bank, INIFAP, CIMMYT, and more recently, the Guatemalan Instituto de Ciencia y Technologia Agricolas (ICTA).

Threats of Extinction

One of the very interesting aspects regarding extinction threats to various teosinte populations, especially those in Guatemala, is the necessity for individuals to sound the alarm. The slow contraction of geographic distribution of the population is the classic example of genetic erosion, but because it is slow, it seldom illuminates a realization of what has happened. People in the country accept and adapt to the changes, but coming from the outside after 10 years, the change for me was obvious and shocking. Long-term familiarity with a region and periodic field observation are the best means to document changes brought about by genetic erosion. The frightening aspect about these observations is the fact that probably four-fifths of all major crop plant wild relatives do not have an advocate with either long-term field studies of the plant or recurrent (three visits per decade) collection trips over a period of time. Given these weaknesses, it is no wonder that there is little concern for the *in situ* conservation of crop plant wild relatives (17). Thirty years ago I did not set out to engage in the genetic conservation of teosinte but now, realizing the implications of what is happening, I am obligated to act.

Teosinte has died out at several sites on Mexico's central plateau in the last 30 years because of more intensive land use (cornfields have become strawberry fields) and pasturage. In general, the distribution of the plant is contracting, and some populations, like the Central Plateau, are losing sites, but the population as a whole is not yet threatened with elimination in Mexico. The chief reasons for the contraction of teosinte populations are (1) intensified land use (which is directly related to the development of roads); (2) genetic swamping of small isolated stands that outcross to maize and thus lose their ability to disperse seed; and (3) cultivation of a cash crop such as short-stature sorghum instead of maize, making the presence of teosinte as a weed more obvious. I estimate that the current distribution of teosinte is about half of what it was in 1900. The disappearance of remaining populations will accelerate as more roads are built and land use is intensified.

Just 30 years ago, a third of the teosinte populations were far from any road, and another third were accessible by dirt roads that could be travelled only with difficulty. Now all are accessible by roads that can easily be negotiated by an automobile. This is especially true for Guatemala. The cluster of localities that makes up the unique teosintes of southeastern Guatemala have dropped to one-fourth of their former distribution (i.e., three out of four are no longer extant or, more bluntly, teosinte is wiped out at three-fourths of the places it used to be found), and in Huehuetenango teosinte is reduced to a stand of approximately one-tenth its former size.

Although it is difficult to define danger of extinction with much precision, some helpful criteria have been established by the Species Survival Commission of the International Union for the Conservation of Nature and Natural Resources (IUCN) in Switzerland. Five categories proposed by the IUCN—extinct, endangered, vulnerable, rare, and indeterminate—can be readily applied to teosinte. Some populations are already extinct and two Guatemala populations are endangered and cannot be expected to persist much longer as naturally occurring populations without *in situ* protection. Most Mexican teosinte populations (the exception is the large populations in the state of Guerrero above the Rio Balsas) can be considered vulnerable; that is, they are declining to such an extent that if nothing is done, they will become endangered. The populations thought to be rare (Nobogame, Durango, and Oaxaca) are single location sites that could be eliminated easily, but they currently appear to be under no immediate threat and are more or less stable (Table 2).

Curbing the Threat

Although teosinte is not likely to provide any means for short-term or intermediate agronomic improvement of maize, it most surely does carry genes not found in maize (18). If teosinte goes there will be no close, living relative of maize to serve as a source of such variation (19). To better protect this vital part of the genetic foundation for the world's maize crop, I received a grant from the U.S. Department of Agriculture's Corn Advisory Committee for collection of teosinte seed. The object of these collections (Fig. 2) will be to put 1-1/2 kg of seed (approximately 100,000+ seed) into long-term gene bank storage (*ex situ*).

For two years I was prohibited from collecting in Guatemala because of a ban on collecting any natural wild plant genetic resources (cultivated maize would not have been considered a national treasure and would not have been prohibited—since teosinte was wild, it was considered a national heritage and therefore collecting was banned). When clearance did come, teosinte populations in Guatemala were discovered to be well on the road to extinction and if actions were not taken quickly, meaning *in situ* conservation, Guatemala's ban on seed collecting would actually ensure the loss of these unique taxa. What follows is an account of the trip to Guatemala and the outcome of the discovery that teosinte is nearing extinction.

Table 2. Teosinte populations in Mexico and Guatemala.

Population and its status as of 1990/91	Common name	Location	Extent	Habitat
Nobogames	*macicillo* (little or lesser corn)	Tarahumare Valley in the Sierra Madre of the state of Chihuahua about 16km northeast of Guadalupe y Clavo.	No more than 50km^2 on the valley floor occuring at a single location.	Along margins of maize fields and (rare) in willow thickets bordering streams.
Durango (rare)		Valley of Guadiana 10km outside Durango in the state of Durango.	No more than 5km^2 occuring at a single location.	Limited to wasteland along irrigation canals.
Central Plateau (vulnerable)	*maiz de coyote* (wild corn)	Isolated populations througout the entire central plateau in the states of Jalisco, Michoacan, and Guanajuato. Largest continuous population is in region north of Lake Cuitzeo.	Once a continuous population covering thousands of square kilometers but now occurs in scattered, isolated pockets rarely larger than 10 km^2.	Occurs in cultivated fields and along field margins or stone walls and in areas protected from grazing.
Chalco (stable) disagreeable)	*accece* or *acece* (unwanted or	Valley of Mexico from Amecameca to Xochimilco, Chalco, and Los Reyes. Isolated populations around Texcoco.	Main populated centered in a 300-km^2 area around Chalco. Estado de Mexico just outside of Mexico City.	Found almost exclusively in maize fields as a "mimic" of the maize but also as a weed along roadsides.
Balsas (stable)	*maiz de huiscatote* (roadrunner) *maiz de pajaro* (bird) *atzitintle*	Hills surrounding Rio de las Balsas Basin. Population distributed discontinuosly, with one located south of Chilpancingo in the state of Guerrero and the other on the northern rim ofof the basin.	Population south of Chilpancingo covers hundreds of sq. km, while the other extends over thousands of sq. km in the states Guerrero, Michoacan, and Estado de Mexico that border Guerrero.	Sometimes observed in maize fields generally in dense stands on hillsides, especially along gullies or other areas where there is rain runoff.
Oaxaca (rare)	*Cocoxle* (ground dove or roadrunner)	San Francisco de Honduras, 5km from San Pedro Juchatengo in the Sierra Madre del Sur of Oaxaca.	No larger than 5km^2, although there may be outlying, isolated pockets in the Sierra Madre del Sur.	Grows on hillsides and in maize fields surrounding the town.
Huehuetenango (endangered)	*milpa de rayo* (where lightning strikes the fields) *salic*	Hills and valleys of the Departamento Huehuetenango around the Guatemalan town of San Antonio Huista, near the Mexican frontier.	Probably not larger than 100km^2 and now very fragmented.	Found along trails, in fields, and on hillsides of deserted milpas, or maize fields.
Guatemala (endangered)	*milpa silvestre* (wild corn) *teocintle*	Distributed discontinuously in south-eastern Guatemala across the hills and valleys of Jutiapa, Jalapa, and Chiquimula.	Was once distributed continuosly covering 500 km^2, but is now fragemented.	Occurs at small, isolated sites along the marginsof fields or in other area protected from grazing.

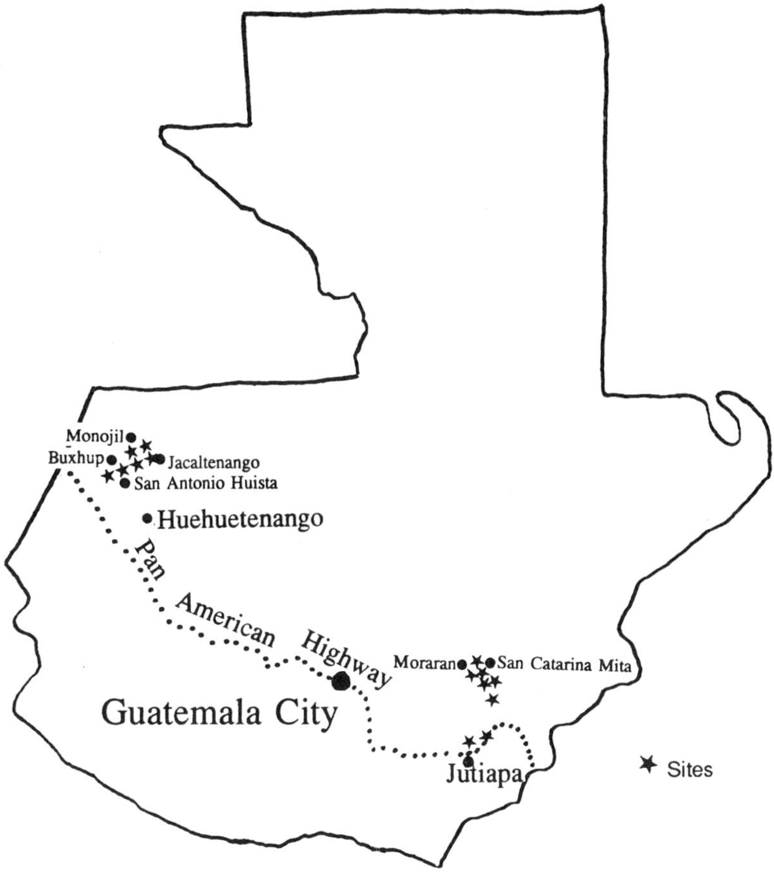

Figure 2. Map illustrating teosinte collection sites in Guatemala referred to in this chapter.

CASE STUDY DIARY: JANUARY 18, 1991

Dr. Suketoshi Taba (Director, CIMMYT maize gene bank) and I flew to Guatemala and met with Francisco Vazques, head of the Genetic Resource Unit, Guatemalan Institute for Science and Technology (ICTA), Francisco Olivet, ICTA Manager; Danilo Gonzales, ICTA Technical Director; and Nory Soto, ICTA Coordinator Maize, to explain the purpose of our trip: to visit the known sites of teosinte populations in the country.

Mutually agreed-upon ground rules for this trip were explained by Francisco Olivet. Collections of seed could be made by the group but the seed was not to leave the country. All seed must remain with the Genetic Resource Unit. In addition, Francisco Vazques was to accompany us at all times during the field work.

Proceeding to the Jutiapa station of ICTA we met with Leroy Gilespie and drove with him down the Pan American Highway to km 128, where teosinte existed on a steep road cut at an elevation of 800 m. This is a local population found only on the margin of three house properties and the road cut—perhaps a thousand plants. After collecting seed and a specimen of an F1 hybrid, we proceeded to another population on the other side of El Progreso on the Jutiapa road at 1,000 m.

I had collected in this area in 1962–63; Gilespie had found plants here in 1976 with Hugh Iltis; and I had collected teosinte here with George Beadle in 1973, but we were not able to locate plants on this trip.

January 19, 1991

We revisited the km 128 Pan American Highway site for photographs and met Cecilio Hernandez, who said he had fields of teosinte on his ranch. We did indeed find heavy populations there, in areas protected from grazing. Hernandez said that teosinte or "teocinte," as the locals call it, was previously widely distributed but was not very common now. He had tried to keep a little forest land on his knolls and thought these areas of "natural vegetation" promoted the survival of teosinte. In fact, his corn fields without fences between the corn and the grazing area (fallow fields) probably did more to protect the teosinte by keeping the livestock out when maize and teosinte was green. When teosinte is dry and the seed mature, it is passed over by livestock. When green, it is sweet and preferred grazing for animals. We collected seed and found the population to consist of tens of thousands of plants.

We took leave of Hernandez at km 129 and proceeded back to El Progreso and the road to Santa Catarina Mita. Around Cuesta de Guayebo, where teosinte used to be common, we found none. In fact, none was seen during the entire trip to Santa Catarina Mita. Admittedly, our survey is anecdotal and can only compare today with what was collected in 1962–63. It is at best a presence/absence measure of known 1962–63 sites. It does not measure total area covered or the density of population stands. Clearly, teosinte is seen less frequently from the road. In fact, no plants were seen until Aqual Blanca, and we would not have seen this population from the road but were told of its existence by a man with two horses loaded with wood on the road Santa Catarina Mita-Aqua Blanca. This population (950 m) of a few hundred plants was typical of fence margin populations in 1962–63.

January 20, 1991

We returned to El Progreso and attempted to find teosinte again, without success. We also searched the Laguna Retana, which had supported a huge population in the past (20), again without success. We were told "teocenta" (teosinte) was up around Moraran just after the second bridge. We located the road, which happened to be the San Manual Chaparron Road and used to be a great locality for teosinte. At San Juan Salamo we inquired about teosinte and, with the help of a young boy, we backtracked to the Finca of Carlos Pinto. Near the stream banks of the Rio Mojarritas, in an area fenced off because of the surrounding maize fields, we found the most well-developed population of the luxurians form of teosinte with the pronounced trapezoidal fruitcase. This is one of the most unique, both morphologically and genetically, of all the teosinte populations in Mexico and Guatemala. In terms of a genetic resource, this is the most valued for preservation. The hypothesis is that this form, which is the least maize-like, is the most primitive form of teosinte and has had the least introgression of maize genes. This was an exciting find because the seed we collected matches the old "Florida Type" teosinte. Florida Type came from Santa Rosa in the nineteenth century and is now extinct in Santa Rosa de Lima (16), since the area is planted to sugarcane.

In total, we have eight stations for teosinte, whereas in 1962–63, I had 37 roadside seed collections. An estimate is that in the last 30 years in the departments of Jutiapa, Jalapa, and Chiquimula, teosinte is approximately 25% of what it used to be. The cause of this decline is increased land use, fewer fallow areas, habitat erosion, and increasing numbers of people and livestock. Increased pressure from grazing livestock is probably the most critical. Repeatedly we were told that teosinte used to be common, but when the plant is green the cattle seek it out because it is sweet. Teosinte is by no means extinct, and careful exploration will locate many additional populations in the region. National programs should undertake such an exploration.

January 22, 1991

On January 21, we crossed Guatemala to Huehuetenango in the Department of Huehuetenango. We left Huehuetenango for the Rio Huista, travelling on the Pan American Highway, and after La Democracia, on a new dirt road that goes though the 800-m plain at the base of the mountains. We turned at four corners for Santa Ana Huista, then climbed to 900 m and found *Tripsacum*, and turned toward San Antonio Huista (1,200 m), which we reached about noon. Javier Gomez, an out-farm agronomist from the Huehuetenango ICTA office, accompanied us. We retraced the road back to Monojil (1,000 m) because we found no teosinte around San Antonio Huista. The previous year, Javier had observed teosinte on the path between Monojil and Buxhap. Toward late afternoon, we returned up the road to San Antonio (for a rest stop) and proceeded on to Jacaltenango for the night. Teosinte was seen in maize fields up to 1,450 m and, in the highest fields, the plants were still green. A review of the day indicates that teosinte is about 10% of what it was in 1962–63.

The absence of teosinte was a surprise to me because it was in such sharp contrast to my observation in 1962–63, and as recently as 1979. To make this change clearer to the readers, I would like to quote from my thesis *Teosinte: The Closest Relative of Maize* (1), page 56:

"Plants (teosinte) are very abundant around San Antonio Huista (1240 m) and the population extends from downstream in the vicinity of Santa Ana Huista (1000 m) and Buxhap (920 m) to the ridge (1650 m) between San Antonio Huista and Jacaltenango. Teosinte is again encountered on the Rio Azul between Jacaltenango and Tzisbaj (1500 m). This population connects with the Rio Huista population at Buxhap.

> "Teosinte is present along trails, in the milpa (maize fields), and covering the hillsides of long-deserted milpas. There are few fences in the region and grazing is controlled by the municipality, which restricts all grazing to a section of the village that is to remain fallow for the year. *Tripsacum* is abundant in the region and, unlike teosinte, is as frequent in the surrounding oak forest as it is in the high grass of deserted milpas. The town of San Antonio Huista (1200 m) lies on the south-facing slope at the vegetation change from oaks to pines. Above 1250 m, teosinte is abundant only on the south-facing slopes. Pines occur below San Antonio Huista on the north-facing slopes, and oaks are abundant above the town on the warmer and drier south-facing exposures.
>
> "The area immediate to San Antonio Huista is seasonally dry four months of the year, and during the rainy season the town is enveloped in clouds of fog and rain. Downstream at Santa Ana Huista (850 m) the climate is much drier and the vegetation is oak scrub. Below, on the plain (750 m) that fans out into Mexico, the habitat is

comparable with the seasonally dry habitat of teosinte in the Bio Blasas (Guerrero, Mexico) and Jutiapa.

"Teosinte is abundant around San Antonio Huista and is present along trails, in the milpas, and covering the hillsides of long-deserted milpas. Maize x teosinte hybrids are most frequently encountered in recently deserted milpas. The habitat varies over very short distances along the Rio Huista due to the steep slopes of the valley, and teosinte reflects these habitat differences in its growth and development. Mid-flowering for teosinte at Buxhap (800–1000 m) is 1–2 weeks earlier than San Antonio Huista (1250 m), which in turn is 2–3 weeks ahead of the population in the fields (1500 m) on the ridge to Jacaltenango. Teosinte is in full flower at San Antonio Huista in early November. Seed is mature in January, but as late as February some plants are still green on the moist upper slopes of the ridge. At the upper elevations, teosinte is limited to the cleared land of cultivated fields, with the naturally occurring wild populations not extending higher than 1300m."

Things have changed considerably in the last 30 years in the region. In 1962, I walked into San Antonio Huista; in 1964, a four-wheel drive road penetrated the town. In 1979, electric power came to the town, and in 1991, telephone service was established. In 1962–63 there was little barbed wire; now all property is fenced. Livestock has become a major source of income in the area, and there is considerable planting of coffee. Milpa is no longer the only land use, and teosinte has almost disappeared except at the margins of distribution where corn is still the major crop.

January 23–25, 1991

We left Jacaltenango after learning that teosinte was still found around La Conception on a road not passable by our station wagon. The road to Tzisbaj also is impassable, so we returned via San Antonio Huista to four corners and Buxhap. Teosinte populations seem to have changed the least around Buxhap (800 m) to Tzisbaj (1,100 m). At the CIMMYT regional office in Guatemala City, S. Taba, Francisco Vasques, and I planned the presentation to ICTA in the afternoon. All the major points were by mutual agreement and Vasques agreed to make the presentation.

In the afternoon, the same group that met with us on January 18 reassembled: Francisco Olivet, Danilo Gonzales, Nory Soto, Francisco Vazques, S. Taba, and myself, plus a new member, Hugo Cordova, from the CIMMYT Central American Program. A summary of our report as presented by Francisco Vazques follows.

Guatemala contains two distinct populations of teosinte: One on the Mexican border that is related to forms found in Mexico (notably teosinte in the Balsas of Guerrero), and another population on the Salvador border that is totally unique and thought to be the most primitive form of annual teosinte. This population, called "Guatemala teosinte" or *Zea luxurians*, had contracted to about 25% of its 1962–63 known collection sites. The other population, known as Huehuetenango teosinte (or *Zea mays* subsp *huehuetenangensis*), had contracted to about 10% of its 1962–63 distribution. Seed was collected from eight sites in the Guatemala teosinte range and six sites in Huehuetenango. This seed is being held by ICTA. The total collection did not exceed one kilo. If these populations were to continue to decline, they would be in danger of extinction. The judicious action would be for Guatemala (ICTA) to undertake conservation measures.

Our first recommendation was that collected seed be grown out and increased on an ICTA station near the zone of collection. The Guatemala type could best be preserved by growing the Carlos Pinto collection, the Cecilio Hernandez collection, and the remaining six collections bulked as three separate and distinct regenerations. This seed in kilo amounts could be put in long-term base storage in CIMMYT, NSSL (National Seed Storage Laboratory), CATIE (Centro Internacional de Agricultura Tropical), and Guatemala. Guatemala (ICTA) would distribute an active seed bank to the Americas and CIMMYT would distribute to the rest of the world but send full records to ICTA. The Huehuetenango type would be grown as a high-elevation collection (La Conception) and an 800 to 1,000 m collection (example Buxhap/Monojil/Tzisbaj) bulked. Again, kilo amounts would be placed in storage and active collection made for distribution. Speaking for ICTA, both Francisco Olivet and Danilo Gonzales agreed to this plan.

Francisco Vazques presented the second recommendation, that being for an *in situ* conservation effort. Without protection and reserved land, teosinte will continue to disappear to hungry cows and goats and alternate land use. We suggested preserving teosinte populations on the ranch of Carlos Pinto and on the small ranch of Cecilio Hernandez. Such *in situ* preservation will require double fences to keep livestock out and to allow teosinte to seed itself. To our great surprise and delight, ICTA pledged support. Our suggestions for Huehuetenango were for the village of Buxhap or Tzisbaj to undertake the preservation in cooperation with ICTA. The same arrangements with the village nearest to the population above Jacaltenango at La Conception was suggested.

This is the first time that *in situ* conservation of an annual wild relative of a major crop plant will be locally managed in contrast to previous efforts relying on central government reserves or parks. ICTA will monitor the *in situ* sites through the genetic resources unit and fully explain to the local conservator the value of its actions, but the commitment to do the job will remain local. This very bold action on the part of ICTA deserves international funding if it is needed. Guatemala has the opportunity to do what has been talked about on paper but never done in the real world.

Restrictions on germplasm collecting in Guatemala were prompted by refusing big corporations and economically powerful governments the right to remove genetic resources and national patrimony from Guatemala. However, denying others access did not solve the indigenous problem that Guatemalans themselves were pushing teosinte to extinction. Once gone, there would be nothing to deny to the international germplasm community. The genetic heritage of Guatemalans would disappear if active conservation was not undertaken to curb the trend of habitat loss and genetic erosion. By assuming direct responsibility for this genetic resource, Guatemalans lay claim to what is uniquely theirs. *In situ* conservation will happen in part because they came to this decision themselves and will thus make it work.

CONCLUSION: PROMOTING *IN SITU* CONSERVATION

Based upon perspectives gained through work in Guatemala, there are actions that the world community can take to enhance workable *in situ* conservation mosaics. I say mosaic because there is no one scheme or way to do *in situ* conservation. This includes actions in three categories: funding, recognition, and networking with researchers in specific crops. The willingness to preserve resources is based upon local or national resolve. Funding is often insufficient to finance full costs of *in situ* preservation. International funding can provide for items such as pollination bags, rodent-proof collection cabinets, and much needed international travel and communication networks.

Recognition is the key to local and national resolve to undertake a long-term commitment to "environmental set asides" for the continued existence of the wild and weedy relatives of the crop plants. If a village is recognized as "holding a national treasure," and it earns respect for its actions from the local agricultural extension officer, the national agricultural research service (genetic resource unit if it exist), and the press, then that *in situ* set aside will work. The press is especially important and could disseminate the message of *in situ* plant genetic resources.

Communicating across the global village is coming at just the right time to focus on regional biodiversity and local or "folk conservation" of a unique genetic resource heritage. Folk conservation (like folklore and folkdress) assures its importance to the world. If *in situ* conservation gains the villages' recognition and importance, there will be pride to carry on a long-term and sustained preservation.

The last area is that of networking by crops. Local *in situ* or *ex situ* conservation is not going to be successful if it is not supported by the national agricultural research bureaucracy (21). So the national agricultural level also needs recognition for its support of *in situ* conservation primarily by the international research community for the particular crop in question. The lack of evaluation of wild relatives is the primary reason they are underutilized and undervalued by the plant breeding community. If there were direct links of targeted, selective evaluation, and the necessary prebreeding to make discovered traits available from national programs with *in situ* conservation, two functions would be served: (1) the national program would not feel that its genetic resources were being stolen (in this age of intellectual property protection) because it would hold the sole distribution handle on the genetic resources, and (2) the profile of the supplying country would be increased and job security of the plant genetic resource personnel enhanced. This is where travel to regional meetings on conservation issues and attendance of crop specific plant breeding/genetics workshops could be financed by the world community. Some of this could come from the private sector, such as crop-specific seed companies and trade associations that have not been paying their freight in the area of plant genetic resources.

Most of the crop genetic resource plants in need of *in situ* conservation are wild or weedy forms of cultivated plants that are of little demonstrated economic value (22). A few plants in need of *in situ* conservation are nondomesticates that have demonstrated economic value such as wild collected drug plants, but I wish to return to the first group and focus on them. These plants possess unique genes not found in the crop, but without extensive evaluation we do not know their value. These plants, many of which are tropical or semi-tropical, do not perform well during the long days of summer in the temperate zone and without prebreeding are just too much trouble for commercial breeders and most public breeding programs. These are not the only reasons wild relatives are undervalued, but it certainly is enough to tip the scale in the minds of practicing plant breeders. This is where a public sector institution or university and/or an individual researcher needs to tie into the *in situ* conservation matrix. This concept of adopt a "crop plant wild relative" brings together the research component and the *in situ* site. In most cases this will be a developed/developing community relationship and the constructive aspect is that they are mutually dependent. The researchers depend on the *in situ* community as a germplasm source and the *in situ* folk conservationist depend on the evaluation/prebreeding to establish the value of their conservation efforts.

A very successful jump-start for agricultural research was accomplished 40 years ago when American universities took on the formation of a sister institution in a developing country. In the same spirit, I would like institution/individuals conducting active research with a specific crop to take on the sister relationship with *in situ* conservation sites for the wild relatives of that crop. If actual individuals are not personally involved and responsible in both the developed and developing world, *in situ* conservation will fail. If a sister institution relationship can develop, then a lot of the suspicion over who owns the "genetic value" that wild relatives can bring to the crop will disappear. By having *in situ* conservation, the country of origin truly establishes claim. The sister institution does the evaluation research and prebreeding, which is then turned back to the country of origin to license or disperse as it sees fit. The evaluation/prebreeding needs to be funded, as does all research, and there are competitive modes to do this. If actual people are responsible for the safety and welfare of crop plant wild relatives, and these people have the attention and respect of the crop-specific research community, then *in situ* conservation for selected crops and sites around the world will work.

I truly hope Guatemalans develop long-term and stable *in situ* preservation for teosinte. They will not be able to do this unless they get recognition from both the maize crop professionals (plant breeders/seed companies/major gene banks) and start-up funds for initial expenses. Global recognition is needed as Guatemalans put these *in situ* conservation actions in place. In this regard, immediate importance should be placed on "media credits" so that the world recognizes that Guatemala is doing something unique and important: *in situ* conservation of teosinte, the closest relative of maize. If this recognition is not forthcoming, other priorities representing more pressing or immediate needs will displace these conservation efforts and teosinte will become extinct.

REFERENCES

1. Wilkes, G. 1966. *Teosinte; The Closest Relative of Maize*, 1–159. Cambridge, MA: Bussey Institution.
2. Wilkes, G. 1972. *Science* 177:1071.
3. Wilkes, G. 1977. *Econ. Bot.* 31:254.
4. Wilkes, G. 1988. In *Recent advances in the conservation and utilization of genetic resources: Proceedings of the Global Maize Germplasm Workshop*, 70. Mexico City: CIMMYT.
5. Benz, B. 1988. In *Recent advances in the conservation and utilization of genetic resources: Proceedings of the global maize germplasm workshop*, 59. Mexico City: CIMMYT; Benz, B., L. Sanchez-Velàsquez, and F. Santana-Michel. 1990. *Maydica* 35:85.
6. Iltis, H. H., J. Doebley, M. Guzmàn, and B. Pazy. 1979. *Science* 203:186; Guzmàn, R., and H. Iltis. 1991. *Diversity* 7:82.
7. Bird, R. 1978. *Taxon*. 27:361.
8. Itlis, H. H., and J. Doebley. 1980. *Am. J. Bot.* 67:994.
9. Wilkes, G. 1985. *Maydica* 30:209.
10. Smith, J., M. Goodman, R. Lester. 1981. *Econ. Bot.* 35:187; Smith, J., M. Goodman, and T. Kato-Y. 1982. *Econ. Bot.* 36:100; Doebley, J., M. Goodman, and C. Stuber. 1984. *Econ. Bot.* 38:97; Doebley, J., M. Goodman, and C. Stuber. 1984. *Syst. Bot.* 9:203; Doebley, J., M. Goodman, and C. Stuber. 1987. *Econ. Bot.* 41:234; Kato-Y, T., and A. Lopez. 1990. *Maydica* 35:125.
11. Doebley, J. 1984. *Ann. Mo. Bot. Garden* 71:1100; Wellhausen, E., L. Roberts, and

E. Hernandez-X, in collaboration with P. Mangelsdorf. 1952. *Races of Maize in Mexico*, 1–223. Cambridge, MA: Bussey Institution.

12. Bukasov, S. 1930. *Bull. Appl. Bot. Genet. Plant Breeding* 47:141.
13. Hernandez-X, E. 1970. *Exploracion Ethnobotanica y su Metodoloia*, 1–67. Privately published by subscription of Friends Rama Botanica Chapingo Edo de Mexico, Mexico.
14. Kato-Y, T. 1976. *Mass. Agricultural Expt. Stat. Bull.* 635.
15. Sanchez, J., and L. Ordaz. 1987. *El Teosinte en Mexico: Systematic and ecogeographic studies on crop genepools*. Rome: IBPGR.
16. Iltis, H. H., D. Kolterman, and B. Benz. 1986. *Econ. Bot.* 40:69.
17. Plucknett, D., N. Smith, J. Williams, and N. Anisletty. 1987. *Gene banks and the world's food*, 1–247. Princeton, NJ: Princeton University Press; Altieri, M., M. Anderson, and L. Merrick. 1987. *Cons. Bio.* 1:49; Hoyt, E. 1988. *Conserving the wild relatives of crop plants*, 1–45. Glands, Switzerland: IBPGR/IUCN/WWF; Wilkes, G. 1991. In *Biodiversity: Culture, conservation and ecodevelopment*, 86. M. Oldfield and J. Alcorn, eds. Boulder, CO: Westview Press.
18. Iltis, H. H. 1987. In *Grass systematics and evolution*, 195. T. Soderstrom, K. Hilu, C. Campbell, and M. Bankworth, eds. Washington, DC: Smithsonian Institution Press.
19. Galinat, W. 1985. In *Breeding strategies for maize production, improvement in the tropics*, 1. A. Brandolini and F. Salamini, eds. Florence, Italy.
20. Kenpton, J.H., and W. Popenoe. 1936. Maize investigations. *Carnegie Inst. Yearbook.* 1936:138–40.
21. Cohen, J.I., J.T. Williams, D.L. Plucknett, and H. Shands. 1991. *Science* 253:866.
22. Goodman, M. 1990. *J. Hered.* 81:11.
23. I wish to thank the staff of the Guatemalan Instituto de y Ciencia y Technologia Agricolas and especially Francisco Vazques of the genetic resources unit of ICTA. As always, I owe thanks to my friend and colleague Suketoshi Taba of the CIMMYT maize gene bank and Hugo Cordova of the CIMMYT Central American Office. I would be remiss not to acknowledge the dialogues and monologues until late in the night with Hugh Iltis in the late 1960s when I taught summer school at the University of Wisconsin. Those conversations have forged some of the ideas in this paper and created a lasting respect for the loyal opposing viewpoint for teosinte in the origin of maize: Thanks, Hugh, for making substantial improvements in this paper.
24. Special thanks to Shauna Lee, who has been loyal in helping me make deadlines while keeping a busy department functioning.

8

Conservation and Utilization of Genetic Resources in Capture and Culture Fisheries

James S. Diana

Development in much of the world, particularly the tropics, has occurred at a pace that alarms conservationists (1). This increasing rate of development has led to the loss of large tracts of tropical rain forests due to clear cutting and other farming practices. Many ecologists are concerned that the loss of tropical rain forest and the diverse flora and fauna of the region is an ecological catastrophe. One element in the effort to reverse deforestation is to promote practices that result in continued exploitation of one tract of land rather than short-term use and abandonment. A second element is the provision of alternative means of subsistence to replace exploitative use of resources as a main source of income.

Aquaculture (the controlled growing of aquatic organisms) is an excellent means for using enhanced and sustainable production cycles of natural organisms for human consumption. In addition, in some areas, aquaculture may be done to treat sewage, since aquatic communities have a large assimilative capacity for natural waste products. While one commonly thinks of high technology systems for rainbow trout or shrimp as aquaculture, most of the world's aquaculture is based on low intensity systems for indigenous fish and even aquatic plants (2). It too uses the aquatic community as a means of food production, and aquaculture ponds might be considered agroecosystems. Aquaculture may provide a unique role in tropical development since it provides an additional means of food production that is sustainable and requires limited land use.

Fish capture from naturally occurring populations has been practiced throughout history as a staple of food production for many indigenous cultures. Major expansion of oceanic harvest by large commercial ships has occurred over the past 100 years. The goal for much of fishery management has been to understand and manage these harvests for maximum yield (3). However, historic capture fisheries were largely artisanal, with local or near shore animal populations exploited on a small scale. Modern fishery management has often ignored or even damaged artisanal fisheries by uneven regulation. The sustainability of artisanal fisheries generally depends on low levels of human effort (4), and much of the inland harvest of fish, especially in the tropics, is by artisanal operators.

The purpose of this chapter is to (1) evaluate and compare fish capture and aquaculture practices, (2) review the future roles of aquaculture and capture fisheries, (3) give two examples of aquaculture systems, and (4) summarize the roles of capture fishing and aquaculture in sustainable development and genetic resource use. This chapter is not intended to be a compilation of aquaculture systems, but rather an example of the role that aquaculture and associated biodiversity may play in sustainable development.

CAPTURE FISHERIES

Catching of fish from natural ecosystems for use as human food has been practiced since prehistoric times. The major utilization of fisheries has occurred since 1850. As human populations have expanded, the catch of fish products has followed. Historic harvests of 1.5 million metric tons (MMT) in 1850 have expanded to about 86 MMT in 1989 (5) (Table 1). Increases in yield have occurred mainly through increased effort, rather than increased fish stocks. This increased effort has come partly from new technology. Historic practices using small vessels and gillnets or trawls produced from natural fibers have been replaced by large vessels with more effective synthetic nets and more sophisticated methods to locate fish.

The future of capture fishery production is controversial. While some authors have predicted stable or declining harvests in the future (6), others believe expansion will continue (7, 8). The harvest of capture fisheries has continued to increase in recent years (Table 1), and the average increase since 1965 has been 2.6% per year. Approximately 30% of the wild fish product is not used for human consumption but for animal feed, fertilizer, or other commodities. This fraction has declined from 42% in recent years. Thus one might estimate the total food production from capture fisheries in 1990 at 60 MMT.

Table 1. World harvests of fish and other aquatic products in millions of metric tons (5, 36).

Year	Freshwater Fishes [1]	Marine Organisms [1]	Capture Harvest	Culture Harvest	Total Yield
1850			1.5		1.5
1900			4.0		4.0
1930			10.0		10.0
1940[2]			21.1		21.1
1950	4.6		20.0		20.0
1960	5.6	34.6	40.2		40.2
1965	7.0	46.2	53.2		53.2
1970	5.8	59.7	65.5		65.5
1975	7.2	60.2	60.3		60.3
1980	7.6	65.8	63.5	6.1	66.4
1985	10.7	76.5	75.8	10.6	86.4
1989[3]	13.8	85.7	86.3	13.2[4]	99.5

[1] Data include aquaculture and capture fisheries.
[2] Data actually from 1938, due to World War II and its effect on fish catch.
[3] Data from FAO[5] that have not yet been reviewed and revised.
[4] Data from 1987 (last available year for aquaculture data).

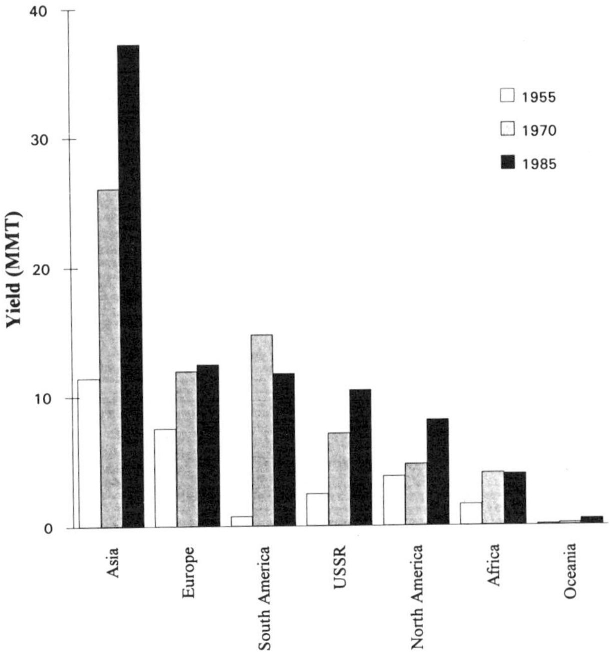

Figure 1. The contribution of different regions of the world to fish harvest from 1955 to 1985. Data from FAO.

The total harvest of fish is dominated by Asia (43%), with similar yields coming from Europe, South America, the former USSR, and North America (Fig. 1).

This regional analysis also indicates that Asia is responsible for most of the increased harvest since 1955, with South America also expanding harvest significantly. These data represent country of landing rather than location of catch. Fleets from many countries, particularly Japan, Korea, the United States, and the former USSR, range widely in search of fish so that their catches reflect some fish taken far from their own waters. The history of harvest for selected countries shows relatively constant harvest in European countries, large increases in Asian countries, and fluctuations in South American countries (Fig. 2). The decimation of the Peruvian anchovy caused a huge reduction in the fishery of Peru. The anchovy was the basis of Peru's fishing economy and in 1970 it contributed 20% of the world's fish harvest. Developing and developed countries have shared the increase in fishery yields since 1955.

Most of the world fish harvest is done by large scale commercial operations on the high seas. Oceanic harvest by these operations provides income for relatively few. Artisanal fishing on a small scale in the oceans or freshwater may add less to the total yields of fish, but adds considerably to the social and economic stability of a region (9). Inland harvest of fish (both artisanal and commercial) has also increased in recent years, but the fraction of total fish harvest from inland waters has remained between 10 and 14%. Harvest statistics for both marine and freshwater animals are problematic in that they include aquaculture yields that are also expanding (5).

The catch by species group (Fig. 3) indicates that much of the increase in yield since 1955 is from harvest of clupeids and cods. Many favorite food fish (salmon, tunas, and flatfish) have stable or decreasing catches in recent years. These changes represent only part of the changing genetic resource picture in the fishery sector.

Figure 2. Changes in fish harvest of various countries from 1955 to 1985. Data from FAO.

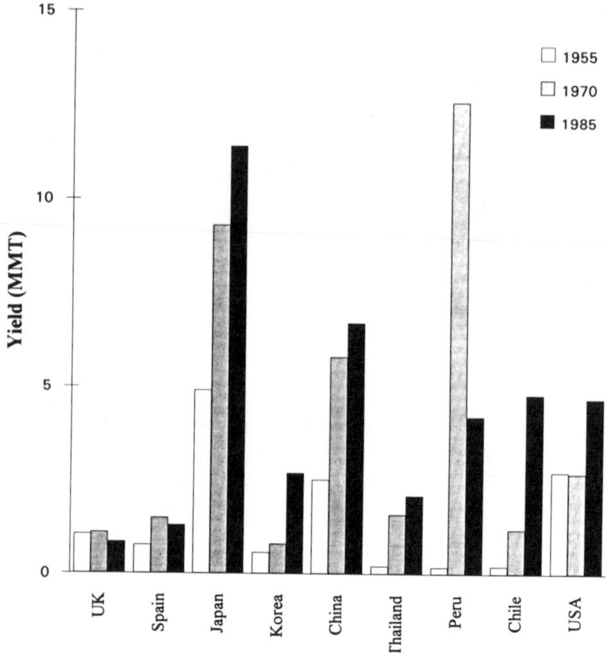

Figure 3. Species groups of aquatic organisms harvested from 1955 to 1985. Data from FAO.

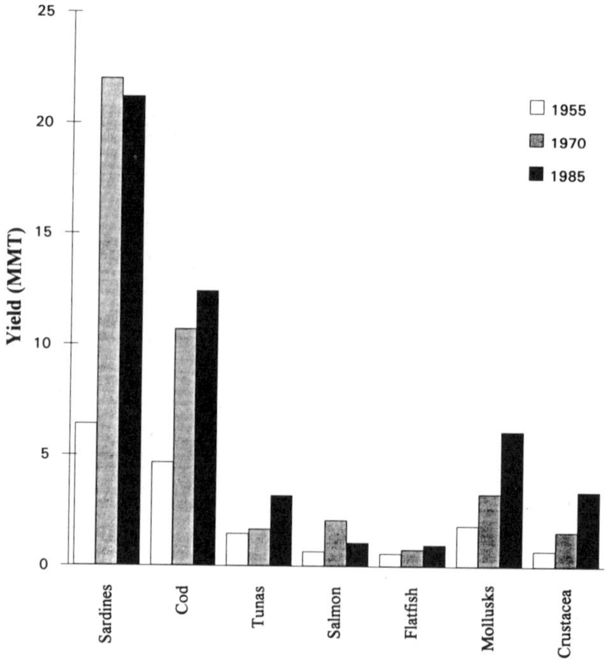

Several species of fish (blue pike in Lake Erie, Peruvian anchovy, California sardine) have become extinct or suffered catastrophic declines in abundance due, at least in part, to over-fishing. While intensive management has been applied to many of these species, over-fishing still remains a major problem. In addition to the extinction of species, local stocks of fish have also been decimated by capture fisheries. For example, sockeye salmon remain abundant in the world, but the Snake River stock is now listed as endangered (10). Similar reductions or losses have occurred in lake trout stocks in the Laurentian Great Lakes (11). Indeed, one of the largest effects of capture fisheries may not be the elimination of species themselves but the changes that have occurred within a species that has been harvested for many years (12). The changes that have been documented include earlier maturation, slower growth, and shorter life (12, 13). The long-term significance of these changes to the fishery itself and to stability of the fish population is uncertain.

Another major problem in capture fisheries is the take of nontarget animals. Regulations often require the return of these species from catches, but the fish commonly die after release. The incidental catch of marine mammals in purse seines set for tuna created such a public outcry that marketers refused to can tuna taken by purse seine in the north Pacific. There are similar concerns for drifting gillnets, which take many nontarget fish even when operated correctly and which, when lost, can continue to kill fish and marine mammals for many months.

AQUACULTURE

Aquaculture presents an opportunity to increase food production in areas, and to do so in a nondestructive manner if reasonable practices are followed. Aquaculture production has increased at 9.68% per year since 1975 (Table 2). The acreage needed to produce fish is much less than otherterrestrial crops. Most of the world's aquaculture production is of animals used for local consumption, so the benefits of aquaculture extend to the regional community. Finfish comprise over half of the world aquaculture products (Fig. 4), and these are dominated by carps, tilapia, and milkfish used mainly for local consumption. High intensity systems such as those used for salmonids or channel catfish may be of interest because of their technological nature, but they contribute relatively little to total world fish production. Some crops, like seaweed and mollusks, are grown in extremely simple systems where organisms are held in the ocean and the main management is to introduce attachment surfaces for the organisms or to seed areas with young propagules. Crustaceans make up a substantial portion of the overall production, and the production systems for some of these are intensive, with products used mainly for export. However, it is clear from Figure 4 that most aquaculture is based on relatively low intensity systems and is used mainly for local consumption.

Table 2. Changes in the relative importance of capture and culture fisheries to the total world fish production.*

Year	Capture Yield (MMT)	Culture Yield (MMT)	% Culture
1975	60.3	6.1	9.2
1980	63.5	8.7	12.0
1985	75.8	10.6	12.2
1987	81.1	13.2	14.0
2000	114.4	38.1	27.7

* Yields for the year 2000 were calculated from current yields and average rates of increase for fisheries (2.6%) and aquaculture (8%) (5, 36).

Figure 4. The contribution of different selected groups of organisms to the total aquaculture production in 1987 (data from 37).

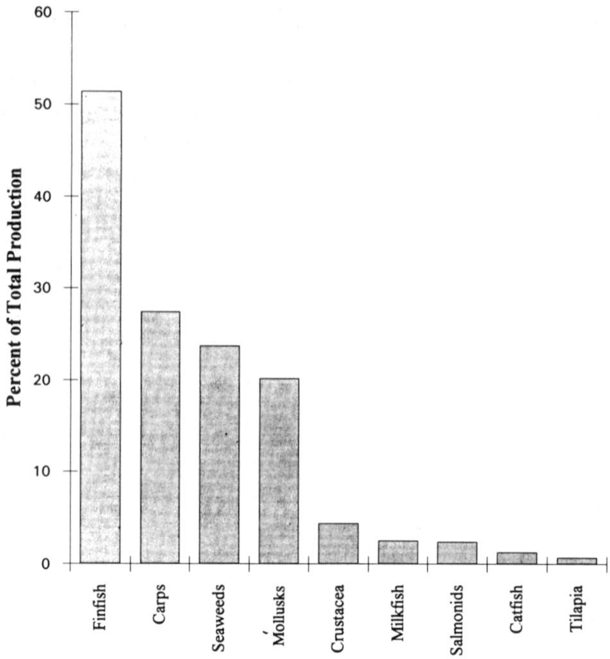

Figure 5. The contribution of different regions of the world to the total aquaculture production in 1987 (data from 37).

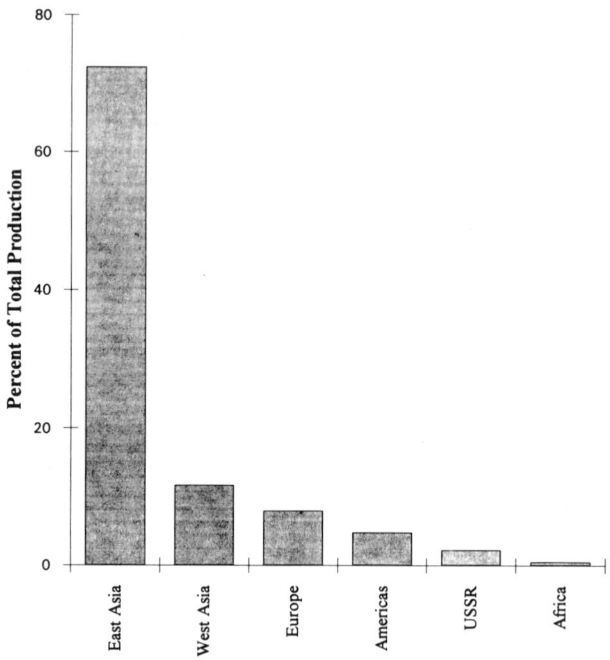

Although aquaculture developed independently in several areas, Southeast Asia might be considered the origin of modern fish culture (14). Carp-silkworm and rice-fish culture systems have existed in China for at least 3,500 years. The first known book on aquaculture was published by Fan Li in the fifth century BC (15). East Asia still dominates total aquaculture production in the world, with West Asia following and virtually all other areas trailing dramatically (Fig. 5). There is obvious potential to increase total aquaculture production in the Americas, in Africa, and other areas.

Aquaculture systems can vary from cages submerged in standing water to small ponds, large ponds, or even the open ocean. Intensive systems may use indoor hatcheries or raceways as well as small tanks or concrete troughs. Polycultures of fish with other crop plants use flooded fields.

The level of management in aquaculture varies from extensive to intensive. Extensive systems are usually ponds with low inputs of materials and relatively low production of crop. A low intensity system for tilapia may use stocking rates of 1 fish/m^2, manure as input, and produce annual yields of 2,500 kg per hectare (16). Intensive systems are on the high input end of the spectrum, with supplemental feeding of fish and water quality management by mechanical devices. These systems may produce 20,000 kg or more per hectare per year in ponds and considerably more in tanks with rapid water turnover. Two special cases of intensive aquaculture are important: polyculture and integrated culture. Polyculture is the production of several species of animals that occupy different trophic levels. These systems may use a benthic carnivore, a planktivore, and an herbivore. Each utilizes different foods to minimize interspecific competition. Integrated culture involves use of the fish culture system with a farm complex (17). In Chinese systems, this may involve using crop wastes (grass, vegetable wastes) as supplemental feed for grass carp, using pig manure and human excrement as nutrient for the pond, and using waste products from animal processing as fish feed. The pond in this system is part of a waste and food processing system for the entire farm (18).

The species cultured are chosen for adaptability to domestication, availability of fish stock, and local market. Aquaculture began with the collection of fry from the wild for stocking into ponds; indeed some believe that the Chinese polyculture systems may have originated from the inability of culturists to distinguish different species of carp fry. In recent years, hormones and semi-natural conditions have been developed to produce fish fry on demand, in some cases on a year-round basis, in spite of natural breeding cycles. In some species, breeding can be promoted by simply providing an environment conducive to spawning. Alternately, fertile eggs and sperm can be collected from the adults at harvest. The ability to produce fry on demand and in large numbers is essential in large-scale aquaculture production. Nonetheless, some of the high volume species (such as milkfish) still rely on fry collection from the wild.

While aquaculture augments food production, it has both positive and negative aspects. Additional benefits of aquaculture are agricultural. Aquaculture can make use of unarable land, such as tropical areas with acidic soils, and since fish feces and pond sediments fertilize soils, land may be brought back into production (7, 17). Fish can be grown with crops in polyculture for a second yield (as in fish-rice systems). They may help control insect pests of a plant crop or other pests like mosquitoes (7). Aquaculture can be used in crop rotation; in fact, this was one of the bases for the chinampa systems of Aztec culture (19). A high fraction of the fish body is usable protein because fish live in a medium where they are naturally buoyant and skeletal development for supporting the body is reduced compared to land animals. Finally, recently discovered health benefits of consuming omega-3 and other unsaturated fatty acids are promoted by fish consumption. These valuable contributions of fish culture extend in various degrees to other aquatic organisms.

The main concerns related to aquaculture are ecological ones. The rapid expansion of acreage in shrimp ponds has resulted in the reduction of mangrove swamps and other sensitive ecosystems in coastal areas, which may affect nutrient cycling, fish nursery areas, and the storm buffering capacity of the near shore environment (20). Nutrient loading of waters receiving aquaculture wastes can cause widespread pollution and eutrophication (20). Probably the largest impact of aquaculture on biodiversity is the importation of exotic animals (7).

Carp were introduced into North America in the late 1800s (because they were a favored food fish in other parts of the world), but never became popular as food. Their population expansion has resulted in reductions of many other fish species and alteration of wetland ecosystems by a feeding behavior that uproots vegetation and suspends sediments (21). This introduction cannot be blamed on the aquaculture industry alone, since the fish were brought in intentionally to be released as targets for commercial or recreational fishing. A more direct impact occurred with tilapia, which were introduced throughout the world as a culture fish (22), in spite of early concerns of their potential negative impact (23). The fish escaped containment and have increased dramatically in natural habitats. Their increase has been followed by the demise of many native species, which are believed to have been driven to extinction by (1) competition with tilapia, (2) diseases brought in with tilapia, or (3) tilapia predation on their eggs and young. Similarly, the widespread importation of tilapia for aquaculture has resulted in hybridization and loss of genetic diversity in the native tilapias in the recipient area.

A COMPARISON OF CAPTURE AND CULTURE FISHERIES

There are large differences in the potential benefits of developing aquaculture or capture fisheries for future food production. Capture fisheries now rely mainly on large vessels and high capital inputs. In contrast, most of the world aquaculture production is accomplished by small operators using relatively low intensity systems. World fish catches are stable or increasing slowly (2.6% per year), while aquaculture production has increased at 9.68% per year since 1975. Management of oceanic fisheries demands increasingly complex regulations and an understanding of the complex ocean environment (climate or upwelling and its effect on fish recruitment). Fish culture is simple and requires little regulation, at least for an individual manager. Some species harvested from the oceans are of limited quality for human consumption, with marginal species comprising 30% of the catch and even valuable species (swordfish, tuna) are often contaminated by pollutants. The species used in aquaculture are valuable and can be controlled to insure quality for human consumption. Both systems of fish production have future importance in food security for humans, but aquaculture can expand this food production using animal husbandry, while capture fisheries are ultimately limited by the natural productivity of the seas.

A comparison of the future potential of aquaculture or capture fisheries indicates that aquaculture will probably contribute 20% of the world's fish production by the year 2000 (Table 2). Fish capture may reach 114 MMT if current growth rates continue, while fish culture is expanding rapidly and could reach 38 MMT by the end

Table 3. World shrimp statistics for 1984 and a comparison to 1987 (24, 37).

	Yield (kg/ha)	Hectares	Total Yield (MT)
		1984	
Asia	400	200,000	80,000
Latin America	605	45,000	27,275
Taiwan	3,600	5,000	18,000
Total			115,000
		1987	
Average	1,000	400,000	454,528

of the century. Aquaculture has already grown at a rate far exceeding predictions of only a few years ago. For example, in 1984, the potential for expansion in shrimp production was considered great (24). It was generally assumed that expansion would occur by adding new acreage to shrimp production. The data available then indicated that most shrimp were being grown under extensive systems (Table 3). Data available for 1987 indicate a huge increase in shrimp culture (from 115,000 MT in 1984 to 454,000 MT in 1987). Much of this increase was due to widespread expansion of the Taiwanese system of intensive culture. For example, in East Java the total acreage in shrimp production increased 265% from 1985 to 1989, mainly through increases in intensive ponds. Average yield per hectare increased 39.5 fold (from 78 to 3,084 kg/ha/yr) (20). On a worldwide basis, a probable doubling of area in shrimp production has occurred, along with a 2.5 fold or larger increase in yield per hectare due to intensification.

TWO EXAMPLES OF AQUACULTURE

As a final aspect of aquaculture, the culture systems for two common fish in Thailand, the walking catfish (*Clarias batrachus*) and the snakeskin gourami (*Trichogaster pectoralis*), will be described. The former is cultured using intensive systems, the latter under extensive polyculture with rice. Both are used for local consumption. These two cases exemplify the sorts of systems that may be useful in the extension of aquaculture into sustainable development in the tropics. This discussion is focused on freshwater fish because I believe they offer the largest potential for ecologically sensitive expansion. Although these two were chosen as examples, other species or systems may be more feasible in areas outside of Thailand.

Clarias is a popular food fish in Southeast Asia. They and many other local fish are air breathers, a common adaptation of fish to the wet-dry seasons of the tropics. *Clarias* are very tolerant of poor water quality and as such are useful in aquaculture. The two main growout systems in Thailand are earthen ponds and concrete tanks.

In earthen ponds, stocking densities vary from 20 to 40 fry per m^2. The ponds are generally stagnant, fed only by rainwater. Supplemental feed (dry pellets or trash fish) is used to grow the fish. Trash fish is a broad category of feed that includes undersized fish taken in commercial nets and sold as fish parts to processing plants. Water quality becomes limited in ponds managed at high fish density with relatively high food input, and the ponds are often anoxic. These conditions result in survival to harvest of only around 30%, and a growth rate of 3% body weight per day (25). Annual yields achieved in such ponds are 20 to 40,000 kg per hectare. The high yields achieved are due to the tolerance of *Clarias* to crowding and low water quality (see Fig. 6).

Figure 6. A *Clarias* pond system near Bangkok. The fish are being fed heads and guts of other fish processed in the plant at rear. Pigs kept in sties along the pond bank are also fed trash fish, and their wastes are washed into the pond. The combination of no water renewal and high fish density often produces anaerobic conditions.

An improved system of growout is in recirculating concrete tanks (Fig. 7). These tanks are usually 9 m in diameter and are stocked at 660 fish/m^3. Water is pumped to the tanks from a large reservoir at a turnover rate of up to three times per day. The fish are fed pellets. Both survival and growth are much higher in this system at 90% and 5% BW/d respectively (26). The Thais believe that this improved yield is due to water quality, but our experiments and simulations have shown that water quality in these tanks is also very low (27). Concrete (in contrast to mud) substrate reduces H_2S and other toxic materials in the water. Limited space may allow more effective utilization of food, which could be the actual reason for improved survival and growth in concrete tanks. Such improvements in yield come at the expense of higher food cost, more electricity for pumping, and more expensive containers.

The fry production system for *Clarias* in Thailand is one of semi-natural spawning. Fields are dug with rows of banks, and small caves are dug in the banks for spawning sites (Fig. 8). Mature fish are stocked in the fields. They occupy the caves and breed there. Every week or so, fry are collected from the caves by dipnet (28). Such a system can produce sufficient fry to easily stock out several farms, and controlled water levels are favorable for fry production in all but the coldest months.

Some variations on these systems are of interest. Dutch scientists have developed growout in cylindrical tubes of water with very high turnover rates and extremely high densities of fish (29). In some of these systems the fish are so crowded that they barely have room to swim, although in nature *Clarias* often group into tight clusters as well. Similarly, scientists in Hawaii have developed methods to produce fry from other species by hormone injection and artificial stripping (30). Hawaiian culture systems are a version of the Thai intensive system but at much lower density and with higher water quality. All of these possibilities work well under different conditions, so that the optimal culture system for *Clarias* may depend largely on

Figure 7. A modern *Clarias* culture facility in Bangkok. The circular tanks are supplied with water from the reservoir (background) at 1–3 turnovers per day. Fish are given processed feeds.

Figure 8. A *Clarias* breeding pond at flooding. Caves in the banks provide nest sites that are occupied by breeding pairs for natural spawning. The fry are manually removed weekly from each cave.

local market price and cost of inputs and land.

Expansion of *Clarias* culture has caused some genetic erosion concerns. In Thailand, the native fishes that were historically cultured (*Clarias batrachus* and *C. macrocephalus*) have been replaced by *C. lazera* (an African species) or its hybrids because of a superior growth rate. However, these exotic fish will undoubtedly escape containment and will likely hybridize or compete with native species. *Clarias* introduced into Hawaii (by Chinese immigrants) and Florida (by the aquarium trade) have expanded dramatically and are considered pests.

The snakeskin gourami is a much lower intensity culture species grown in Asia. The gourami is generally grown in combination with rice fields, although around Bangkok acid sulfate soils have sometimes excluded rice growing and the fish are grown alone in a similar setting. The rice field is surrounded by a peripheral canal, so that as the field is drained the fish are concentrated into the deepest water for harvest. Adult fish are stocked at 40 to 220 kg/ha and allowed to spawn naturally in the fields. Since natural spawning is practiced, the density of fry in the field is variable and so is growth. Organic fertilizer (green manure, which is simply vegetation cut and placed in ponds to decompose) can be applied to increase yields (Fig. 9). A growout cycle of six to nine months produces 450–1,700 kg/ha (31, 32). Improvements on this system involve stocking controlled densities of fry, shorter growouts, and addition of chicken manure, but since the value of these fish is relatively low, such improvements are not often economically feasible. Improved yields may be more than 2,000 kg/ha. The combination of gouramis with rice may produce not only fish but also better rice yields, as the fish consume insect pests.

There appear to be fewer concerns with genetic resources in rice-fish culture. The relatively limited yields and control of the rice farming system for fish culture have led to use of a variety of local species. Most rice fields also contain significant populations of other native fishes. The importation of new species, as well as the use of hybrids or genetically altered fishes, is not common in rice-fish culture. However, pesticide use may be problematic in the suitability of these fish for human consumption.

Figure 9. A snakeskin gourami farm near Bangkok, converted from an old rice field. The peripheral canal is barely visible. Workers are cutting reeds that are left in the pond to compost and serve as green manure.

CONCLUSION

The future of capture fisheries is unclear. Large-scale commercial operations have resulted in the reduction or extinction of many fish species, from sardines throughout the world to blue pike in Lake Erie. Small scale commercial or artisanal fisheries may provide a more reasonable contribution to the local economy (33). However, artisanal fisheries require special regulations compared to large commercial fisheries (34).

Aquaculture is well developed in the tropics of Asia. The provision of a less damaging system for food production and local economic development through aquaculture may have significant impact in the neotropics, where the success of management to alleviate rain forest destruction depends on providing alternative means for survival. Expansion of aquaculture to the neotropics should be promoted with ecological sensitivity to avoid the problems previously encountered.

One key development in the extension of aquaculture is the choice of species to be cultured. The Peace Corps and other development agencies have attempted to extend Asian fish culture to the Americas. At present, the most successful aquaculture program in South America is for shrimp, mainly because it is based on large scale programs with international companies and export markets. It probably has little significance for the lives of peasants in the interior. There are also large environmental impacts as described earlier.

The use of exotic species for interior aquaculture may fail economically in addition to causing ecological problems. One major ingredient in success of aquaculture is market potential of the crop. In South America, artisanal fisheries for local species are common, and potential for sale of these fishes is very good. However, if one visits a fish market there, the last fish sold are often exotic species like tilapia, and they fetch a very low price. The low market value of tilapia is a concern throughout the world (7). To overcome this social bias, and to promote more sensitive ecological interactions in the region, there is a strong need for research and extension of fish culture techniques to locally endemic fish (35). Since few aquaculture species have been domesticated to any degree, the use of new wild stocks should not prove particularly difficult. There are a variety of fish in the neotropics with high potential for success in controlled culture, but a paucity of research on them. We need to reorient our research programs from using standard animals and trying to export them, to using local animals which may be less understood but hold higher potential both in the market and in the local environment.

A promising situation for developing aquaculture in tropical rain forest regions is the use of seasonal wet-dry cycles to resupply water. Goulding (34) has documented large variations in discharge for the Amazon River, and the importance of these variations on riverine and flood-plain fisheries. There are a number of possibilities for aquaculture in flood plains, including damming of small channels, constructing drainage ponds, using natural backwaters, using cage culture in rivers, or normal intensive aquaculture (36). Typical aquaculture operations may require modification to function in areas with large seasonal fluctuations in water depth, and research should be targeted to studying this sort of modification.

Both capture and culture fisheries provide considerable food for human consumption, and it appears they will continue to do so in the future. Expansion of aquaculture in ecologically sensitive ways can dramatically increase protein availability in South and Central America, as well as other wet regions of the world. The environmental impacts of fish culture may be much lower than livestock production, and the efficiency of production much higher. Extensive aquaculture systems, with low material input and high biodiversity, should be able to provide large benefits for economic development in many countries.

REFERENCES

1. Cunningham, W.P., and B.W. Saigo. 1990. *Environmental science.* Dubuque, IA: William C. Brown Publishers.
2. Nash, C.E., and C.B. Kensler. 1990. A global overview of aquaculture production in 1987. *World Aquaculture* 21(2):104–112.
3. Sissenswine, M.P. 1981. An overview of some methods of fish stock assessment. *Fisheries* 6(6):31–35.
4. Larkin, P.A. 1988. The future of fisheries management: Managing the fisherman. *Fisheries* 13(6):3–4.
5. Food and Agriculture Organization of the United Nations (FAO). 1990. FAO yearbook of fishery statistics, Vol. 66 (and preceding years). Rome.
6. Robinette, H.R., J. Hynes, N.C. Parker, R. Putz, R.E. Stevens, and R.R. Stickney. 1991. AFS position statement on commercial aquaculture. *Fisheries* 16(1):18–22.
7. Pillay, T.V.R. 1990. *Aquaculture principles and practices.* Oxford: Fishing News Books.
8. Wise, J.P. 1991. North American fish landings on the rise. *Fisheries* 16(4):53.
9. Larkin, P.A. 1988. The future of fisheries management: Managing the fisherman. *Fisheries* 13(6):3–4.
10. Anonymous. 1992. Salmon recovery efforts accelerate. *Fisheries* 17(3):45–46.
11. Goodier, J.L. 1981. Native lake trout (*Salvelinus namaycush*) in the Canadian waters of Lake Superior prior to 1955. *Canadian J. of Fisheries and Aquatic Sciences* 38:1724–37.
12. Ricker, W.E. 1981. Changes in the average size and average age of Pacific salmon. *Canadian J. of Fisheries and Aquatic Sciences* 38:1636–56.
13. Diana, J.S. 1983. Growth, maturation, and production of northern pike from three Michigan lakes. *Transactions of the Am. Fisheries Soc.* 112:38–46.
14. Ling, S.W. 1977. *Aquaculture in southeast Asia: A historical review.* Seattle: University of Washington Press.
15. Lam, T.J. 1982. Fish culture in southeast Asia. *Canadian J. of Fisheries and Aquatic Sciences* 39:138–42.
16. Diana, J.S., C.K. Lin, and P.J. Schneeberger. 1991. Relationships among nutrient inputs, water nutrient concentrations, primary production, and yield of *Oreochromis niloticus* in ponds. *Aquaculture* 92:323–41.
17. Little, D., and J. Muir. 1987. *A guide to integrated warm water aquaculture.* University of Stirling, Stirling, Scotland: Inst. of Aquaculture.
18. Lin, H.-R. 1982. Polycultural system of freshwater fish in China. *Canadian J. of Fisheries and Aquatic Sciences* 39:143–50.
19. Coe, M.D. 1964. The Chinampas of Mexico. *Scientific American* 211(1):90–98.
20. Chamberlain, G. 1991. Shrimp farming in Indonesia 1: Growout techniques. *World Aquaculture* 22(2):12–27.
21. Scott, W.B., and E.J. Crossman. 1973. Freshwater fishes of Canada. Bulletin 184. Fisheries Research Board of Canada, Ottawa.
22. Pillay, T.V.R. 1992. *Aquaculture and the environment.* Oxford: Fishing News Books.
23. Bardach, J.E., J.H. Ryther, and W.O. McLarney. 1972. *Aquaculture—the farming and husbandry of freshwater and marine organisms.* New York: John Wiley and Sons.

24. Rhodes, R.J. 1986. Status of world aquaculture: 1985. *Aquaculture Magazine Buyer's Guide* 1986

25. Diana, J.S., S.L. Kohler, and D.R. Ottey. 1988. A yield model for walking catfish production in aquaculture systems. *Aquaculture* 71:23–35.

26. Tarnchalanukit, W., W. Chuapoehuk, P. Suraniranat, and U. NaNakorn. 1983. *Pla duk dan culture in circular concrete ponds with water recirculating system.* Department of Agriculture, Faculty of Fisheries, Kasetsart University, Bangkok, Thailand.

27. Diana, J.S., and A.W. Fast. 1989. The effects of water exchange rate and density on yield of the walking catfish, *Clarias fuscus*. *Aquaculture* 78:267–76.

28. Areerat, S. 1987. *Clarias* culture in Thailand. *Aquaculture* 63:355–62.

29. Huisman, E.A., and C.J.J. Richter. 1987. Reproduction, growth, health control, and aquaculture potential of the African catfish, *Clarias gariepinus* (Burchell 1822). *Aquaculture* 63:1–14.

30. Young, M.J.A., A. W. Fast, and P.G. Olin. 1989. Induced maturation and spawning of the Chinese catfish *Clarias fuscus*. *J. World Aquaculture Soc.* 20:7–11.

31. Bardach, J.E., J.H. Ryther, and W.O. McLarney. 1972. *Aquaculture*. New York: John Wiley and Sons.

32. Diana, J.S., W.Y.B. Chang, D.R. Ottey, and W. Chuapoehuk. 1985. Production systems for commonly cultured freshwater fishes of southeast Asia. International Programs Report No. 7, Great Lakes and Marine Waters Center, University of Michigan, Ann Arbor, Michigan.

33. Goulding, M. 1980. *The fishes and the forest.* Berkeley: University of California Press.

34. Munro, J.L. 1980. Stock assessment in tropical fisheries: Past and present practices in developing countries. In *Stock assessment for tropical small-scale fisheries*, P.M. Roedel and S.B. Saila, eds., 27–34. International Center for Marine Resource Development, University of Rhode Island, Kingston, Rhode Island.

35. Welcomme, R.L. 1985. River fisheries. FAO Fisheries Technical Paper 262. Rome: FAO.

36. Food and Agriculture Organization of the United Nations (FAO). 1989. Aquaculture production (1984–1987). FAO Fisheries Circular No. 815, Revision 1. Rome.

37. Nash, C.E., and C.B. Kensler. 1990. A global overview of aquaculture production in 1987. *World Aquaculture* 21(2):104–12.

9

Mangrove Ecosystem Biodiversity and Conservation in Ecuador

Robert R. Twilley, Alejandro Bodero, Donald Robadue

Mangroves are one of the dominant features of coastal ecosystems of the tropics. While ecologically they have been considered an important component of coastal watersheds, they are continually exploited for aquaculture, agriculture, charcoal, urban development, and other economic activities. Attention has been focused on the utilization of mangrove habitat for the expansion of shrimp ponds, particularly in Ecuador, where in some watersheds such as Rio Chone over 80% of the mangroves have been reclaimed for mariculture. The success of this economic activity in Ecuador is quickly spreading to other developing countries in Central and South America where there is concern for sustaining the important linkages of ecological and social systems. Management of coastal resources such as mangroves is limited by the complex nature of their various ecological functions that are poorly understood. These ecological functions include the more commonly discussed issues of habitat quality, water quality, and shoreline stabilization. Some of the less tangible functions of mangroves, such as supporting biodiversity of coastal regions, have not been directly addressed as part of the habitat quality that mangroves provide. This chapter will describe a case study of mangrove utilization in Ecuador to demonstrate the importance of genetic resource management as a property of tropical coastal ecosystems.

In August 1986, a workshop was organized in Guayaquil by the University of Rhode Island coastal resource management project to evaluate the shrimp farming industry in Ecuador, which had recently suffered loss of production due to the lack of available postlarvae for ponds. Total production and export of shrimp was down in 1984 and 1985; only half of the 75,000 ha of shrimp ponds constructed in the coastal provinces were in operation. Several factors had been associated with the decline in postlarvae in the estuaries along the coast of Ecuador, including lower water temperatures (back to normal temperatures following an El Niño event), loss of mangrove habitat, decline in water quality (increased occurrence of red tide, pesticides, and heavy metals), and indiscriminate overfishing of available wild

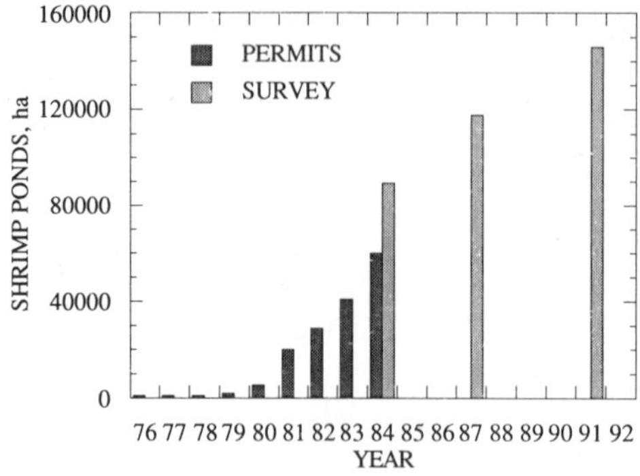

Figure 1. Index of the rate of shrimp pond construction in Ecuador based on permits issued by the Ministry of Agriculture, compared with direct delination of shrimp pond area using remote sensing (1983, 1987, and 1991).

stocks. From 1980 to 1991 nearly 15,000 ha of shrimp ponds were authorized for construction annually, increasing the total to 143,676 ha by 1991 (Fig. 1); most of these ponds had been constructed in the intertidal zone. There was immediate concern that the unregulated growth of this industry had destroyed the ecological processes of coastal ecosystems, threatening the sustainability of shrimp farming in Ecuador. Declines in both the recruitment and survival of wild postlarvae demonstrated the susceptibility of this industry to the environmental quality of Ecuador. The constraint of the shrimp farm industry by the limited supply of postlarvae is considered an example of negative feedback from commercial activities on the ecological function of natural resources.

The first commercial shrimp operations began in 1969 (1), nearly 400 years following the Incas' practice of closing off lagoons that were temporarily flooded with seawater and penaeid shrimp larvae. Ecuadorean farmed shrimp production rose dramatically from 1979 to 1984 and by 1989, the year of the highest production on record, shrimp ponds produced over 50,000 metric tons while production from the trawl industry remained at 7,500 metric tons (Fig. 2). The value of production from shrimp ponds increased from U.S. $56.9 to $287.9 million from 1980 to 1986 (Fig. 2). The export value of the 1991 crop increased to U.S. $482 million, ranked second only to petroleum as an export commodity for Ecuador (2, 3). The cash generated by this mariculture activity is more important to the economy of Ecuador than bananas and cacao combined, and twice as important as coffee (3). The tremendous growth of this industry has made Ecuador the second leading farm shrimp producer in the world, providing nearly 16% of the world market (4). Nearly all of this market is consumed in the United States.

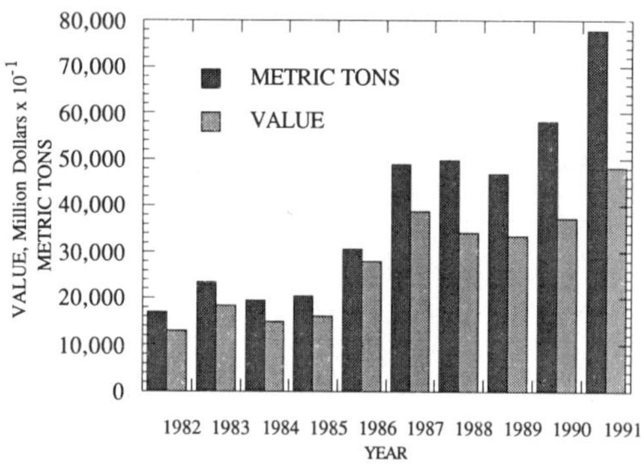

Figure 2. Productivity of the shrimp pond industry from 1982 to 1991 based on United States dollars and mass of shrimp sold per year.

The expansion of the farmed shrimp industry has been largely confined to the two southern coastal provinces of Guayas and El Oro (Table 1). This coastal region handles 95% of the country's imports and 50% of its exports, and its coastline includes the most populated city in Ecuador, Guayaquil (5). The population in Guayaquil has doubled to 2.5 million inhabitants in the last 10 years, and nearly 82% of the city discharges its waste directly into coastal waters (6). In addition, over 50% of the agricul-

Table 1. Land use patterns in the four provinces along the coast of Ecuador (areas in ha, percents follow).

Land Use	Guayas	El Oro	Manabi	Esmeraldas	Total
Mangroves	121,464	24,489	7,973	21,293	175,219
	69.4	14.0	4.4	12.2	100.0
Shrimp Ponds	52,912	26,484	8,377	1,596	89,368
	59.2	29.6	9.4	1.8	100.0
Salinas	17,340	2,520	164	4	20,028
	86.6	12.6	.8	.0	100.0
Total	191,715	53,493	16,514	22,893	284,615
	67.4	18.8	5.8	8.0	100.0

ture activity for the four coastal provinces of Ecuador occurs in the Guayas River basin. The Daule-Peripa dam was recently constructed to provide water for expansion of agriculture, particularly rice, by 17,000 ha during the initial phases of the project. This dam will also limit the discharge of freshwater into the Guayas River estuary. This coastal zone region also maintains an extensive area of intertidal communities including nearly 83% of all mangroves in Ecuador (Table 1). It is the exploitation of mangrove resources, together with their diverse nature, that results in complex issues of coastal resource management. The multi-use nature of mangrove ecosystems is attributed to the diverse ecological processes that they provide, including biomass production, habitat quality, sediment retention, and water quality (Fig. 3). The biodiversity of mangrove ecosystems is also an issue in developing management plans for the sustainability of the shrimp pond industry.

The biodiversity of mangrove ecosystems is unique forested wetlands that can inhabit salt water, and their location in the intertidal zone links these habitats with both upland watersheds and coastal ecosystems. The plant diversity of mangals has been estimated at approximately 90 species, depending on definition of mangrove-dwelling species (7). There is a great incongruity in number of species between the eastern and western world, and it is not certain why fewer than nine species have colonized the new world compared to over 30 in the old world. Although species diversity of true mangroves is lower in the eastern hemisphere compared to the western hemisphere, there does not seem to be a similar contrast in the diversity of fauna that utilize mangals. The most extensive work on faunal composition of mangroves is that of Macnae (8), who described not only the species

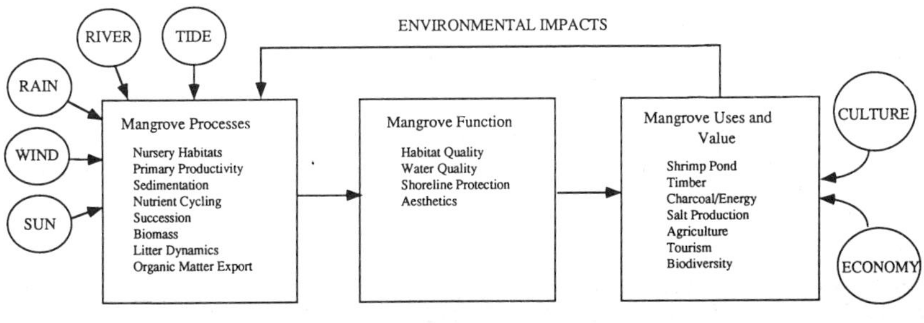

Figure 3. Diagram of the linkages among the environmental setting, ecological processes, functions, and uses of mangrove ecosystems (from 18).

composition, but also the habitat function of organisms inhabiting mangroves in the Indo-West-Pacific region. Tomlinson (7) cites that Saenger *et al.* (9) lists 95 species of mollusca, 65 determined species of crustacea, 97 species of worms, and 242 species of birds in an Australian mangrove and tidal salt marsh. Odum *et al.* (10) summarized existing information on the diversity of mangrove ecosystems in Florida; and for many groups of organisms, such as invertebrates, species diversity of mangroves is intermediate between salt marsh and rocky intertidal communities among coastal ecosystems. Extensive surveys of a mangrove island in the Caribbean are discovering the immense diversity of reef mangrove ecosystems, particularly within prop root communities and the number of insects that utilize mangrove habitats (11). These surveys lead to the conclusion that the combination of intertidal habitat in the mud and terrestrial habitat in the canopy, together with the transition from sea to land, result in a wide variety of enviroments for fauna to colonize (7).

The life cycle of shrimp demonstrates the important linkages of mangroves with other coastal ecosystems that are required to support the shrimp farm industry. Adult shrimp reproduce in the coastal ocean and release eggs that develop into initial larval stages while floating in pelagic waters. The microscopic larvae are part of the plankton communities of the coastal ocean and utilize phytoplankton during their development into postlarvae. Tides and currents provide the transport mechanisms to recruit postlarvae into nearshore and estuarine waters, where they receive protection and food in mangrove habitats. It is the combined processes of reproduction, maturation, and recruitment that supply the estuarine waters with postlarvae. Mangroves produce detritus that is used as food by shrimp, and prop roots provide refuge for protection from predators, creating a nursery for the critical stages of shrimp life cycle. Juvenile shrimp mature to adults and return to coastal ocean, again aided by tidal currents. The adults reproduce and resupply the estuary with a new generation of shrimp. The shrimp life cycle is representative of the pattern of offshore and inshore migrations of crabs and several species of fish. It is particularly important in Ecuador because of the unique oceanographic processes found in this Pacific coastal region of South America.

REGIONAL SETTING

The coastal zone of Ecuador (1°N to 3°S 20'S) consists of four coastal provinces (Esmeraldas, Manabi, Guayas, and El Oro) situated in 284,000 km^2 of lowlands between the Pacific Ocean and the Andean highlands (Fig. 4). There are three climatic zones along the coast: a moderately wet climate in the south with abundant fresh water from runoff around Guayaquil; an arid central province with very sparse vegetation; and in the north near Esmeraldas, a more humid, tropical zone with abundant rainfall and runoff. More than 95% of the annual precipitation falls during the wet season from January to May (12) and varies from less than 500 mm in the central provinces and the coast of the southern provinces to over 3,000 mm at Santo Domingo de las Colarados in the north (13). Annual mean temperatures (from 24.2 to 27 C) vary little along the coast; thus, potential evapotranspiration is about 1,300 mm/yr.

The two major river and estuarine ecosystems of the coast are Esmeraldas River estuary in the north and Guayas River estuary, which flows into the Gulf of Guayaquil in the south (Fig. 4). The Gulf of Guayaquil receives runoff from some 20 rivers with a watershed of 51,230 km^2 and is the largest estuarine ecosystem on the western Pacific coast of South America (14). The major source of freshwater is

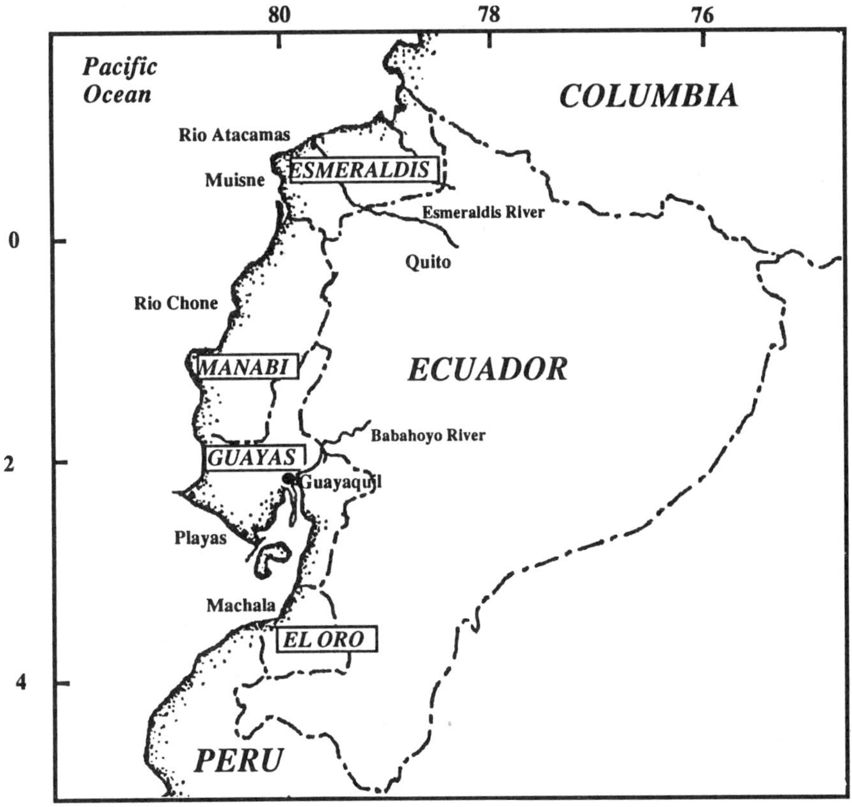

Figure 4. Map of Ecuador showing the four coastal provinces of Esmeraldas, Manabi, Guayas, and El Oro.

the Guayas River, which forms 60 km upstream at the confluence of Rio Daule and Rio Babahoyo. The mean discharge of 1,143.7 m^3/s for the Guayas River is the highest among the 30 rivers in the coastal zone of Ecuador, representing 39% of the total discharge from this lowland region. Discharge is strongly seasonal, ranging from 200 m^3/s during the dry season to 1,600 m^3/s in the wet season with an average amount of precipitation (Fig. 5). Mean precipitation in the Guayas River drainage system north of Guayaquil is 885 mm/yr, which may range from less than 400 to more than 1,800 mm during any one year. The huge variation in precipitation and discharge indicates that normal conditions are very arbitrarily defined and masks the importance of freshwater to coastal processes, such as the life cycle of estuarine organisms (Fig. 5).

The management of shrimp farming depends on oceanographic processes near Ecuador because of the strong influence that currents have on the availability of postlarvae. Coastal Ecuador is a transition zone, or equatorial front, between southerly flowing tropical water from the Panama Bight and northwardly flowing Humboldt Current from Peru. The mixing of these two water masses occurs between Manta and Punta Santo Elena along the coast of Ecuador and gradually moves southward into the Gulf of Guayaquil. The dominance of the Panamanian Current occurs during the summer, causing an increase in sea water temperature, and initiates the onset of the rainy season (14). Years of abnormally warm water

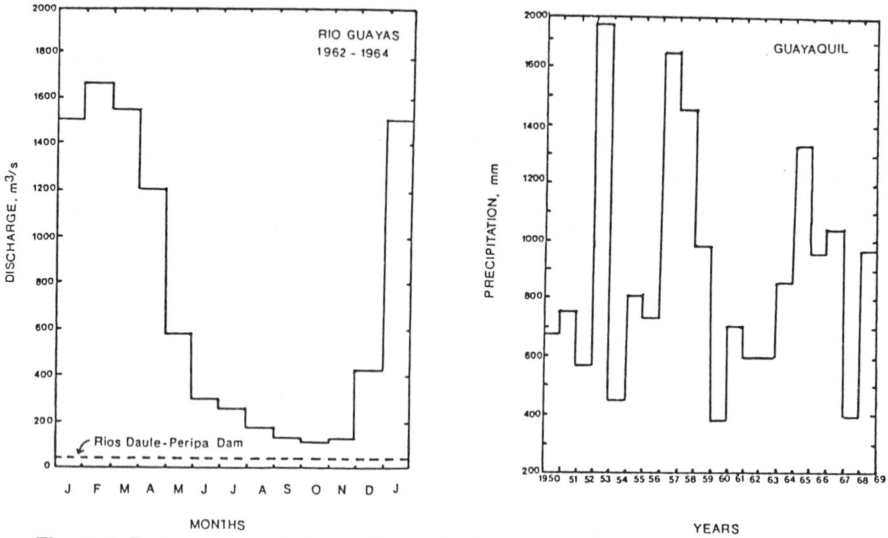

Figure 5. Forty-year record of annual precipitation in Guayaquil (left panel). Average monthly discharge of the Guayas River from 1962 to 1964 (right panel) [Both figures from (12)].

temperatures and high rainfall are associated with El Niño climate patterns due to the influx of unusually warm surface water in the southeast Pacific Ocean. The warmer offshore waters have resulted in the explosive populations of white shrimp off the coast of Ecuador from enhanced spawning, maturation, and recruitment. In the last century, major El Niño events were recorded in 1925, 1929, 1939, 1941, 1953, 1957–58, 1965, 1972–73, 1976, and 1982–83 (15). The high availability of postlarvae that supported the expansion of the shrimp industry in 1983 and 1984 has been associated with the latter El Niño event. The unpredictable nature of oceanographic events and their influence on the ecological processes of upland and coastal watersheds contributes to the complex nature of coastal resource management in Ecuador.

MANGROVE EXPLOITATION

Eight species of mangroves are distributed along a relatively narrow band of the intertidal zone of Ecuador, with a nonvegetated area called salinas in more inland intertidal areas (Table 2). Narrow zones of mangroves result from high rates of evapotranspiration relative to precipitation that occurs in the southern regions of Ecuador. The existence of 121,464 ha of mangrove in the Guayas province has been attributed to extensive river flow from the Guayas river basin (13). High tidal frequency and river discharge create conditions suitable for mangrove forest structure in the northern provinces with a tree density of 185/ha and a basal area of 62.4 m^2/ha. This forest structure is greater than for mangroves in Venezuela, Columbia, Malaysia, and Puerto Rico (16).

The most observable exploitation of mangroves along the coastal zone of Ecuador is the construction of shrimp ponds, particularly in the southern provinces of El Oro and Guayas. A survey from the Center for Remote Sensing (CLIRSEN) (17) cites that these two southern provinces have 79,396 ha or 88.8% of the total area of shrimp ponds that are constructed in the intertidal zone (Table 1). The close

Table 2. Mangrove species found along the coast of Ecuador.

Rhizophoraceae
Rhizophora harrisonii (Leachm)
Rhizophora racemosa (GFM Mayer)
Rhizophora mangle (L.)
Rhizophora samoensis
Pellicieraceae
Pelliciera rhizophorae (Planchon and Triana)
Avicenniaceae
Avicennia germinans (L.)
Combretaceae
Laguncularia racemosa (L.) Gaertn f.
Conocarpus erectus (L.)

proximity of ponds to the shore lowers costs associated with supplying water and larvae to the ponds. Initially, shrimp ponds were constructed in the salinas areas that are basically void of any vegetation. As the more inland areas became scarce, more of the ponds were constructed in mangrove habitats resulting in the increased loss of this natural resource from the coastal zone. The loss of mangroves by 1984 was estimated at 10.6% (21,587 ha) of the 1969 estimate of 203,695 ha of total mangrove area in Ecuador (Fig. 6). Urban expansion contributed to a loss of 1,200 ha, and the remaining loss was associated with construction of shrimp ponds. By 1984 there were 89,367 ha of ponds, suggesting that 23% of shrimp ponds had reclaimed mangrove areas. Recent surveys of mangrove areas by CLIRSEN (17) estimate that 14.4% of the national mangrove resources has been lost from coastal zone of Ecuador from 1967 to 1989. Based on a total area of 114,000 ha of shrimp ponds, only 26% of the ponds have been constructed in mangrove habitat; the remaining apparently have used more upland areas of the intertidal zone. Surveys of land use in the coastal zone in 1991 estimate that pond construction increased to 150,000 ha, responding to the available postlarvae during the 1988 El Niño. The loss of mangroves continues to increase, demonstrating the susceptibility of this natural resource to the availability of postlarvae along the coast.

The average annual loss of mangroves between 1969 and 1984 was 1,434 ha/yr, compared to 2,618 ha/yr from 1984 to 1987 (Fig. 6) (12). While the national average of mangrove loss is about 14%, the range for the four coastal provinces was from 8.4% for the Guayas province to 30.5% for Manabi. In Manabi, the loss of mangroves was nearly 6% per year from 1984 to 1987. By late 1988 the destruction of mangrove habitat in some estuaries was virtually complete, such as in Rio Chone estuary, a large estuary in the province of Manabi. From 1974 to 1988 mangrove area along this estuary declined from 3,973 to 600 ha, and nearly all of the mangrove loss was associated with construction of shrimp ponds. In Rio Atacames, there are only 50 ha of mangroves remaining of the 578 reported in the 1970s, representing a loss of 90.1% of the mangrove resources. In the southern province of El Oro, the Machala–Puerto Bolivar area lost over 50% of a very productive mangrove system. Thus, the perspective of mangrove loss in Ecuador is both national and regional, and the significance of mangrove loss is site specific. Historical documentation of mangrove loss along the coastal zone of Ecuador remains one of the key issues

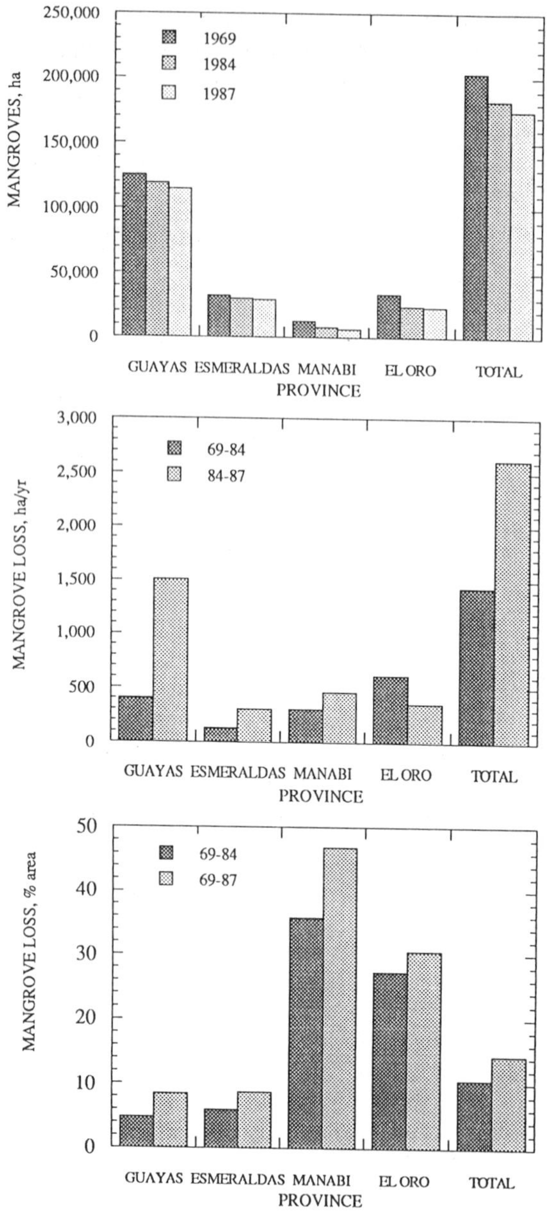

Figure 6. Loss of mangroves in Ecuador from each of the four coastal provinces and the analysis based on change per year and percentage of the total managrove area in each province.

associated with resource management.

In the northern coastal province of Esmeraldas, mangrove timber is used for the production of charcoal in the local region. Mangrove wood produces 4,500 kcal per kg of wood and is considered an economical supply of energy in the rural coastal areas of Ecuador. This is the only location in Ecuador that uses mangrove wood extensively for charcoal, although much of the woody debris from clear cutting mangrove forests for shrimp pond construction in the southern provinces was also converted to charcoal. Much of the cheap mangrove timber that supplied the charcoal industry in the northern provinces was from clearing the woody debris from initial stages of pond construction. Once pond construction declined, particularly following El Niño periods when postlarvae were scarce, wood from pond construction was limited and the charcoal industry lost a cheap and adequate supply of mangrove timber. Now there is a strong demand for mangrove timber to supply an industry that expanded along with the shrimp pond industry. In Esmeraldas, it is estimated that 2,000 m^3/yr of mangrove wood is needed to supply the current demand, which would require 20 ha of mangrove forests per year. However, mangrove silviculture is not commonly practiced as a form of mangrove management, and even the minimum rotation of 20 ha of timber per yr to supply the charcoal industry is a problem.

There is also a natural loss of mangroves in the southern coastal province of Guayas associated with excessive leaf herbivory. A massive defoliation of mangroves has been documented in the Churute Ecological Preserve, in the Guayas province, caused by the bagworm, *Oiketicus kirbyi* Guilling (18). Initial observations of the impact of the insect larvae were made in February 1989, and it was estimated that nearly 1,000 ha of mangroves had been defoliated. Leaf defoliation of complete tree canopies was most extensive in the fringe zone, dominated by *Rhizophora mangle*, where nearly 75% of the trees in infected areas were damaged by the bagworm. However, the insect larvae also defoliated about 10% of three other mangrove species, *Avicennia germinans*, *Conocarpus erectus*, and *Laguncularia racemosa*. Most of the damage was located in the vicinity of Isla Churitillo, but by 1990 the insect damage had spread into the less saline regions of the Guayas River estuary. This area of the Guayas River estuary is the most extensive area of mangrove defoliation caused by insect herbivory observed along the coast of Ecuador. It is postulated that the change in production of secondary compounds (such as tanins) in response to stress causes mangroves to be more susceptible to herbivory. Another factor that may influence the outbreak of these larvae is the loss of natural predators, such as parasites and birds.

The influence of natural resources on the supply of postlarvae in the coastal zone of Ecuador are complex. Both the oceanographic processes and mangrove habitat provide an excellent source of shrimp through optimal spawning, maturation, and recruitment processes. The extreme temporal variation of these processes limits the ability to predict the nature of available resources. Thus, the relative role of offshore temperature, rainfall, and mangrove habitat in sustaining genetic resources is complicated by the continuous change of each with time. In addition, the coast of Ecuador provides a very diverse set of environmental settings, from drowned river estuaries with abundant river discharge to dry conditions dominated by beach processes. The temporal and spatial characteristics of the coastal zone of Ecuador complicate the effect of resource utilization, such as shrimp farming, on the sustainability of genetic resources.

IMPACTS ON SPECIES AND GENETIC RESOURCES

Loss of Mangroves and Habitat Quality

The ecological function of mangroves is attributed to providing habitat and water quality in coastal ecosystems (19, 20). Mangroves may provide food and habitat to a variety of trophic levels, as well as influence nutrient and sediment concentrations in estuarine waters. The specific ecological function of mangrove ecosystems may be related to environmental settings or forcing functions of the coastal zone (Fig. 3) (19). Environmental settings include the geomorphological characteristics and the geophysical energies such as river discharge, tidal amplitude, and precipitation, and together these factors may be important in determining the structure of mangrove forests. Thom (21) classified the diverse nature of terrigenous coasts into six groups based on the magnitude of hydrological and geomorphological characteristics of the coastal environment. Each type of environmental setting was proposed to control the structure of mangrove forests and influence forest development. It is less clear if the properties of mangrove ecosystems are also linked to the forcing functions of coastal environments (19).

There is evidence that leaf litter productivity and export (22–24) and sedimentation (25) vary according to the hydrological characteristics of the coastal zone. The environmental settings of mangroves in the four coastal provinces of

Ecuador vary from high tidal amplitude and river discharge (Esmeraldas) to arid environments with minor tides and little freshwater input (Manabi). Based on the assumptions described above, the function of these wetlands in water and habitat quality may also vary among each watershed. Thus, the application of ecological concepts on the function of mangroves in coastal ecosystems has to account for the diverse coastal regimes that occur in Ecuador. This is important to understanding the ecological significance of mangrove loss and genetic resources in Ecuador.

The loss of mangroves from tropical estuaries may have direct consequences on food webs by representing a loss of habitat and organic matter. The contribution of mangroves to sustaining secondary productivity of coastal ecosystems is dependent on the fate of leaf litter, including production in the forest, transport to the estuary, and use by marine food webs. Production and transport of litter are seasonal within a system, and variation among systems seems to be related to the environmental setting (19). Those systems in areas with more hydrologic energy, as represented by river discharge, tidal amplitude, and precipitation, have more leaf production and transport (23, 24). There are also microtopographic factors that influence productivity and transport of organic matter within an estuarine watershed. Thus different types of mangroves, such as fringe and basin forests, may contribute different quantities of organic matter to adjacent estuaries. In Rookery Bay, Florida, fringe mangroves export twice as much organic matter per unit area than more inland basin forests. Yet, mass balance calculations of total organic matter contribution of each type of mangrove, taking into account the areal coverage of each habitat, resulted in equal loading rates of detritus to the estuary (6, 24, 27). In this case, the relative value of mangrove forests cannot be associated with distance from shoreline, but must also account for the distribution of the resource.

Litter productivity in three riverine forests in Ecuador range from 8 to 12 Mg ha^{-1} yr^{-1}. These mangrove forests have a 3-m tidal amplitude, and leaf litter on the forest floor is absent except for three months of the year (28). This may be associated with greater export owing to the effect of tides on the transport of leaf litter from the forest. Yet observations in the mangroves suggest that most of the leaf litter on the forest floor is harvested by the mangrove crab, *Ucides occidentalis*, and transported to sediment burrows (28). During September and October, when the crab aestivates, the standing crop of leaf litter increases on the forest floor. The levels of leaf litter during these two months are still much lower than expected based on daily rates of leaf fall, suggesting that leaf export is significant. The influence of mangrove crabs on litter dynamics has been described in other mangrove ecosystems with high geophysical energies and rates of litter turnover above 5 yr^{-1} (29, 30). It is uncertain if the crab burrows in mangrove sediment also provide other important functions to the productivity of mangroves by channeling nutrient-rich estuarine water to mangrove roots.

Three approaches have been used to establish utilization of mangrove detritus in coastal ecosystems including (1) correlations of fishery yields with habitat area, (2) habitat surveys of fauna density and diversity, and (3) food web analyses. Associations exist between the production rate of shrimp and the extent of mangrove area (31, 32, 33) such that one hectare of mangroves can yield without management more than 600 kg/yr of shrimp and 100 kg/yr of fish (32). Based on an approximate loss of 25,000 ha of mangrove, the reduction in shrimp production from the coastal zone in Ecuador would be 15,000 mt/yr. This is equivalent to about 30% of the 1987 yield from shrimp ponds. Although these statistics do not bear causal relationships, they do point out that wherever a productive postlarvae fishery exists, there is the presence of a mangrove habitat, as has been observed in Malaysia.

Habitat surveys indicate that the loss of mangroves from estuaries in Ecuador may influence the genetic resources of economically important fisheries, as well as many other less documented artisinal fisheries. Zimmerman and Minello (34) have found that *P. vannamei* and *P. stylirostris* inhabit areas in the mangroves, but it is not known whether these habitats enhance the survival or growth of these and other marine organisms in the Estero Salado. During periods of warmer temperature of offshore waters, there is a significant increase in frequency of Penaeus sp. throughout the mangrove habitats. Zimmerman *et al.* (35) found that recruits of three species, *P. californiensis*, *P. vannamei*, and *P. stylirostris* were abundant and used the nursery habitats associated with mangroves, while juveniles of *P. occidentalis* and *P. brevirostris* occurred infrequently and were not associated with mangroves. *P. vannamei* was more abundant during years with higher rainfall, particularly during the 1987 El Niño event, while *P. californiensis* was more abundant during drier years. This multi-year study demonstrated that tropical estuaries vary annually in habitat suitability as shrimp nurseries, depending on oceanographic processes and available habitat.

Confusion remains over the relative role of offshore processes and inshore destruction of mangrove to the decline in abundance of postlarvae in the last decade. Nationally there has been a loss of about 14% of the mangrove resources from the coast, but in some watersheds the loss of mangroves is greater than 90%. The cumulative impacts of mangrove loss may be site specific, particularly in regions where mangrove loss is high, and other estuaries must provide habitat to sustain the natural genetic stock. During periods of high recruitment, the impacts of habitat loss may not be significant. Yet, during the more normal oceanographic conditions, the negative impacts of habitat loss may be more pronounced along the coast. The life cycle of penaid crustaceans links the physical processes of the coastal oceans with the ecological processes of mangrove estuaries. The combination sustains this genetic resource.

Mangroves also support estuarine food webs that have more resident life histories and a more local impact on economies. In Ecuador, the mangrove crab, *Ucides occidentalis*, is harvested for 10 months and sold locally. There are no reports of the value of this fishery to the local economy, but it is listed as an important management issue in the Esmeraldas, Manabi, and Guayas provinces. A similar economic use of mangrove crab occurs in Sabah, Malaysia, where *Scylla serrata* earns U.S. $1.2 million annually in the local economy.

Mangroves also provide habitat for a variety of organisms that do not contribute directly to the activities of tropical countries. There are very few estimates of the genetic resources associated with mangrove ecosystems in Ecuador. Although there have been extensive surveys of the offshore environments in the Galapagos Islands, no surveys exist of the genetic diversity of mangrove ecosystems along the coast. Ortiz (36) observed that over 40 species of birds were associated with mangrove ecosystems in Rio Chone, Manabi. The high number of organisms that are associated with mangrove habitats is apparently due to mangrove proximity to both upland and marine environments (7).

Shrimp Pond Management and Genetic Resources

Shrimp ponds represent managed ecosystems that are linked to the ecological processes of several coastal ecosystems. Methods of shrimp mariculture in the intertidal zone are grouped into three classifications based on the densities of juvenile shrimp stocked in the ponds. Extensive mariculture uses a stocking density of 10,000 to 20,000 juveniles per ha and relies little on further supplements from seawater exchange via pumping or from artificial fertilization. Predators are present and annual yields are relatively low at 100 to 400 kg/ha. An increase in stocking rates to 50,000 to 60,000 juveniles/ha is a semi-extensive system that requires some supplemental feeding for an enriched supply of food and exchange of seawater to control for water quality problems such as decreased levels of dissolved oxygen. Production rates more than double with this program. Semi-intensive operations stock ponds at 100,000 juveniles/ha. Feed is supplied or ponds are fertilized to increase sources of food. Water exchange with the estuary is higher and annual production rates increase to 1,000 to 1,800 kg/ha. The shrimp producers' association estimates that 60,000 ha, or nearly 60%, of operational shrimp ponds use extensive management (estimate for 1989). Semi-extensive and semi-intensive operations include 25,000 and 15,000 ha, respectively. However, since the semi-intensive operations are much more productive, they produce nearly the same amount of shrimp as the extensive operations.

The dramatic expansion of the farmed shrimp industry and increased levels of pond management stimulated the development of a new fishery to provide postlarvae and seed shrimp for stocking mariculture ponds. Industry sources estimated that up to 90,000 artisinal fishermen were involved in the 1983 harvest and in 1984 numbers of fishermen working along the coast were even higher (4). Seed fishing is concentrated in areas of significant freshwater discharge along the coastline, such as El Oro and Esmeraldas, with the highest effort occurring in the Guayas province. The catches of these fishermen are nonselective, with small fish, penaeid postlarvae, and juvenile shrimp including a mixture of *P. vannamei*, *P. stylirostris*, *P. occidentalis*, and *P. californiensis*, as well as some freshwater Carid species. Since only the former two species survive best in mariculture ponds, owners pay according to the proportion of the stock that is *P. vannamei* and *P. stylirostris* (4). Selection is a postharvest process and therefore less-valued species are lost from the estuary. The peak of the seed fishing season is from December to March, when fishermen may take up to 40,000 postlarvae a day at a size ranging from 7 to 10 mm. The annual demand for postlarvae is estimated at 16.5 billion based on 120,000 ha of shrimp ponds using mostly extensive pond management (1989 estimates). Only about half of the postlarvae collected along the beaches is *P. vannamei*, requiring a total harvest of 33 billion postlarvae. This demand for *P. vannamei* represents a potential impact on the genetic stability of nontargeted organisms that use the coastal environments during their life cycle.

The abundant supply of postlarvae during the El Niño events created an excessive demand for construction of shrimp ponds from 1985 to 1987. The natural supply of postlarvae during the more normal years of recruitment could not stock the existing ponds, such that by 1985 nearly 50% of the shrimp ponds were not in operation. During this period, there was a major emphasis to produce postlarvae with hatcheries and acclimate these shrimp to growout ponds. In 1986 there were only 12 hatcheries in operation. By 1989 there were 106 laboratories in operation and another 60 planned for construction. Nearly half of these hatcheries were located in Guayas province and 25 were constructed along the beaches of Manabi. The anticipated production of these hatcheries was nearly 10 billion larvae, yet less than 10% of these will survive under pond conditions. Most hatcheries produce larvae from wild gravid females because they have better survival in growout ponds.

Thus, the hatchery industry, while quickly responding to the demand for larvae, is presently not replacing the natural genetic resource needed to sustain the shrimp pond industry.

Eutrophication of Coastal Waters

Some preliminary evidence indicates that mangroves may be a sink of nutrients in coastal waters. This may seem to contradict the outwelling concept of mangroves as a source of detritus to estuarine ecosystems (37, 22, 23). One explanation is that net nutrient uptake may be a balance between inorganic nutrient input and organic nutrient export. However, the net balance of nutrient exchange has seldom been investigated for mangroves (38–41).

Sediments suspended in the water column are deposited in mangroves during flooding, enriching mangrove soils. The extensive root system of mangroves enhances this trapping process and retards the forces of erosion along the shoreline (26, 42). Although this function has been overstated to the extent of calling mangroves "walking trees," they do contribute to the sedimentary processes that exist in estuarine ecosystems (26, 43).

The replacement of mangrove ecosystems with shrimp pond ecosystems may influence the eutrophication of adjacent coastal waters. Shrimp farming contributes to the degradation of habitat and water quality of coastal ecosystems by loading nutrients (fertilization) and increasing freshwater loss (pumping). Pumping and fertilization are an integral part of pond management to improve the productivity of shrimp in growout ponds by decreasing mortality and increasing growth rates. The benefits of pond management from these practices must also be evaluated in the context of their possible negative feedback on shrimp production through their deleterious effects on water quality. The extent of this potential impact has not been evaluated and is important in understanding the costs of replacing natural ecosystems that serve as nutrient sinks with ponds that are nutrient sources in estuarine watersheds. Twilley (6) estimated that the exchange of water with 50,000 ha of shrimp ponds in the Guayas province is equivalent to half the peak freshwater discharge of the Guayas River during the wet season. The effects of shrimp pond management on coastal resources include not only the loss of mangroves and their ecological functions, but also the replacement or alteration of these natural functions with those of shrimp ponds (6).

Changes in water quality have been associated with the occurrence of phytoplankton blooms that discolor the water along the coast of Ecuador, commonly known as red tides. Although their cause is not positively known, these types of blooms are a common occurrence in the Gulf of Guayaquil and in the inland waters of the Guayas River estuary (44–46). There have been 28 reportings of red tides in the coastal zone of Ecuador, and these blooms vary in species composition, density of cells, and duration. The effect of red tides may be direct by producing toxic substances that impact higher trophic levels. Phytoplankton blooms also exclude more common and diverse phytoplankton communities that support food webs that change in response to new producers. These indirect effects cause shifts in the higher trophic levels and potentially impact the economically based fisheries of an area.

The most direct influence of red tides on the estuary is fish kills caused by the presence of toxic organisms such as *Gonyaulax catenella* and *Gymnodium breve*. *Gonyaulax monilata* occurred in the upper portion of the Gulf of Guayaquil in April

1980 and in March 1986 along the coast of Manglaralto. The 1980 bloom resulted in high fish mortality (45), while the 1986 bloom caused significant mortality of shrimp postlarvae in eight hatcheries, interrupting operations for 30 to 45 days (46). Other red tides in the Guayas River estuary include *Gyrodinium stratum* in September 1982, *Mesodinium rubrum* in August 1984, and *Prorocentrum maximum* from February 1985 to February 1986, and a recent bloom of *Nitzxchia* sp. (47). These blooms caused high mortality in shrimp ponds when phytoplankton-contaminated waters were pumped from the estuary.

MANAGEMENT AND CONSERVATION OF MANGROVE RESOURCES

The three central elements of Ecuadorean institutional policy for the regulation of the shrimp farming industry include shrimp farm site selection and construction, fisheries management, and conservation of mangrove ecosystems (47). The government's policy of granting royalty-free 10-year concessions for the use of land on which to build and operate shrimp ponds led to extensive farming practices and mangrove destruction in the intertidal zone. While there is an existing government policy to provide the framework for site selection of ponds and prevent mangrove destruction, it is seldom used or consistently enforced. Land in the beach and bay zone, which is essentially the intertidal zone, is leased from the Merchant Marine and Coastal Directorate for Fisheries (DIGMER), while vacant upland can be purchased from the National Institute for Agrarian Reform (IERAC). The site acquisition process is very complex and may involve as many as seven different agency departments. As a result, few of the ponds constructed during the expansion of the industry from 1980 to 1987 had complete authorization. It was estimated that only 10% of the ponds in 1985 were operating with completion of all necessary permits. In addition, the authority between DIGMER and IERAC lacks delineation between intertidal and upland sites that could be used for shrimp pond construction. There have been several decrees and laws that forbade the construction of shrimp ponds in mangrove areas, including those of 1978 and 1985. However, these laws have not been enforced owing to the lack of clear definition of agency responsibility and the lack of personnel resources to manage such regulation. In 1990, a presidential decree forbade the destruction of mangrove forests for any exploitation.

The government has also helped establish the development of conservation areas such as the Churute Ecological Preserve along the Guayas River estuary. This conservation area was established by the Direccion Nacional Forestal (DINAFOR) as part of the 14 forestry reserve areas in the country. The preserve includes 30,000 ha of coastal and freshwater wetlands serving as important feeding and habitat for fish, waterfowl, and other wildlife. However, even within the preserve there has been much shrimp pond construction—demonstrating the problem of economic pressure even in areas designated as ecological preserves. In 1985, DINAFOR accused IERAC of granting leases for shrimp farms in the national forestry domains. In many cases IERAC approved the lease of lands as upland sites in mangrove areas that were within its jurisdiction. The development of shrimp ponds in the Churute Ecological Preserve demonstrates the inadequate control of mangrove destruction by shrimp farming in Ecuador.

Coastal resource management was established in 1986 through a convenio among the U.S. Agency for International Development (USAID), the University of Rhode Island (URI), and the Government of Ecuador (GOE). Complementary financing is provided by the USAID Mission in Ecuador and the Government of Ecuador. The Coastal Resources Management Project (Proyecto Manejo Recursos Costeras—PMRC) was established by the government of Ecuador through Executive Decree 375 in January 1989. The PMRC is implemented by the Technical Secretary to the National Coastal Resources Management Commission. The convenio was modified in 1991 to extend the project through October 1994 and to make the Secretaria de Administration Publica the counterpart agency to URI. The Secretaria Technica and URI share central offices in Guayaquil, Ecuador. The PMRC initially established three working groups to develop management and research activities and serve as conduits for information gathering and synthesis. The Mangrove Working Group was one of the initial efforts of PMRC, followed by the Water Quality Working Group and Fauna Working Group.

The ZEMs

Coastal resource management is focused in five specific coastal watersheds referred to as Zonas Especiales de Manejo (ZEM). Four of these zones represent some of the more acute problems in mangrove management in Ecuador. The ZEM projects are designed to coordinate integrated development and environmental management plans that will solve coastal area use conflicts of mangroves and provide the framework for a variety of development initiatives that are based on sustainable use of coastal resources. Each ZEM plan deals with a unique combination of issues, resource users, and opportunities for development. The coordinator of each ZEM plays a crucial role in providing the background information needed to address each issue, identifying and working to gain the cooperation of the individuals who use or are affected by the condition of the coast. The boundary of each ZEM is defined by the coastal watershed unit (or hydrologic unit) that must be carefully delineated and mapped. In addition, the ZEM must also recognize the larger regional planning zone for consideration of factors that could affect the coastal resources of the ZEM. The watershed concept is central to utilizing information on ecosystem properties in development of resource management. This approach also provides a method of accounting for the complex linkage of mangrove estuaries with coastal and upland ecosystems.

Area management work plans for each ZEM are designed to make the best use of available information in setting out the resource management challenges faced in each ZEM. The objective is to develop a policy framework for specific actions and to interpret the overall ZEM plan in terms of its significance for particular communities. The Atacames-Sua-Muisne ZEM is an example of the management plans that have recently been adopted for the four ZEMs. These areas represent coastal watersheds that are typical of rural Ecuador, containing a number of small villages that depend on the coastal environment for income and food. The continuing degradation of the environment and loss of genetic resources are particularly significant to these more rural regions of the coast. The objectives of the ZEM plan for the Atacames-Sua-Muisne region include the following:

1. Halting mangrove cutting and establishing community-based reforestation, sustained use, and protection measures;
2. Protecting public health, especially for tourists, by improving sanitary conditions in popular beach areas and coastal communities;

3. Shifting the focus of tourism activities toward those that emphasize the high quality natural environment and spur local economic interests to lead efforts to protect and restore key sites, including information and training as well as carefully planned facilities to manage visitors;

4. Creating a local capacity for controlling land use through development and implementation of shore line and coastal land zoning, regulations, and decision procedures;

5. Reducing the stress of estuarine and nearshore fisheries, including those involving shrimp larvae and egg-bearing adult shrimp, caused by overfishing, poor handling of shrimp larvae, and habitat degradation, through training, fisheries assessments, and experimental management initiatives.

The more urbanized ZEMs, such as Playas-Posorja and Machala, represent a different set of problems facing rapidly growing urban coastal centers, such as Guayaquil. The ZEM management plan is working to resolve conflicts among tourists, fishermen, and residents at the beaches, control fishing in nearshore waters, build local capability to control shore and land use, and undertake environmental enhancement and public sanitation initiatives. The challenge in the urban areas is also to define an agenda that matches the likely resources available for implementation and fits closely with the dominant role played by municipal and provincial officials.

Implementation of the ZEM management plans to conserve natural resources will require coordinated efforts of the working groups concerned with mariculture, mangroves, water quality, and shoreline processes. The issues of mariculture management include appropriate technology for alternative mariculture species, new practices in mariculture, and the development of mariculture policy. Work in this area includes development of educational materials on the proper handling of wild shrimp postlarvae, and a survey of the shrimp postlarvae fishery to document its size and socioeconomic characteristics, fishing techniques, and the impact of handling techniques on yield. This assessment is important to evaluate the effect of mariculture activities on genetic stocks of natural populations.

The Mangrove Working Group

The mangrove working group is dominated by officials from the government forestry sector; recently, however, participation of the fishery groups has been encouraged. It has worked with DINAFOR to review coast-wide assessment of mangrove management problems and options in a document entitled Strategy for the Conservation and Management of Mangroves in Ecuador (48). This document is aimed at gaining national acceptance of management concepts now being tested in the ZEMs. The group has also been responsible for the delineation of mangrove loss and identifying areas suitable for mangrove restoration projects through efforts of CLIRSEN. The mangrove working group is also presently developing technical training in mangrove reforestation and development of sustainable mangrove plantations for the remediation of coastal areas and provision of timber for the charcoal industry. In addition, the group is working on mangrove protection by supporting the Ranger Corps, a group working to improve enforcement of existing coastal resource management laws and regulations. The Ranger Corps is organized according to the seven port captaincies that have provided training exercises in the respective ZEMs. Members of the corps include the Merchant Marine, the Tourism Agency, the Sub-Secretary of Fisheries, and the Ministry of Agriculture. There are two levels of activities in these troops, the decision makers in the institutions and the field inspectors charged with enforcement.

The Water Quality Working Group

The water quality working group is involved in characterization of coastal and estuarine waters in the ZEMs, although the availability of funds has limited field surveys to the Santa Rosa and Rio Chone estuaries. The objective of the working group is to determine what water quality parameters represent threats to the living resources of the coastal zone. Of particular interests are heavy metals, pesticides, nutrients, and dissolved oxygen (49). The objective of the shoreline process and mapping working group is to provide information on changes in coastal landscapes, particularly related to potential impacts of mangrove deforestation and shrimp pond construction on sedimentation. These maps are important tools for planning ways to stabilize coastal resources.

Private foundations, such as Fundacion Pedro Vicente Maldonado, are important participants in the development of programs for the conservation of coastal resources in Ecuador. These organizations provide leadership in public education of coastal environments and provide assistance in mangrove management. Television and press coverage of coastal issues and PMRC activities, including quarterly publication of *Costas*, are used to inform the public.

In May 1992, the National Coastal Resource Management Commission formally adopted the five Special Area Management Plans, which, according to recently signed Presidential Executive Decree 3399, are binding on Ecuadorian agencies. This greatly strengthens the ability of the PMRC to move forward in its community-based mangrove management techniques, coordinated enforcement of existing laws, and public education programs.

SOCIOECONOMIC COSTS AND BENEFITS

Mangroves are a common access resource (Fig. 3). Since different people benefit from the various functions of mangroves, the market substantially undervalue its existence. Underestimates of the tradeoffs between the various goods and services provided by mangroves lead to a poor understanding of how best to manage them, for whatever objectives society might have. For example, without a knowledge of the opportunity cost of increasing timber production from mangroves, we cannot know whether such a policy makes sense, or who benefits or loses from such a policy. Similarly, lack of knowledge of the total impacts of mangrove removal inhibits any careful examination of this issue.

The net production of ecosystems is utilized by humans to support goods and services, such as fisheries (19). There has been much effort to establish the value of services that mangroves and other wetlands provide to humans (50–53). For mangroves, these include indirect use such as providing food and habitat for economically important fisheries and water quality for aquaculture. Direct uses of mangroves include timber harvest and charcoal production, and nonuse value functions of mangroves to maintain biodiversity (19). The link between the properties of ecosystems and the goods and services they provide is an important concept in natural resource management. Valuation techniques and land-use management schemes should identify those ecological processes that control the functions of ecosystems that sustain their social and economic value. One of those values is the genetic resource they provide.

The valuation techniques of wetlands have generally used either neoclassical economics or energy analysis (54, 55). Authors using the former have tended to value various goods and services such as timber, wastewater treatment, and fish production in isolation, using a variety of techniques (54, 56). Some authors in this group have attempted to determine the overall value of wetlands by valuing the current levels of goods and services and summing (50, 51, 57). These studies measure consumer and producers' surplus, or they use less precise proxies. However, few authors have addressed the fact that ecological goods and services are interrelated. For instance, using wetlands for wastewater treatment can affect their ability to serve as a genetic stock for commercially important fisheries. Although some authors such as Gupta and Foster (58) have attempted to value different mixes of system goods and services, few have attempted to value a natural system as an asset that can provide a variety of interrelated functions. Similarly, none seem to address the problems that arise when two natural systems are linked together in a larger system such as a watershed. In other words, they do not consider the valuation problems arising from the choice of boundaries among coupled ecosystems (53).

Dixon (52) estimated that the value of mangroves ranges from \$25 ha^{-1} yr^{-1} to over \$1,000 ha^{-1} yr^{-1}, depending upon the extent of goods and services included in the analysis (58). Whereas some on-site or off-site products or services are marketed, others are not. Consequently, many goods and services provided by mangroves are omitted from the valuations, and most of the values included fail to subtract out the costs of production, such as harvesting, or the effects of negative feedbacks, and thus substantially overstate net benefits. Valuation techniques for mangrove ecosystems have several shortcomings, including failure to analyze (1) the tradeoffs and the "elasticities" between the various goods and services produced; (2) the value of wetlands as part of a larger system; and (3) the distinction between the average (per unit) value of a hectare of mangrove and its marginal value, the amount by which the value of all remaining hectares of mangrove (total wetland value) would decrease should a hectare of mangrove be reclaimed (53).

Energy analysts utilize energy flows in natural systems as a measure of the total work performed by those systems. This holistic method is designed to account for the variety of functions that ecosystems provide to support goods and services. This method converts energy flows to dollars by assuming proportionality between energy inputs and GNP (50, 51, 59). Odum (60) analyzed the ecological economics of the shrimp pond industry in Ecuador to demonstrate the use of the EMERGY method for evaluating environmental contributions to assist in natural resource management. The EMERGY method is a scientific-based measure of wealth, which puts raw materials, commodities, goods, and services on a common basis. According to the theory, the pattern that maximizes EMERGY contributes more wealth. Economic design and management alternatives can be evaluated by maximizing for EMERGY, which can generate wealth according to an area's potential. Evaluations of design are based on comparing the EMERGY investment in purchased inputs to those that are provided free by the environment.

This analysis emphasizes the importance of feedback that encourages the sustainable use of the free services provided by the environment. The ratios indicate the huge investment of environmental inputs to support international monetary exchange for Ecuador. Given the lack of feedback to support the genetic and environmental quality of the coastal systems that provide these environmental inputs, there will be some eventual shift in designs of the economic systems as these free services decrease. The total annual macroeconomic monetary value for the ecological processes of coastal ecosystems (out to the 100-m contour of the continental shelf) is about $21.4 billion per year (60). The costs of the degradation of these environmental resources is compared to the realized income from shrimp exports to Ecuador. Discounting the contribution of environmental inputs and the loss of these natural resources to the subsidy of the shrimp farming industry can be addressed with the following recommendations: (1) decrease channelization; (2) return *P. vannamei* produced in the shrimp ponds to the estuary to improve larval stocks; (3) reforest mangrove areas; and (4) manage the Daule-Peripa dam to allow for seasonal discharge. In addition, the report recommended that shrimp pond management be less intensive, shift some ponds to other more artisinal fisheries, and discontinue exchange of shrimp with developed countries.

CONCLUSION

The environmental quality of mangrove ecosystems is important to sustaining the shrimp farm industry by providing wild stock of postlarvae, a productive pond environment, and minimum water quality problems such as turbidity and dissolved oxygen. Mangroves and tides provide the shrimp industry with clean water and vital habitat that enchance wild postlarvae supply and shrimp production in ponds. Thus, successful shrimp pond managment is tightly coupled to the environmental quality of the estuarine ecosystem.

Community support and participation—especially by key user groups—provide the pressure to develop coastal zone management. The success of the ZEM concept will be evaluated by the following indicators: increased understanding of the importance of coastal resources by community members; awareness among users of natural resources that ecosystem processes are important in sustaining their livelihoods; the level of participation of resource users in ZEM program activities; and development of specific actions aimed at resource protection and sustainable use.

The government's lack of control of mangrove destruction is one of the critical weaknesses of present coastal zone management. The granting of royalty-free concessions encourages short-term planning in the development of mariculture in Ecuador (61, 62). Once the free services of the environment sustaining the profitability of shrimp farming have diminished, operations will have to rely more on purchased inputs, such as postlarvae and feed. As the quantity and quality of the genetic resources decline, the industry becomes susceptible to disease and other perturbations that can drastically reduce productivity. Existing policy that bans the cutting of mangrove relies on routine enforcement and, as a result, has been completely unworkable. Until recently, it has been hard to justify more activist approaches to mangrove management because of lack of information on the full range of environmental and social values lost.

There are several questions that arise from this case study of coastal forested wetlands that are not normally considered in the conservation of upland forested resources. What size of conservation areas are needed to maintain genetic resources in coastal environments? The processes of recruitment and habitat utilization of the diverse life cycles in coastal ecosystems have to be considered. This has been discussed for some marine habitats such as coral reefs, but few analyses have been made of river-dominated estuarine ecosystems. An important consideration is the proper wetland area to water area ratio that is needed to maintain biodiversity of coastal ecosystems.

What are the linkages of upland, coastal, and offshore environments to sustain genetic resources? This is particularly important for organisms whose life cycle includes a transition among oceanic, estuarine, and freshwater environments. It also includes the importance of fluctuating salinities in estuaries that control the competitive advantages of certain organisms and food webs.

How can policy and management goals be implemented for coastal resources that are under the jurisdiction of several government agencies? The use of mangroves in navigation, fishery resources, and timber include many institutional guidelines, some with contradictory goals for resource utilization.

Successful widespread adoption of site-specific mangrove management practices that foster community participation, user group agreements, and local enforcement will require careful evaluation of the pilot projects that are now under way in some of the special area management zones. Through the continued analysis and development of research and management information, we hope to provide models to assist in the analysis of other tropical coastal ecosystems.

Funding for the mangrove research in Ecuador was from the U.S. Agency for International Development Program in Science and Technology Cooperation (Grant No. DPE-5542-G-SS-8011-00) and the University of Rhode Island/AID Coastal Resource Management Program. Support from the University of Southwestern Louisiana includes LEQSF (1988-94-GF-15) from the Board of Regents, Faculty Research Awards, Graduate Student Organization, and Department of Biology.

REFERENCES

1. Siddall, S.E., J.A. Atchue III, and R.L. Murray, Jr. 1985. Mariculture development in mangroves: A case study of the Philippines, Ecuador and Panama. In *Coastal resources management: Development case studies*, J.R. Clark, ed. Renewable Resources Information Series, Coastal Management Publication No. 3. Prepared for the National Park Service, U.S. Department of the Interior, and the U.S. Agency for International Development. Columbia, SC: Research Planning Institute, Inc.
2. Olsen, S., and L. Arriaga. 1989. Establishing a sustainable shrimp mariculture industry in Ecuador. University of Rhode Island, Technical Report Series TR-E-6.
3. Aiken, D. 1990. Shrimp farming in Ecuador, an aquaculture sucess story. *World Aquaclture* 21:7–16.
4. McPadden, C.A. 1985. A brief review of the Ecuadorian shrimp industry. Instituto Nacional de Pesca/Overseas Development Administration, UK.
5. Ochoa, E., W. Macas, and J. Marcos. 1987. Ecuador: perfil de sus recursos costros. Part III. Guayas, pp. 135–211. Guayaquil, Ecuador: Funacion Pedro Vicente Maldonado.
6. Twilley, R.R. 1989. Impacts of shrimp mariculture practices on the ecology of coastal ecosystems in Ecuador. In *Establishing a sustainable shrimp mariculture industry in Ecuador*, Stephen Olsen and Luis Arriaga, eds., 91–120. University of Rhode Island, Technical Report Series TR-E-6.
7. Tomlinson, P.B. 1986. *The botany of mangroves*. New York: Cambridge University Press.

8. Macnae, W. 1968. A general account of the fauna and flora of mangrove swamps and forests in the Indo-West-Pacific region. *Advances Marine Biology* 6:73–270.

9. Saenger, P., M.M. Specht, R.L. Specht, and V.J. Chapman. 1977. Mangal and coastal salt-marsh communities in Australasia. Chap. 15 in *Ecosystems of the world*. Vol. 1, Wet coastal ecosystems. Amsterdam: Elsevier Scientific.

10. Odum, W.E., C.C. McIvor, and T.J. Smith III. 1982. The ecology of the mangroves of south Florida: A community profile. Fish and Wildlife Service/Office of Biological Services, Washington, DC: FWS/OBS-81/24.

11. Ruetzler, K., and C. Feller. 1988. Mangrove swamp communities. *Oceanus* 30:16–24; C. Feller, personal communication.

12. Stevenson, M.R. 1981. Seasonal variations in the gulf of Guayaquil, a tropical estuary. Inst. Nacional de Pesca. *Bol. Cient. Tec.* 4:1–133.

13. Schaeffer-Novelli, Y. 1983. Inventario de los biorecursos del manglar en la costa ecuatoriana. Informe Final. Mision de Consultoria Sobre el Tema.

14. Cucalon, E. 1984. Temperature, salinity and water mass distribution off Ecuador during an El Niño event in 1976. *Rev. Cien. Mar. Limn.* 2:1–25.

15. Cucalon, E. 1986. Oceanographic characteristics off the coast of Ecuador. In *Establishing a sustainable shrimp mariculture industry in Ecuador*, Stephen Olsen and Luis Arriaga, eds., 185–94. University of Rhode Island, Technical Report Series TR-E-6.

16. Cintron, G. 1981. El manglar en la costa Ecuatoriana. Hoja de Documentacion. Departamento de Recursos Naturales.

17. CLIRSEN. 1984, 1988. Applicacion de sensores remotos en el estudio del levantamientos integrados de recursos naturales por sensores remotos. Center for Remote Sensing.

18. Sarango, A. 1990. El gusano de canasta (Leridoptero: Psychidiae) en los manglares Ecuatorianos de Churute. Proteccion Forestal.

19. Twilley, R.R. In press. Energy signature and properties of mangrove ecosystems. In *Maximum power*, C. Hall, ed.

20. Lugo, A.E., and S.C. Snedaker. 1974. The ecology of mangroves. *Ann. Rev. Ecol. Syst.* 5:39–64.

21. Thom, B.G. 1982. Mangrove ecology—a geomorphological perspective. In *Mangrove ecosystems in Australia*, B.F. Clough, ed., 3–17. Canberra: Australian National University Press.

22. Twilley, R.R. 1985. The exchange of organic carbon in basin mangrove forests in a southwest Florida estuary. *Estuarine, Coastal and Shelf Science* 20:543–57.

23. Twilley, R.R., A.E. Lugo, and C. Patterson-Zucca. 1986. Litter production and turnover in basin mangrove forests in southwest Florida. *Ecology* 67:670–83.

24. Twilley, R.R. 1988. Coupling of mangroves to the productivity of estuarine and coastal waters. In *Coastal-offshore ecosystem interactions*, B.O. Jansson, ed., 155–80. Berlin: Springer-Verlag.

25. Lynch, J.C. 1989. Sedimentation and nutrient accumulation in mangrove ecosystems of the Gulf of Mexico. M.S. Thesis, University of Southwestern Louisiana, Lafayette.

26. Lynch, J.C., J.R. Meriwether, B.A. McKee, F. Vera-Herrera, and R.R. Twilley. 1989. Recent accretion in mangrove ecosystems based on ^{137}Cs and ^{210}Pb. *Estuaries* 12:284–99.

27. Twilley, R.R. 1982. Litter dynamics and organic carbon exchange in black mangrove (*Avicennia germinans*) basin forests in a southwest Florida estuary. Ph.D. dissertation. University of Florida, Gainesville.

28. Twilley, R.R., R. Zimmerman, L. Solorzano, V. Rivera-Monroy, A. Bodero, R. Zambrano, M. Pozo, V. Garcia, K. Loor, R. Garcia, W. Cardenas, N. Gaibor, J. Espinoza, and J. Lynch. 1990. The importance of mangroves in sustaining fisheries and controlling water quality in coastal ecosystems. Interim Report,

U.S. Agency for International Development, Program in Science and Technology Cooperation, Washington, DC: USAID.

29. Robertson, A.I. 1986. Leaf-burying crabs:their influence on energy flow and export from mixed mangrove forests (*Rhizophora* spp.) in northeastern Australia. *J. of Experimental Marine Biology and Ecol.* 102:237–48.
30. Robertson, A.I., and P.A. Daniel. 1989. The influence of crabs on litter processing in high intertidal mangrove forests in tropical Australia. *Oecologia* 78:191–98.
31. Macnae, W. 1974. Mangrove forests and fisheries. FAO/UNDP Indian Ocean Programme. IOFC/DEV/7434.
32. Turner, R.E. 1977. Intertidal vegetation and commercial yields of penaeid shrimp. *Transactions of the Am. Fisheries Soc.* 106(5):411–16.
33. Jothy, A. A. 1984. Capture fisheries and the mangrove ecosystem. In *Productivity of the mangrove ecosystem: Management implications*, J.E. Ong and W.K. Gong, eds., 129–41. Unit Pencetakan Pusat, University Sains Malaysia, Penang, Malaysia.
34. Zimmerman, R., and T.J. Minello. 1986. Recruitment and distribution of postlarval and early juvenile penaeid shrimp in a large mangrove estuary in the Gulf of Guayaquil during 1985. In *Establishing a sustainable shrimp mariculture industry in Ecuador*, Stephen Olsen and Luis Arriaga, eds., 233–45. University of Rhode Island, Technical Report Series TR-E-6.
35. Zimmerman, R.J., R. Garcia, T.J. Minello, and T.J. Baumer. 1991. Variability in juvenile shrimp abundances related to location and rainfall in the gulf of Guayaquil estuary, Ecuador. Abstract, Estuarine Research Federation Conference, San Francisco.
36. Ortiz, D. undated. Comparacion estructural de dos bosques de manglar cercanos a poscinas camaroneras, y estudio socio-economico de la comuna salinas de Bahia, en la provincia de Manabi. Quito: Universidad Catolica.
37. Odum, W.E., and E.J. Heald. 1972. Trophic analysis of an estuarine mangrove community. *Bull. Marine Science* 22:671–738.
38. Walsh, G.E. 1967. An ecological study of a Hawaiian mangrove swamp. In *Estuaries*, G.H. Lauff, ed., 420–31. Washington, DC: AAAS.
39. Nedwell, D.B. 1975. Inorganic nitrogen metabolism in a eutrophicated tropical mangrove estuary. *Water Res.* 9:221–31.
40. Sell, M.G., Jr. 1977. Modelling the response of mangrove ecosystems to herbicide spraying, hurricanes, nutrient enrichment and economic development. Ph.D. thesis. University of Florida, Gainesville.
41. Clough, B.F., K.G. Boto, and P.M. Attiwill. 1983. Mangroves and sewage: A re-evaluation. In *Biology and ecology of mangroves*, H.J. Teas, ed. The Hague: Dr. W. Junk Publishers.
42. Scoffin, T.P. 1970. The trapping and binding of subtidal carbonate sediments by marine vegetation in Bimini Lagoon, Bahamas. *J. of Sedimentary Petrology* 40:249–73.
43. Nixon, S.W., B.N. Furnas, V. Lee, N. Marshall, O. Jin-Eong, W. Chee-Hoong, G. Wooi-Khoon, and A. Sasekumar. 1984. The role of mangroves in the carbon and nutrient dynamics of Malaysia estuaries. Proceedings Symposium on Mangrove Environments—Research and Management, 534–44. Kuala Lumpur: University of Malaya.
44. De Arcos, T.V. 1982. Mareas rojas en aguas ecuatorianas. *Revista de ciencias del mar y limnologia* 1:115–25.
45. Jimenez, R. 1980. Marea roja en el Golfo de Guayaquil en abril de 1980. *Bol. Informativo Inst. Nac. Pesca* 11–13.
46. Jimenez, R. 1986. Red tide and shrimp activity in Ecuador. In *Establishing a sustainable shrimp mariculture industry in Ecuador*, Stephen Olsen and Luis Arriaga, eds., 179–84. University of Rhode Island, Technical Report Series TR-E-6.
47. Perez, E., and D.D. Robadue, Jr. 1986. Institutional issues of shrimp mariculture in Ecuador. In *Establishing a sustainable shrimp mariculture industry in*

Ecuador, Stephen Olsen and Luis Arriaga, eds., 71–88. University of Rhode Island, Technical Report Series TR-E-6.

48. Mangrove Working Group. 1991. Strategy for the Conservation and Management of Mangroves in Ecuador. PMRC.
49. Armijos, M.M. 1992. Estudio de la calidad del aqua costera ecuatoriana. Proyecto de Manejo de Recursos Costeras (PMRC).
50. Farber, S., and R. Costanza. 1987. The economic value of wetlands systems. *J. of Environmental Mgmt.* 24:41–51.
51. Costanza, R., S.C. Farber, and J. Maxwell. 1989. The valuation and management of wetland ecosystems. *Ecological Economics* 1(4):335–61.
52. Dixon, J.A. 1989. Valuation of mangroves. *Tropical Coastal Area Mgmt.* 4(3):1–6.
53. Gottfried, R.R. The value of a watershed as a series of linked multiproduct assets. *Ecological Economics*.
54. Bell, F.W. 1989, June. Application of wetland valuation theory to Florida fisheries. Report No. 95. Florida Sea Grant College. (University of Florida, Gainesville).
55. Pearce, D.W., and R. Kerry Turner. 1990. *Economics of natural resources and the environment*. Baltimore: Johns Hopkins University Press.
56. Hufschmidt, M.M., D.E. James, A.D. Meister, B.T. Bower, and J.A. Dixon. 1983. *Environment, natural systems, and development: An economic valuation guide*. Baltimore: Johns Hopkins University Press.
57. Gupta, T.R., and J.H. Foster. 1975. Economic criteria for freshwater wetland policy in Massachusetts. *American J. of Agricultural Economics* 57(1):40–45.
58. Hamilton, L.S., and S.C. Snedaker, eds. 1984. Handbook for Mangrove Area Management. Honolulu: U.N. Environment Programme, and Environment and Policy Institute, East-West Center.
59. Gosselink, J.G., E.P. Odum, and R.M. Pope. 1974. The value of the tidal marsh. Center for Wetland Resources, Baton Rouge: Louisiana State University. LSU-SG-70-03 (unpubl).
60. Odum, H.T., and J.E. Arding. 1991. Emergy analysis of shrimp mariculture in Ecuador. Working Paper, Center for Wetlands, University of Florida, Gainesville.
61. Meltzoff, S.K., and E. LiPuma. 1985. The social economy of coastal resources: Shrimp mariculture in Ecuador. *Culture and Agriculture* 28:1–10.
62. Bailey, C. 1986. Shrimp mariculture development and coastal resources management: Lessons from Asia and Latin America. In *Establishing a sustainable shrimp mariculture industry in Ecuador*, Stephen Olsen and Luis Arriaga, eds., 45–67. University of Rhode Island, Technical Report Series TR-E-6.

10

Primate Conservation and Ecotourism in Africa

William Weber

Perhaps the most common image of Africa is that of abundant wildlife. The East African savanna, in particular, with its seemingly endless landscapes and countless herds of large mammals, presents an unmatched natural spectacle. Even greater biological riches have more recently been revealed within the continent's vast rain forests. Yet this image of biological wealth stands in stark contrast to another image of Africa: that of chronic economic poverty and underdevelopment.

In this context of contrasting wealth and poverty, many see in Africa an ideal laboratory in which to test new ideas to reduce conflicts between conservation and development. Common to most of these approaches is a mix of traditional protection efforts and sustainable use options, either within or around a core nature reserve. The objective of these use systems should be to generate a steady flow of benefits to local populations, without degrading the natural ecosystem.

This chapter will examine the concept of ecologically sensitive tourism—or ecotourism—as an effective means of integrating conservation and development interests in Africa. Rather than look at traditional safari tourism in East African grasslands, however, the focus will be on the distinctively different attractions of African primates and their rain forest habitat. Still, the fundamental question will remain: can ecotourism help maintain biological wealth, while also improving economic well-being?

THE AFRICAN RAIN FOREST ENVIRONMENT

Rain forests cover more than two million square kilometers across equatorial Africa (Fig. 1). This total area includes the largest single block of contiguous rain forest outside of the Amazon: the more than one million km^2 forest of the central Congo Basin. In West Africa, the forest is more fragmented, though several major blocks remain. This trend is even more exaggerated at the eastern edge of the biome, where an archipelago of remnant forest "islands"—few of which exceed 1,000 km^2—is dispersed across the highlands of east-central Africa.

Biological Diversity

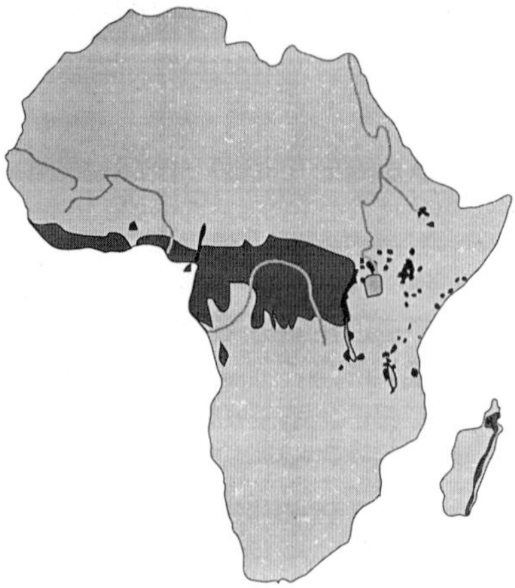

■ MONTANE FOREST
■ LOWLAND FOREST

Figure 1. Rain forest distribution in Africa. [Adapted from Hamilton, 1976 (50)].

As in other tropical regions, rain forests are the most diverse of Africa's many biogeographic zones (1). Millions of species are found there, the vast majority of which remain unknown to science. Even the smaller, generally less diverse forests at the edges of the biome are often characterized by a high percentage of rare and endemic species. Thus, while wide-open savannas may attract the greatest tourist interest, the forests increasingly command greater scientific attention.

The staggering diversity of rain forest biota consists primarily of plant and invertebrate animal species. Yet it is large mammal fauna that distinguish the African forest from other tropical rain forest regions. More than 225,000 elephants are estimated to inhabit this forest, or roughly one third of the entire remaining African elephant population (2, 3). Though most of these are the smaller forest elephant (*Loxodonta africana cyclotis*), their ecological importance is increasingly evident. Other large mammals that share this habitat include the forest buffalo (*Syncerus caffer nana*), the okapi (*Okapia johnstoni*), the bongo (*Tragelaphus euryceros*), and more than 11 species of duiker (4).

Up to now, however, greatest attention has been given to the forest-dwelling primates, especially the great apes. The African rain forest is home to three of the four anthropoid apes—gorillas (*Gorilla gorilla*), chimpanzees (*Pan troglodytes*), and bonobos (*Pan paniscus*)—which are our closest living relatives, and which generate considerable popular and scientific interest. An additional 63 species of primates share the continental rain forest (5), most of which are arboreal Cercopithecines and Colobines, while 30 species of lemuriformes (1, at p. 126) are endemic to the island of Madagascar off the southeast coast of Africa.

The Socioeconomic Context

African wildlife does not exist in a void. The relevant ecosystem includes those people who live in and around the forest, as well as forces that originate in distant cities and foreign lands. At the macro level, few indicators are moving in a positive direction. Per capita GNP, food production, and export:import account balances are down in most countries (6, 7), while population growth rates and debt service ratios are rising (6, at 60 and 70; 7). Among the complex and interrelated results of these trends is a generalized increase in environmental degradation, whether in the form of desertification across the Sahel or accelerated clearing of the rain forest for both subsistence and export needs.

In the case of the forest, local people are often both agents and victims of degradation, pursuing short-term benefits at the expense of long-term well-being. This results in part from the pressures of outside market forces to exchange forest resources for material goods on an unequal basis. It is also a function of the political marginalization of such groups and their relative inability to control outside forces, such as intensive commercial logging—or even the establishment of new parks and protected areas.

Dealing with deforestation and other environmental problems is further complicated by the fact that most African nations are quite recent creations. As a result, institutional foundations—legal, administrative, and technical—are not yet fully developed. Worse, the human resource base needed to staff those institutions is even less developed, with critical shortages of well-trained upper- and middle-level technicians in most ministries. Finally, the relative youth of modern African nations can be seen in the political instability that many are now experiencing. Political change is essential if the continent is to move beyond the stagnating grip of corrupt dictators, but the potential for violence and disorder in the process provides an uncertain context for progress in other sectors.

Threats to African Primates and Forests

Deforestation and the loss of wildlife may be more easily understood in the context described above, but understanding does not diminish some hard facts:

- Some of the highest rates of forest clearing in the world are found in West Africa (8). The primary cause is commercial logging, though the opening of roads has also facilitated the establishment of permanent agricultural settlements throughout the former forest zone.

- As commercial exploitation has depleted primary hardwood stocks in West Africa, logging interests have turned their attention to the rich forests of Gabon, Cameroon, and Congo. Though related settlement is rare, hunting pressures are generally intense along logging roads, even in remote areas, due to a combination of subsistence and urban market demand for bushmeat (9, 10, 11). Among those species hunted are the endangered lowland gorilla and drill, as well as many other primates, although subsistence staples include more common species such as porcupines, cane rats, and a variety of duikers. Elephants are hunted for both ivory and their meat.

- In the Afromontane region, logging and settlement are constrained by the generally small size and rugged terrain of the remnant forest blocks, most of which enjoy some form of protected status. Rising human population densities—often reaching several hundred per km^2—have placed great pressures on these reserves, however, as sources of food, fuel, and fodder.

The combined effects of deforestation, degradation, fragmentation, and hunting have led to much greater awareness of the need to conserve critical forest habitats and their component species. More than 80% of the existing parks and protected areas in Africa, however, are found in savanna and semi-arid environments, while barely 5% of the continent's rain forests have any comparable status (7, at 124). The creation of several new forest reserves in recent years has helped to redress this situation, but far more needs to be done.

At the same time, a consensus has emerged that sees the greatest challenge in the integration of conservation and development interests outside of traditional parklands. Applying this approach, core protected areas are surrounded by progressively more intensive zones of managed use. Utilization practices, however, must meet criteria for biological and ecological sustainability, and economic benefits must flow to designated local groups as well as to appropriate regional and/or national recipients. Under this system, tourism is clearly one of the lower impact forms of forest exploitation, compared with hunting, farming, and commercial logging. Tourism is also believed to have considerable potential to generate both local employment and foreign revenue.

Whether tourism can meet these high expectations in African rain forests is the subject of the remainder of this chapter. The following case studies focus on apes and other primates as tourism "flagships." The locations of the studies are shown in Figure 2. Special attention is given to the Afromontane region for several reasons: it is the area where primate tourism has been most fully developed for the longest period; it is the primary source of reliable documentation and data; and it is where the author has the greatest personal experience. The question of that region's relevance to other rain forest environments is then given a critical assessment, along with other issues, in the concluding section.

Figure 2. Primate tourism sites in Africa: 1. Tiwai Island, Sierra Leone; 2. Korup National Park, Cameroon; 3. Dzanga-Sangha Reserve, Central African Republic; 4. Tongo, Zaire; 5. Virunga National Park, Zaire; 6. Kibale Forest, Uganda; 7. Volcanoes National Park, Rwanda; 8. Gombe Stream Park, Tanzania; 9. Nyungwe Forest, Rwanda; 10. Kahuzi-Biega National Park, Zaire; 11. Perinet Reserve, Madagascar. [Adapted from Oates, 1986 (5).]

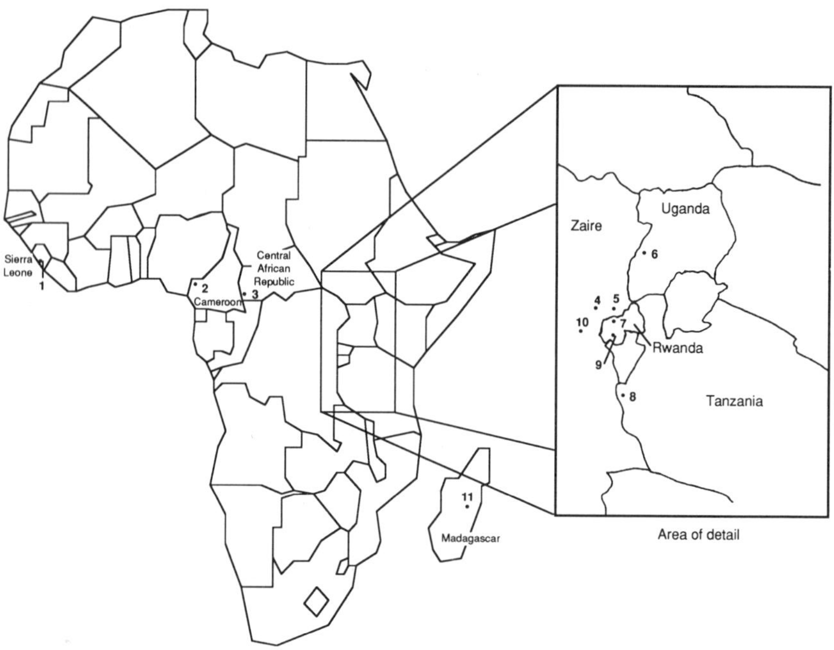

CASE STUDIES

The Mountain Gorilla Project: Rwanda

The best-documented example of primate-centered tourism in Africa is that of the Mountain Gorilla Project (MGP) in Rwanda (12–18). More than a dozen years have passed since its initiation, permitting a reasonable assessment of its accomplishments.

The mountain gorilla (*Gorilla gorilla beringei*) occupies a very limited range in the Virunga volcanoes of east-central Africa. Rising to heights of more than 4,500 meters, this 65-km chain forms part of the international boundaries between Rwanda, Zaire, and Uganda. The area was set aside as the first national park in Africa in 1925, but administrative responsibility was then divided among those nations at the time of Independence.

The Virunga ecosystem lies within the Afromontane biogeographical region. A majority of the 420 km^2 reserve consists of montane rain forest, while Ericaceous and subalpine formations predominate above 3,200 m. While less species-rich than most lowland rain forests, the Virungas nevertheless provide a refuge for several rare and endangered species. Foremost among these is the mountain gorilla.

By the mid-1970s, much of the developed world was aware of the mountain gorilla, due to the pioneering study of George Schaller (19, 20) and subsequent long-term work by Dian Fossey (21, 22) and others at Rwanda's Karisoke Research Centre. Yet just as popular articles, books, and films brought these special creatures and their complex social lives into the general consciousness, it also became clear that their very survival was at risk. Between 1960 and 1973, the Virunga population of mountain gorillas crashed dramatically from roughly 450 individuals to no more than 260 (13).

Research background

The MGP stemmed primarily from a project initiated in 1978 to better understand the causes of this decline and to recommend corrective action (12–18). The research phase of this effort took an interdisciplinary approach, combining traditional conservation biology with assessments from the socioeconomic and political spheres. The first step was to determine the current status of the gorilla population by means of a census. This showed some stabilization of the total population, at a level of 268 individuals, but also revealed a continuing decline in the percentage of young gorillas as a direct result of poaching and other disturbance factors (13, 23). Whether or not the remnant forest habitat could sustain even the existing population was the focus of a complementary study of gorilla feeding ecology and habitat utilization. In this case, results showed that while significant gorilla resources had been lost due to forest clearing, sufficient high-quality habitat remained to support population growth, possibly to the 1960 level of 400 to 500 gorillas (13, 24).

Whereas the demographic and ecological viability studies indicated some potential for reestablishment of the population, research conducted outside of the reserve yielded less encouraging results (14, 15). The fertile volcanic soils of the region had attracted human settlement at densities of more than 400/km^2 by 1978, and population growth was continuing at an annual rate of 3.7%. The result was a reduction of average farm size to barely one acre (0.5 ha) per family and a shortening of the fallow cycle. Intensifying this pressure on the land was a lack of nonfarm employment in a rural economy where 95% of the population lived by farming, and family income averaged less than $200 per annum.

Surveys of local attitudes and awareness also demonstrated that there was little support for continued protection of the Volcanoes Park (14, 15). Few local farmers perceived any benefits of forest conservation, and barely one third of the population saw any nonconsumptive value associated with wildlife. Among those who did cite a value, most referred to tourism and its associated employment potential. This was especially interesting, since the park at that time generated only minimal revenues and employment (14, 15, 25).

Given the lack of land, employment alternatives, and perceived conservation values, it is not surprising that a majority of survey respondents believed that the park should be abolished and opened for use and settlement. Faced with this popular interest, and the lack of off-setting foreign revenues, it was also understandable that the government of Rwanda supported this conversion. Thus, for the third time in 20 years, a major section of the Volcanoes Park was slated for clearing in early 1979: this time, one third of the remaining forest—including the richest gorilla habitat—was to be converted for a cattle-raising scheme supported by international development agencies (14).

Project development

Armed with the information outlined above, and faced with the immediate threat of continued poaching and conversion, the action phase of the Mountain Gorilla Project (26) began in August 1979. A three-pronged approach was taken, with activities focused on (1) improving park security, (2) creating conservation education programs, and (3) developing a tourism program centered on the gorillas. While the first two of these programs achieved considerable success in their own right (12, 14, 15), the flagship of the MGP has been the tourism component.

The first step in establishing a tourism program in the Volcanoes Park was to incorporate the idea that this would not be "tourism as usual." The term "ecotourism" did not exist at that time, but the intent was quite similar with regard to an interest in ecologically sensitive tourism. In particular, impacts on the gorillas were to be minimized by limiting the number of visitors to six per gorilla group each day; making no more than one visit to each group per day; timing that visit to roughly coincide with regular late-morning gorilla rest periods; and excluding direct physical contact with the gorillas. In this way, the primary focus was placed on the conservation and well-being of the focal species; it also served to provide a high-quality experience for visitors.

Habituation of the gorillas to tourists was obviously central to the project's success. Feeding, other inducements, and physical restriction of the gorillas were excluded. Instead, two wild groups were selected from the 1978 census, then followed on a consistent, nonthreatening basis until they accepted a human presence. This process proceeded more rapidly than expected, and regular tourist visits were being made to the first group after only three months. Eventually, a total of four gorilla groups were habituated in a similar manner.

Another key element was the training of local personnel as guides. While these men knew the forest and animals well, they had no experience dealing with tourists. They therefore needed to learn how to both inform and control visitors, including the acquisition of appropriate foreign language skills. The training process required regular hands-on experience with expatriate advisors in the first few years, but the vast majority of daily visits have been conducted independently by the Rwandan guides for the past 10 years.

Beyond continued fund-raising to support its operations, the primary responsibility of the MGP in its final years was to provide technical advice to the Rwandan Office of Tourism and National Parks. This advice focused on how to manage the Volcanoes Park, its personnel, and the gorilla tourism program on a long-term, sustainable basis. By 1989, the MGP and the Rwandan government agreed to phase out all but a limited advisory role (27).

Results

Tourism has been a key factor in the success of the Mountain Gorilla Project over the past 13 years. Visitation rates rose steadily through the early 1980s to reach an average level of 5,000 to 6,000 per year (Fig. 3). With rising entry fees failing to affect this demand, annual revenues in 1989 surpassed $1 million for direct entry fees alone. Since the vast majority of these visitors are nonresident foreigners (Fig. 4), it is estimated that an additional $3 to 5 million per year enter the national economy through tourist expenditures on lodging, rentals, and purchases of various kinds (28).

At the local level, tourism revenues have also generated a significant increase in employment, albeit on a limited scale. Park guard numbers increased from fewer than 20 to more than 70 over the life of the MGP. These guards, the majority of whom come from the Virunga region, receive annual salaries which average more than $1,200, or more than four times the national average. Guides make even more as a result of tips, which can surpass 50% of base income.

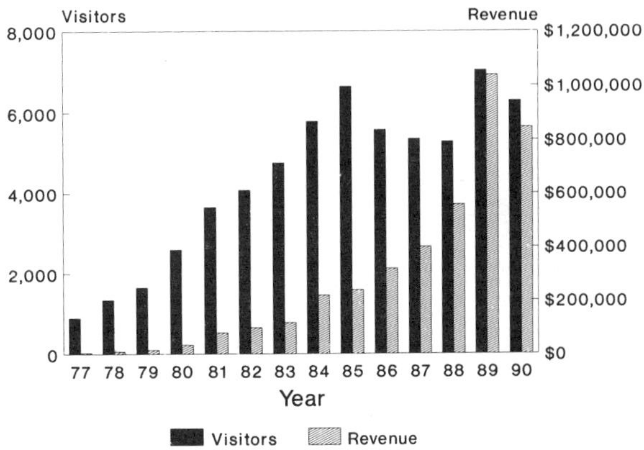

Figure 3. Visitation and revenue: Gorilla tourism at Volcanoes National Park, Rwanda (51).

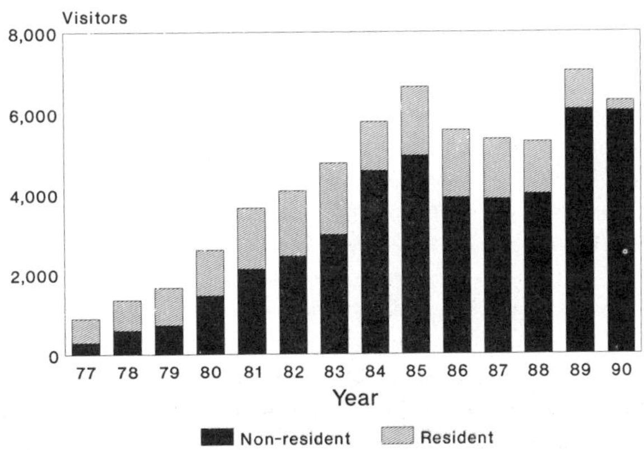

Figure 4. Visitation by origin: Gorilla tourism at Volcanoes National Park, Rwanda (51).

Although studies have not yet been done to assess the overall economic impact of this revenue flow on local communities, there is clear evidence of a positive effect. In particular, follow-up surveys in the project area have shown an increased appreciation of wildlife values, as well as a marked decrease in those opposed to the park, from

51% to fewer than 20% (14). Furthermore, the combination of foreign revenue and local employment has created a strong foundation of political support for gorilla conservation within Rwanda: support that can be seen in conservation policies and practices.

In wildlife conservation, the bottom line for any endeavor is its impact on the species and/or ecosystem in question, and here again the results of the MGP have been highly encouraging. In the 10 years following the project's inception, total gorilla numbers increased from barely 260 to roughly 320 in 1989 (23). Tourism revenues were again the key: paying for increased guards whose surveillance cut gorilla poaching to zero between 1984 and 1992, and justifying government resistance to all further park conversion schemes. As an added bonus, the four gorilla groups visited regularly by tourists were among the largest and had the highest percentage of young in the 1989 census (23).

As a final comment on the MGP, it should be noted that the project's successes served to generate increased international interest and support. Thus, while the Rwandan government was able to apply tourism revenues to the cost of park operations and personnel, additional technical and financial assistance to the project was provided by international NGOs and agencies that were attracted by earlier successes. The effect of this positive feedback cycle even extended to agroecosystems outside of the park, as discussed in a later section.

Other Gorilla Projects

Although the Mountain Gorilla Project has received more attention than any other effort, it is not the only attempt to develop gorilla-based tourism. Nor was it the first. A predecessor project and subsequent MGP spin-offs are briefly described below.

Kahuzi-Biega (Zaire)

The first successful effort to habituate gorillas for tourism purposes took place in the Kahuzi-Biega National Park in eastern Zaire. Extending over 5,000 km^2 of premontane and montane rain forest, this park provides habitat for a population of more than 1,000 Grauer's gorillas (*G. g. graueri*): one of three gorilla subspecies and the only one endemic to Zaire. The Belgian warden of Kahuzi-Biega, Adrien Deschryver, began the habituation process in 1973 and was able to guide paying visitors by the end of that same year. As word spread, the number of visitors increased to several thousand per year by the late 1970s.

No effective controls over visitation had ever been put in place, however, and groups of as many as 30 to 40 tourists were sometimes taken out on a single visit. To accommodate this number, guides would hack down the vegetation around the gorillas, provoking displays and alarm calls and clearly disturbing the animals. The untrained and underpaid guides were also known to stage sit-down strikes in the middle of the forest for more pay, or simply not show up for work for days at a time. At the same time, corrupt park managers were more interested in diverting profits than in disciplining guides. Still, international tour agencies kept bringing visitors to one of the only sites in the world where one could see wild gorillas.

In retrospect, certain caveats should be applied to any assessment of the Kahuzi-Biega experience. First, the primary emphasis was on generating income for the park, not necessarily on the well-being or conservation of gorillas. Second, there were no relevant standards or models for forest-based tourism when this project was launched nearly 20 years ago. Finally, those who eventually set up the MGP learned

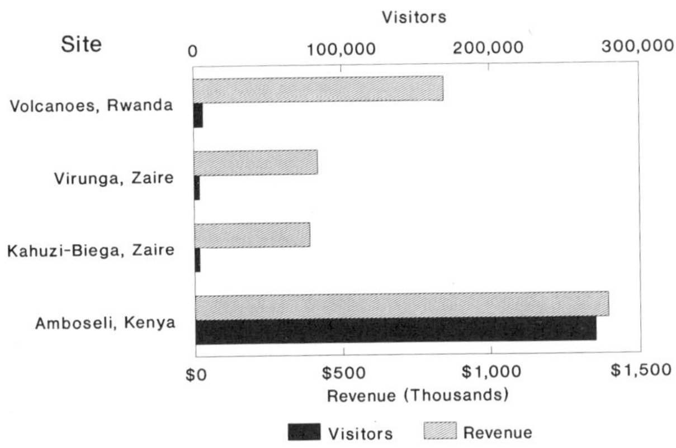

Figure 5. Tourism visitation and revenue at selected sites in Africa in 1990 (51–54).

several important lessons of what not to do in Rwanda by observing first-hand the problems faced by Kahuzi-Biega (29)—and in turn set an example for Kahuzi-Biega itself in later years.

In the mid-1980s, a German technical assistance program began working with the Zairian Institute for Nature Conservation to improve tourism in Kahuzi-Biega. Learning from the MGP, the project instituted guide training programs and limits on visitation. While the transformation has been imperfect in some regards, it is nevertheless considered to be a great improvement by international tour operators and their clients. The result is increased visitation and a willingness to pay higher entry fees: a combination that generated nearly $400,000 in direct park entry fees in 1990 (Fig. 5).

Virunga (Zaire)

The Zaire Gorilla Conservation Project (ZGCP) began its activities in the Virunga National Park in 1984. Founded by former MGP workers, this project focused on the critically important Zairian sector of the Virunga chain—home to more than half of the region's remnant gorilla population.

In approach, the ZGCP was modeled closely on the Rwandan MGP (30). With regard to tourism, this meant that three families of gorillas were eventually habituated to daily visits by groups of up to six tourists. Although the first official tourists did not arrive until 1985, the number of annual visitors rose rapidly into the thousands (Fig. 6), fueled in part by pent-up demand from Rwanda. With visitation relatively unaffected by price increases, entry fees from 3,728 tourists surpassed $418,000 in 1990.

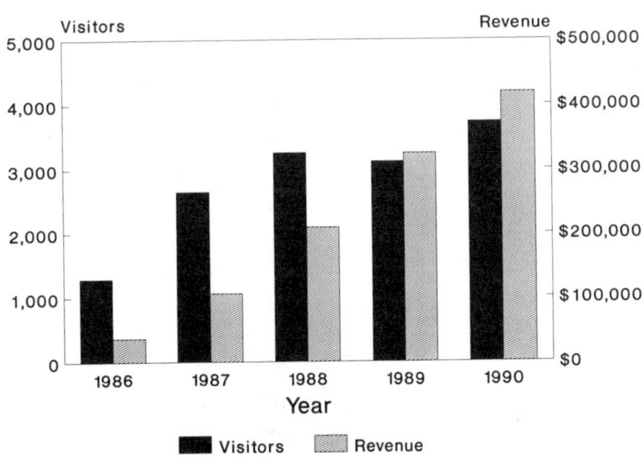

Figure 6. Visitation and revenue: Gorilla tourism at Virunga National Park, Zaire (52).

While the Zaire Gorilla Conservation Project can be seen as a mirror image of the MGP in many respects, there are notable differences. In particular, the project's experience with earmarked funds for local benefit and complementary rural development efforts will be addressed in the discussion section.

Chimpanzee Tourism

Very few sites exist where even an adventuresome ecotourist can view wild chimpanzees. The two projects considered here therefore stand out as much for the rarity of their offering as for their differences in history and approach.

Gombe Stream (Tanzania)

World famous as the site of Jane Goodall's long-term research on chimpanzee behavior, Gombe Stream has attracted a small but steady stream of visitors over the past 10 years (Fig. 7). In the 1990–91 fiscal year, 1,118 tourists paid a little more than $17,200 to visit this relic forest and its famous chimps.

Most commonly, the chimps are first seen around the Gombe station, where many of the original research subjects and their offspring continue to congregate. The reason for this is that they were first habituated by feeding, and those who became dependent on this practice continue to be fed, although the "banana ration" has been steadily reduced over the years. While this concentration around the camp facilitates viewing, it is not compatible with the ecotourist's demand for a more natural experience. Fortunately, this too is possible at Gombe, and most visitors also spend one or two days trekking in search of the more reclusive, free-ranging chimps, who still inhabit this isolated mountain woodland (31).

Tongo (Zaire)

Perhaps no more ideal conditions currently exist for viewing wild chimpanzees than at Tongo, in eastern Zaire. Adjoining the Virunga Park described above, this 8 km² patch of forest is surrounded by open lava fields, creating a naturally restricted reserve with a healthy population of chimpanzees.

Beginning in 1987, ZGCP personnel began habituating the Tongo chimps as a complement to the on-going gorilla tourism program (32). This process took roughly one and one-half years to complete and Tongo was quickly added to many tour group itineraries. Nearly 450 visitors, 78% of them nonresident foreigners, paid more than $14,000 in entry fees in 1990 (Fig. 8). Basing their decision in part on a 98% "success rate" in tour groups actually seeing chimps, Zairian authorities increased nonresident fees from $40 to $60 per person in 1991. Data are not yet available, however, to indicate the

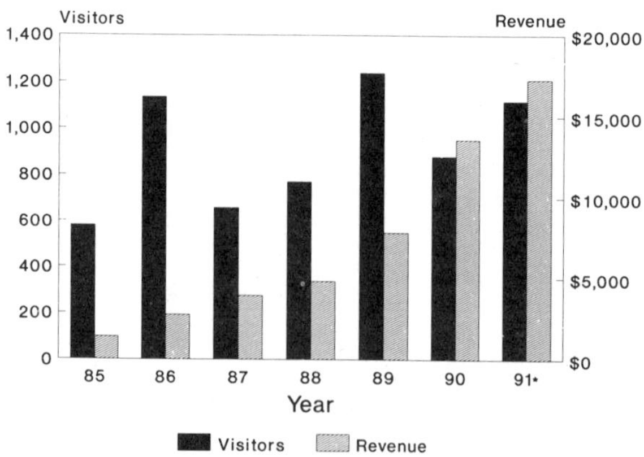

Figure 7. Visitation and revenue: Chimpanzee tourism at Gombe Stream National Park, Tanzania (55).

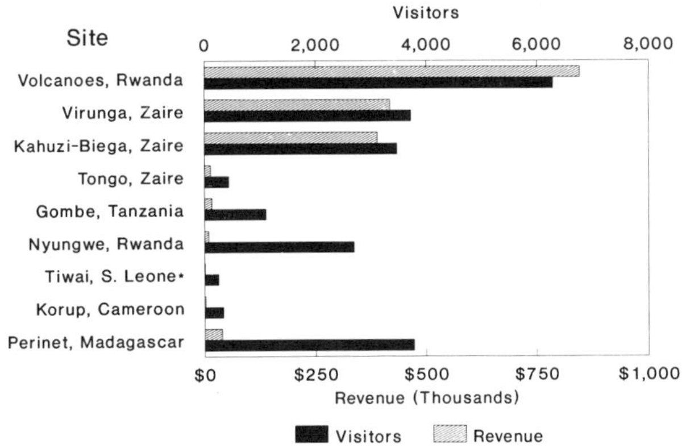

Figure 8. Tourism visitation and revenue at selected sites in Africa in 1990. * = 1988 data (51–53, 55–60).

effect of this change on overall visitation and revenues.

The most direct social benefits of the Tongo project can be seen in the creation of 25 new jobs. Fifteen of these positions were initiated by the project and then taken over by the park service; the other 10 were created by a private businessman to staff a new lodge built specifically to serve chimp tourism. In addition, a local clinic has been constructed, access roads have been improved, and complementary rural development activities have been initiated, all as a direct or indirect result of the chimpanzee project.

Other Afromontane Projects

All of the projects described above are situated in the highlands surrounding Lakes Tanganyika, Kivu, and Edward in east-central Africa. In addition to chimps and gorillas, this Afromontane vegetation zone supports some of the most diverse, concentrated, and visible primate communities on the continent, as well as a distinctive avifauna (1, at 33; 33). It is also a region of spectacular landscapes and moderate climate.

Nyungwe (Rwanda)

The most successful attempt to develop forest-based ecotourism without the magnet of the great apes has been in the Nyungwe Forest of southwestern Rwanda. This 970 km^2 montane rain forest is home to 13 different kinds of primates and more than 270 bird species (34). It is not a national park, however, but instead is managed as a multiple-use reserve by the Rwandan forest service. It was to demonstrate the potential for nature tourism as a low-impact use of the forest that the Nyungwe tourism effort began in 1988 (35–37).

Although the Nyungwe Forest supports a healthy population of several hundred chimpanzees, rugged terrain combines with high chimp mobility to severely restrict their potential for habituation. Given this situation, multiple attractions were highlighted in Nyungwe. Chief among these was the presence of spectacular associations of more than 300 black-and-white colobus monkeys (*Colobus angolensis ruwenzorii*). Two of these groups had already been habituated for research purposes, and another two were added later. In addition, groups of blue monkeys (*Cercopithecus mitis*) and grey-cheeked mangabeys (*Cercocebus albigena*) were also habituated. In 1990, half of all tourists paid almost $6 apiece to make a guided visit to view the colobus and other monkeys. Most other visitors came simply to hike parts of a 60-km network of trails, with the virtual certainty of incidental primate sightings along the way. This trail

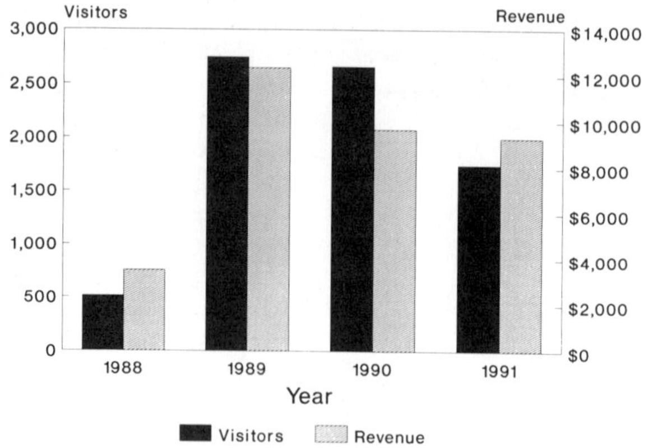

Figure 9. Visitation and revenue: Primate tourism at Nyungwe Forest, Rwanda (57).

network takes visitors through a wide variety of habitats, from marshes to exposed ridges, and has been very popular with both tourists and school groups (38). Bird-watchers have also appreciated the trail system and several tour operators have been interested in expanding this specialty market.

Total visitation for the Nyungwe Forest reached 2,659 in 1990, generating direct revenues of almost $10,000 (Fig. 9). This relatively low sum reflected intentionally low entry fees during the project's start-up phase. Prices have since been doubled, and revenues are expected to increase proportionally—though never to approach the premium rates for gorillas. By far the greatest economic benefit from the Nyungwe program, however, comes from indirect sources. Two-thirds of all visitors in 1990 were nonresident foreigners who also rented cars, stayed in hotels, and purchased other local goods and services. Although no detailed studies have been made of these expenditures, total revenues would probably range between $540,000 and $1 million for 1990 (39). With regard to employment, the Nyungwe project currently supports a staff of 19 Rwandans, all from the local area, while the park service pays the salaries of 20 guards assigned to patrol the entire reserve.

Kibale Forest (Uganda)

The most recent addition to the highlands ecotourism network was initiated in late 1991 in the Kibale Forest of western Uganda (40). The Kibale Reserve consists of a unique mosaic of dense forest and open grasslands, spread over 540 km^2 of rolling terrain. Along with a highly diverse avifauna, this habitat supports the world's highest recorded density of primates, including chimpanzees and an endangered subspecies of red colobus (*Colobus badius tephrosceles*). Long the focus of scientific research, a decision was made to exploit this primate resource through controlled tourism as an alternative to a proposed logging venture.

Drawing on experience from Nyungwe and Tongo, a program has been developed that places its initial focus on the establishment of a trail network to facilitate primate and bird viewing. A parallel effort is also being made to habituate chimpanzees; yet while early results are promising, no formal visits will be made by tourists until the process permits a reasonable certainty of viewing. Meanwhile, visitors have already expressed their satisfaction with the existing focus on hiking and primates, and several tour agencies have indicated an interest in increasing their visits to Kibale.

Finally, it should be noted that local participation has been a central element in the Kibale tourism project. Neighboring communities were consulted on key issues, and almost all of the guides and support staff have been hired from the region. Furthermore, a mechanism has been established to direct 75% of all tourism profits for the use of local villages and rural communities, with the remainder to go to the Forest Department to pay for guards and other management costs (40).

Lowland Forest Sites

As described in the introduction, Africa's 2 million km^2 of lowland forests contain the continent's highest levels of biological diversity. They also support more than 60 primate species, including the vast majority of the remaining gorilla and chimpanzee populations. Yet while the highland forests and their primates have succeeded in attracting a significant share of the African ecotourism market, tourism in the lowland rain forests remains virtually nonexistent. The following brief descriptions illustrate the situation.

Korup (Cameroon)

The Korup National Park was created in 1987 to preserve roughly 1,200 km^2 of extremely rich rain forest in southwestern Cameroon. Among the park's many attributes is the highest number of primate species (17) of any African forest. Korup has also been the subject of a widely acclaimed wildlife film of the same name, and it has probably received more international publicity than any other African rain forest.

A comprehensive management plan for Korup was elaborated in the late 1980s, which placed its primary emphasis on two areas: the establishment of a park infrastructure, and the reduction of human pressures on the park through an extensive rural development program (41). Tourism was also addressed in this plan, but intentionally as a lower priority in the initial five-year phase. Perhaps as a result of this deemphasis, or perhaps because of natural limitations, the total number of tourists who visited Korup in 1990 did not surpass 350 (Fig. 8)—and most of them were hardy individuals who managed to reach the park without any of the standard support services. Once at the park headquarters, they did find a trained guide at their disposition, and most set off to look for primates: but no primate groups have yet been habituated and viewing them in the high canopy is problematic.

Tourism in Korup was to be reassessed in 1992 and it is possible that more emphasis will be put on this sector. Still, despite considerable publicity among audiences with high ecotourism potential, visitation remains minimal—and the total entry revenues of roughly $2,800 for 1990 would not even cover the annual running costs of a single park vehicle.

Tiwai Island (Sierra Leone)

Tiwai Island, in southeastern Sierra Leone, has an even longer history of primate-centered tourism than Korup and an even lower level of visitation. Situated in the Moa River, Tiwai is covered with a mix of primary and secondary forest that provides habitat for 11 species of primates in an area of only 12 km^2. These include chimpanzees, red colobus (*Procolobus badius*), and the spectacular Diana monkey (*Cercopithecus diana*).

To help support long-term research and conservation activities on the island, as well as to generate income and employment for local communities, tourism was initiated on Tiwai in 1987. Trails were established, an overnight visitor center was built, several primate groups were habituated for viewing in the wild, and local guides were trained. While visitors have appreciated this support structure and the quality of their experience, annual visitation has never surpassed 260 individuals (Fig. 8). The barely $1,000 in revenues from this level of tourism has supported the visitor center's operating costs and salaries for personnel and even generated a small surplus for local communities. Whether or not this amount can offset any future alternative-use values is another question, but that was not the original objective of the project (42, 43).

DISCUSSION AND CONCLUSIONS

There is no question that controlled nature tourism, or ecotourism, can be an extremely important factor in favor of conservation. It is also clear from the preceding examples that African primates and rain forest environments can compete with that continent's more traditional wildlife attractions to generate significant revenues and support conservation. At the same time, it must also be recognized that there are very real limits to this kind of ecotourism.

Ecotourism Successes and Benefits

Nature tourism has grown steadily over the past decade (44), fueled both by a fascination with exotic locales, such as rain forests, and by a concern for the plight of endangered species. Yet whereas most scientists see the latter with reference to millions of bugs, beetles, and other biota, the concerned ecotourist is much more likely to see one of the great apes or some other threatened forest primate as the face of extinction. This concern combines with an existing fascination with our closest wild relatives to give primates high ecotourism appeal.

The results of this appeal can be translated into dollars, jobs, and political support for conservation. Gorillas, in particular, generated nearly $2 million in combined entry fees at three different reserves in 1990 (Fig. 5). The Rwanda project alone would have surpassed $1 million, as it did the preceding year, if civil disturbance had not shut down tourism in the last two months of the year (see below). As is, 1990 gorilla revenues in Rwanda were more than half of the $1.4 million earned by the Amboseli National Park in Kenya: a park visited by more than 270,000 tourists—430% more than in Rwanda—with attendant environmental impacts.

Tourism focused on chimpanzees and other primates has thus far earned considerably less than gorillas, but even those low-level inputs represent a significant increase over previous revenue levels for forest reserves. In addition, each of the tourism programs described above created jobs for local people in remote areas where salaried employment opportunities were either rare or nonexistent. The multiplier effect of this cash infusion into marginal, nonmarket economies is certainly significant, although supporting research is lacking on this important subject.

In many of the cases under consideration, even greater funding and employment opportunities have come from complementary activities supported by international development agencies. Most of these projects attempt to integrate forest conservation with more comprehensive approaches to rural development in surrounding areas. The scale of funding for these initiatives ranges from roughly $100,000 per year in the case of Nyungwe, to several million dollars per year for the Virunga region in Zaire. In most instances, this has included at least some direct support for tourism and reserve management, as well as significant employment outside of the reserves. In all cases, but especially in places like Korup, where visitation is low, this supplemental funding can be extremely important with regard to local perceptions of conservation.

The issue of complementary funding notwithstanding, it is important to note that the majority of in-country tourism benefits do not stay in the project area, but instead go to central government agencies and urban-based businesses. Some of this money may recycle locally, as in Rwanda where the park service uses revenues to pay increased guard salaries and other local operating costs. In general, however, park revenues go into national treasuries—or individual pockets—as do most indirect revenues from hotels and car rental agencies within the centralized economies that still predominate in Africa. Private ventures are on the rise, however, and to the extent that these businesses are not foreign-owned, this money at least stays in the national economy.

Still, the greatest share of tourism benefits accrues to parties outside of the countries concerned. International airlines alone make at least $10 million per year from the Rwandan gorilla project (46); European companies own a share of most African hotels and car rental agencies; and American travel agencies make an attractive profit from package tours.

From a conservation perspective, benefits and success must be measured not in revenues and jobs, but in the status of the species and ecosystems concerned. In fact, most of the primate populations considered here appear stable—or increasing, in the case of the mountain gorilla—and hunting and encroachment are generally on the decline at sites visited regularly by tourists (47). Nevertheless, it is the combination of revenue and employment that is generating essential political support for conservation at the local, regional, and national levels.

Ecotourism Problems and Limits

This generally positive assessment of primate-centered tourism in rain forests must be tempered with a recognition of some serious problems as well. Some of these problems constrain tourism growth at existing sites, while others appear to preclude its extension to other forest areas.

Problems at existing sites

A fundamental concern for any tourism program is that it do no harm. Bringing humans into close contact with primates requires strict controls to prevent potentially serious problems. This is especially important for the various mountain gorilla tourism projects.

Gorillas are capable of contracting virtually all human diseases, yet may have little if any immunity to their effects. Bringing six new people to within 10 meters of a gorilla family almost every day therefore entails obvious risks. Set against the certain extinction of the population in the face of hunting and habitat loss, however, these risks were determined to be worth taking, under the strict controls described earlier. In general, guides in Rwanda appear to have maintained the strongest commitment to control such contact, while recent reports indicate a breakdown of supervision in Zaire. One factor influencing this behavior in Rwanda was a suspected outbreak of measles among the gorillas in 1988 (37). Though never confirmed, this scare has had a lasting effect. And while gorilla numbers have climbed steadily since the tourism program began, concern about disease transmission remains strong, with prevention efforts reinforced by a long-term veterinary presence at the park.

Quality control is another serious problem for any ecotourism project. Individuals who pay premium prices to visit wild gorillas, or incur high travel costs to see other species, will not continue to do so if they perceive a qualitative decline in the experience. In forest environments, visitors seem sensitive to the condition of trails, the abilities of the guide, and the availability of information in the form of maps and guidebooks. What visitors appear most sensitive to, however, is tourist group size during in-forest visits. This holds true across primate types, but is particularly important for groups visiting gorillas. In part, this is a function of the dense ground vegetation in montane forests. Constraints on tourists' ability to see gorillas—and gorillas' ability to see and thereby be comfortable with visitors—were the primary reason for setting a limit of six visitors per group in the original Rwandan project. Over time, however, it became clear that six was also an optimal number for visitors' appreciation of the restricted, or 'select' nature of their experience. This was highlighted when Rwandan authorities decided to increase daily group size to eight, and complaints about jostling, displacement from the guide, and limited visibility began to increase (27).

The problem of expanded group size resulting in lowered vistor appreciation is one that could become worse. As decision makers see visitation and revenues level off under current controls, they are likely to respond in one of two ways. They can raise prices, which has worked up to a point with gorillas, but is much less likely to succeed with other primates. They can also loosen limits on visitation. In this case, however, they may alter the demand equation by lowering the quality of the experience, thus attracting fewer visitors in the end.

A third problem that has arisen at existing sites is that of economic equity. As mentioned earlier, the vast majority of benefits from tourism flow out of the immediate project area. Though this is tempered by increased local employment, there remains a fundamental imbalance in that those communities asked to give up the most for conservation, in the form of foregone benefits of forest use, receive the least in return.

Several of the projects described above have attempted to redress this imbalance, with limited success. The greatest failure in this regard has been the Mountain Gorilla Project. Though several proposals have been made to earmark a percentage of entry fee revenues for local use or redistribution, these have been systematically rejected by the Rwandan government. The initial justification for this rejection was the need to cover park operating costs, but subsequent refusals came at times of high net profits. At the proposed level of a 5% earmark, this has meant a lost opportunity to channel $40,000 to $50,000 per year from tourism receipts into health, education, or other uses by local communities—and thereby to generate additional support for conservation (14).

Across the border, the Zaire gorilla project did succeed in obtaining government approval for a limited earmark system. Under this approach, a small percentage of tourist entry fees was used to supplement the extremely low official salaries for guards and guides working in the tourism sector. This fund was also used to build and maintain a health clinic for parks personnel and their families. When the Tongo chimpanzee project began as an offshoot of this program, the precedent was set for a comparable system of revenue-sharing. In both cases, it should be noted that direct employees were the principal beneficiaries, though surrounding communities benefited from related rural development activities.

The project that has given the highest share of revenues to local populations is the Tiwai Island initiative in Sierra Leone, where all tourism profits are divided among the local chiefdoms. That the central government has permitted this is to its credit. The fact that total annual profits do not exceed a few hundred dollars, however, may make such government beneficence easier. The proposed 75% earmark for local communities in Uganda's Kibale Forest project will be a much more important test of a government's willingness to share, if projected visitation and revenue levels are realized.

Current models for ecotourism development correctly insist that projects address this issue of revenue-sharing. At the same time, it should be recognized that most of the projects described in this chapter were initiated long before the concept of ecotourism even existed. They were conceived primarily to help save wildlife by generating foreign revenues that would influence central government policies and practices in favor of conservation. That much said, their experience with revenue-sharing has contributed greatly to the current debate on economic equity. Second, even under the best of conditions, some benefits from tourism must accrue to the central government institution responsible for reserve management. If not, there may be no alternative source of support and conservation may suffer greatly. Benefits to the national treasury may also help to offset alternative use options, such as the proposed cattle-raising scheme in Rwanda. Finally, it must be recognized that while foreign individuals and NGOs play an important advisory role in most projects, decisions about revenue-sharing are ultimately in the hands of sovereign governments.

The last problem arising from on-going projects concerns the degree of uncertainty inherent in ecotourism. Whereas overall trends are positive, visitation may drop off rapidly at any particular site as a result of several factors. International tourism in general is subject to ebbs and flows related to economic conditions in the source countries of the North. International political conditions also affect travel, and tourism has suffered at different times from perceived threats of terrorism arising from events in the Middle East, North Africa, and Iraq.

Of more direct concern to African travelers, however, is the wave of political unrest and civil disturbance now sweeping across sub-Saharan Africa. Nowhere is this more apparent than in the most successful of the projects under consideration: the Mountain Gorilla Project. In October of 1990, Rwandan exiles launched an attack on their former country from Uganda. Failing to control any land, the rebels retreated into the rugged refuge of the Volcanos National Park, from where they have continued to harass Rwandan forces for more than a year. In this state of hostility, both sides have issued statements that they will not harm the gorillas—a stand that they appear to take seriously, and surely a first in international bio-diplomacy (48). Still, tourism has suffered dramatically, as seen in the fall-off in visitation in 1990 (Fig. 3). Unofficial estimates for 1991 indicate an even more dramatic drop, with total visitation less than one-third of its previous level. Furthermore, current visitors include a much higher percentage of residents and backpackers: groups that tend to spend much less than the tourists who came before. The resulting decline in revenue has not led to any visible change in the official government policy toward conservation. Still, limited funds are far more likely to go to fight guerillas than to protect gorillas; and tourists are likely to stay away until there is calm.

Rwanda is not alone in dealing with the effects of political instability. In Zaire, the drawn-out collapse of the Mobutu regime has led to sporadic uprisings in cities, along with almost total economic devastation. As a result, fewer tourists are coming, and those who do find little in the way of support services. They also find parks personnel who have not been paid for months. The result, according to some recent accounts, is a marked increase in corruption, with tourists shaken down for more money at almost every step of their visit to see the chimps and gorillas. In addition, these same reports indicate that group size limits have been abandoned and as many as 20 tourists at a time are now being taken out to see the gorillas.

It is ironic that of the countries under consideration, Uganda stands out in 1992 as one of the most peaceful. Perhaps no other African country has suffered as much from civil war and instability; none have had a worse external reputation. That Uganda is now considered a good site for tourism development is a hopeful sign for its neighbors. That it has taken so long to overcome its notoriety should be taken as a warning.

Limits to the expansion of forest primate ecotourism

First, it must be recognized that not all primates have equal touristic value. Apes apparently rank higher than monkeys, and gorillas are best of all from a tourist's perspective. Not only do many more people come to see gorillas, but visitors have demonstrated a willingness to pay increasingly higher fees for the experience. The Nyungwe experience may indicate that equally high numbers can be attracted to view certain monkeys as part of a more comprehensive forest experience, but not under a comparable pricing structure. Reinforcement for this possibility comes from Madagascar, where 3,782 visitors came to view lemurs in the Perinet Forest in 1990, generating more than $40,000 in entry fees and much more in indirect revenues (Fig. 8). While $40,000 is an interesting sum of money, it is only a fraction of what is earned by gorillas and the major savanna parks. If decision makers make comparisons with either of the latter, they may find other forms of primate-based tourism wanting and try to raise prices. If they do, they may find that demand declines.

Second, it appears that tourists from the temperate North are much more attracted to highland and montane environments in the tropics. The most successful projects presented in this chapter are concentrated in the highlands of east-central Africa, and tourism is certainly helped by the cooler climate and decreased incidence of disease at these sites.

That temperate highlands are more attractive for tourism has helped to provide financial support for some critically important conservation efforts. Mountain gorillas and several other primate species are totally dependent on small islands of forest habitat to sustain their populations, and tourism revenues can cover many of their maintenance costs. Saving important primate habitat in such instances should not be confused with saving highly diverse ecosystems, however. The montane forests of east-central Africa contain many rare, endemic, and endangered species, but they are generally far less diverse than lowland rain forests.

Which leads to the third—and potentially most critical—constraint on tourism development in African rain forests: possible limits to its extension into the lowlands. Visibility is a very important issue here. While lowland forests are undeniably rich in biological terms, the vast majority of animal species are invertebrates living either in the subsoil and ground litter, or far up in the forest canopy. Most primates, too, tend to remain in the canopy and generally out of sight, though frequently heard. Lowland gorillas spend more of their time on the ground, but have proven to be extremely elusive and difficult to habituate: partly because of the underbrush, partly because of hunting pressure in many areas, and partly because of their shy nature. The other large mammals that are most distinctive in the African rain forest—the various duikers, bongo, okapi, buffalo, and elephants—are also quite difficult to see. The one exception is the forest elephant population of southwest Central African Republic (C.A.R.) and northern Congo. In the Dzanga-Sangha Reserve of the C.A.R., more than 1,400 individual elephants have now been identified frequenting a single clearing (49): a site also being developed for tourism purposes. The Dzanga elephants, however, are the exception that proves the rule.

All of this should not rule out the potential for tourism development in lowland forests. Thousands of people each year are now seeking the experience of visiting rain forests across the tropics. Most leave with a much greater appreciation of a fascinating environment. Yet it will not take too many sites to satisfy even the most optimistic estimates of demand, and this level and form of tourism will never generate revenue flows approaching those for mountain gorillas in Rwanda. Nor will tourism be able to fully offset the potential short-term revenues from some other forms of use, such as logging. With this in mind, it should be clearly understood that ecotourism is not going to provide the economic justification for saving large expanses of lowland rain forest. And, as a result, the major share of the earth's richest biome will require other arguments and approaches if it is to maintain its biological values and ecological integrity.

CONCLUSION

The identification of serious constraints to ecotourism development should permit us to recognize it as a very useful tool with limited geographic—and perhaps taxonomic—applicability. At the same time, one must avoid the tendency to discard or denigrate the tool when it does not work in all situations. The conclusion that ecotourism is not a universal solution only reflects poorly on those who have oversold its utility. What is necessary now is to move forward with the testing of other approaches that may hold more hope for conserving primates and other threatened biota in lowland rain forests. In some instances, the traditional combination of fences and force may work. More likely, we will have to come up with far more creative and flexible solutions to the highly complex problems that confront the forest and its diverse inhabitants. Many of those solutions will undoubtedly involve more intensive use and cause more serious impacts than the soft path of ecotourism.

It is sobering to recognize the degree to which we are now conducting an extensive set of experiments to determine the fate of the forest, and even more so to realize how little empirical knowledge is being generated and disseminated in the process. This feeling was strongly reinforced in the course of research for this manuscript by the paucity of field-based data. Visitation and revenue data were hard enough to come by; published accounts of the evolution of projects were rare; and more detailed studies of social and economic impacts were virtually nonexistent. As a conservation practitioner, I am well aware of the constraints on research and publication when faced with the immediate pressures of saving species and their habitats. As a final conclusion to this chapter, however, I would highlight a fundamental need for more applied research in this and other key areas where planners and practitioners are now making critical decisions in an informational void.

REFERENCES AND NOTES

1. Stuart, S. N., and R. J. Adams with a contribution from Martin D. Jenkins. 1990. *Biodiversity in sub-Saharan Africa and its islands: Conservation management and sustainable use*. Occasional Papers of the IUCN Species Survival Commission.

Occasional No. 6. Gland, Switzerland: International Union for the Conservation of Nature.

2. Barnes, R.F.W. 1989. *The status of elephants in the forests of central Africa: Results of a reconnaissance survey*. New York: Wildlife Conservation International (unpubl.).
3. The African Elephant Conservation Coordinating Group. 1991. *The African Elephant Conservation Review*. Oxford: AECCG Environment and Development Group.
4. Nowak, R. M. 1991. *Walker's Mammals of the World*. Vol. II, p. 1434. 5th ed. Baltimore: Johns Hopkins University Press.
5. Oates, J. F. 1986. *Action plan for African primate conservation: 1986–1990*. IUCN/SSC Primate Specialist Group.
6. McNamara, R. S. 1990. *Africa's development crisis: agricultural stagnation, population explosion, and environmental degradation*. Address to the Africa Leadership Forum, June 21, at Ota, Nigeria, pp. 51 and 59.
7. World Resources Institute in collaboration with The United Nations Environment Programme, The United Nations Development Programme. 1991. *World Resources 1990–91*, pp. 248–49. Oxford: Oxford University Press.
8. Repetto, R., and M. Gillis, eds. 1990. *Public policies and the misuse of forest resources*. Sponsored by World Resources Institute. Cambridge: Cambridge University Press.
9. Stromayer, K.A.K., and A. Ekobo. 1991. *Biological surveys of southeastern Cameroon*. New York: Wildlife Conservation International. (unpubl.).
10. Wilkie, D. S., J. G. Sidle, and G. C. Boundzanga. In press. *Mechanized logging, market hunting and a bank loan in Congo*. Boston: Conservation Biology: The Journal for the Society of Conservation Biology.
11. Lahm, S. A. 1988–1990. Unpublished material from wildlife surveys in Gabon. New York: Wildlife Conservation International.
12. Vedder, A.L., and A.W. Weber. 1990. The mountain gorilla project (Volcanoes National Park). In *Living with wildlife: Wildlife resource management with local participation in Africa*. A. Kiss, ed. Washington, DC: World Bank.
13. Weber, A.W., and A.L. Vedder. 1983. Population dynamics of the Virunga Gorillas: 1959–1978. *Biological Conservation* 26:341–66.
14. Weber, A.W. 1989. Conservation and development on the Zaire- Nile divide: An analysis of value conflicts and convergence in the management of afromontane forests in Rwanda. Ph.D. diss. University of Wisconsin, Madison.
15. Weber, A.W. 1987. Sociologic factors in the conservation of afromontane forest reserves. In *Primate conservation in the tropical rainforest*, Clive W. Marsh and Russell A. Mittermeier, eds., 205–30. New York: Alan R. Liss, Inc.
16. Weber, A.W. 1981. Conservation of the Virunga Gorillas: A socioeconomic perspective on habitat and wildlife preservation in Rwanda. Masters thesis. University of Wisconsin, Madison.
17. Weber, A.W. 1979. Gorilla problems in Rwanda. *Swara* 2:29–32.
18. Vedder, A. L. 1989. In the hall of the mountain king. *Animal Kingdom* 92:30–43.
19. Schaller, G. B. 1963. *The mountain gorilla: Ecology and behavior*. Chicago: University of Chicago Press.
20. Schaller, G. B. 1964. *The year of the gorilla*. Chicago: University of Chicago Press.
21. Fossey, D. 1970. Making friends with mountain gorillas. *National Geographic Magazine* 137:48–67; Fossey, D. 1971. More years with mountain gorillas. *National Geographic Magazine* 140:574–86.
22. Fossey, D. 1983. *Gorillas in the mist*. Boston: Houghton Mifflin.
23. Vedder, A. L. 1989. Ecology and conservation of the mountain gorilla (*Gorilla gorilla beringei*). Ph.D. diss. University of Wisconsin, Madison.

24. Vedder, A. L. 1984. Movement patterns of a group of free-ranging gorillas (*G. g. beringei*) and their relationship to food availability. *American J. of Primatology* 2:73–88.

25. Administration Géneral de la Cooperation au Développement. 1980. Evaluation du projet Belgo-Rwandais: tourisme et parcs nationaux. Brussels: AGCD.

26. The research leading up to the creation of the MGP, as described in this chapter, was supported by Wildlife Conservation International (WCI). The consortium formed to implement the MGP, however, consisted of the African Wildlife Foundation, the Fauna and Flora Preservation Society, and the World Wildlife Fund. WCI has continued to support regular censuses of the gorilla population and attitudinal surveys.

27. C. Sholley, personal communication.

28. Vedder, A. L. 1988. Technical analysis: Biodiversity, 10. Natural Resources Management Project, USAID, Rwanda.

29. In fact, the approach taken by the MGP in Rwanda was first designed to be implemented in Kahuzi-Biega by this author and A. Vedder as an initiative under the Peace Corps–Smithsonian program in 1974. Funding and administrative problems in Zaire, however, forced the change of venue to Rwanda.

30. Aveling, C., and R. Aveling. 1989. Gorilla conservation in Zaire. *Oryx* 23:64–70.

31. E. Williamson, personal communication.

32. Lanjouw, A. 1989. Chimpanzee Project, Zaire. In *Proceedings from the First International Workshop for the Conservation and Management of Afromontane Forests*, Cyangugu, Rwanda, June 19-23, 1989. Nyungwe Forest Conservation Project, Rwanda.

33. Collar, N.J., and S.N. Stuart. 1988. Key forests for threatened birds in Africa. *ICBP/IUCN, Monograph No. 3.* Cambridge: ICBP.

34. Offutt, K., comp. 1990. *Nyungwe Forest Reserve*. Nyungwe forest conservation project. Cyangugu, Rwanda.

35. Vedder, A. L. 1988. Conservation of the Nyungwe Forest Reserve. New York: Wildlife Conservation International. (unpubl.).

36. Clausen, R.M. 1989. Nyungwe forest conservation project: progress report #1. New York: Wildlife Conservation International. (unpubl.).

37. Weber, A. W., and A. L. Vedder. 1984. Forest conservation in Rwanda and Burundi. *Swara* 7:32–35.

38. Williamson, E. 1992. Nyungwe Forest conservation project: progress report 1991. New York: Wildlife Conservation International. (unpubl.).

39. This calculation is based upon each foreign visitor spending \$300 to \$500 in Rwanda during a 2- to 3- day stay (foreign visitation in 1990 was 1,800 people).

40. Johns, A. 1992. Annexe 5: Tourism development. In *Conservation and Management of the Kibale Forest, Uganda: Phase II*. Wildlife Conservation International, New York, and USAID, Uganda. (unpubl.).

41. The Korup Project: Plan for developing the Korup National Park and its Support Zone. 1989. Republic of Cameroon: Ministry of Plan and Regional Development.

42. Todd, A. M. 1988. Tiwai Island Wildlife Sanctuary: Kikihun Nature Center final report. New York: Wildlife Conservation International. (unpubl.).

43. J. Oates, personal communication.

44. Lindberg, K. 1991. Policies for maximizing nature tourism's ecological and economic benefits. Washington, DC: World Resources Institute.

45. Calculations based on official statistics for Amboseli National Park, Kenya, provided by the Kenya Wildlife Service.

46. Calculation based upon average airfare of \$1,800 round-trip for nonresident visitors.

47. The one notable exception is in Korup, where there is a minimal tourist presence and where the hunting of certain primates for market may be causing a decline in their numbers.

48. The first known gorilla casualty of this conflict occurred as this manuscript was in preparation, when the dominant male of a Rwandan tourism group was shot and killed by unidentified assailants in May 1992.
49. Fay, J. M. and A. Turkalo, personal communication.
50. Hamilton, A. C. 1976. The significance of patterns of distribution shown by forest plants and animals in tropical Africa for the reconstruction of upper Pleistocene palaeoenvironments: A review. *Paleocolgy Africa* 9:63–97.
51. Office Rwandais du Tourisme et des Parcs Nationaux, official statistics 1977–90 for Volcanoes National Park, Rwanda.
52. Institute Zairois pour la Conservation de la Nature, official statistics for Virunga National Park, Zaire. Revenues estimated from official visitation rates and park fees.
53. Institute Zairois pour la Conservation de la Nature, official statistics for Kahuzi-Biega National Park, Zaire. Revenues estimated from visitation rates and park fees.
54. Kenya Wildlife Service, official statistics for Amboseli National Park, Kenya.
55. Tanzania National Parks, official statistics for fiscal years 1985–91 for Gombe National Park, Tanzania. The 1991 revenues estimated from visitation rates and park fees.
56. Lanjouw, A. Tongo Chimpanzee Project, unpublished material for Tongo, Zaire.
57. Nyungwe Forest Conservation Project, unpublished material for Nyungwe Forest, Rwanda. Revenues estimated from official visitation rates and park fees.
58. Tiwai Island Visitor Center, unpublished material for Tiwai Island, Sierra Leone.
59. Gartlan, S. The Korup Project, WWF-Cameroon, unpublished material. Revenues calculated from estimated visitation rates and park fees for Korup National Park, Cameroon.
60. Service des Eaux et Forêts, Madagascar, official statistics for Perinet National Park, Madagascar.

11

Ecologically Sustainable Forest Management in the Peruvian Amazon

Gary S. Hartshorn and William Pariona A.

LOSS OF TROPICAL FOREST BIODIVERSITY

Tropical deforestation is generally viewed as a primary causal agent in the loss of biodiversity. Because of the increasing rate of loss of tropical forests (1), habitats of untold numbers of species are being destroyed, causing an inexorable reduction in remaining forest habitats, as well as more fragmentation of the remaining forest patches. Tropical deforestation and fragmentation of the remaining forests not only contribute to loss of species, but also decrease the genetic diversity of those species that survive habitat destruction. Though our knowledge of species richness and diversity in tropical forests is woefully inadequate, the most recent and conservative estimate (2) is that tropical deforestation is causing the loss of 74 species per day.

Despite rampant deforestation in virtually all countries with tropical forests (3), there are several regions and countries that still have appreciable amounts of tropical forests, e.g., New Guinea, Borneo, the western Congo Basin, the Amazon Basin, and the trans-Andean Chocó of Colombia. Most countries rich in tropical forests have set aside significant areas for protection as national parks and equivalent reserves (4); however, unprotected forests may include up to 10 times the total area designated as conservation units. Unprotected tropical forests are usually viewed by colonists and politicians alike as opportunities for development. Traditional rural development in tropical America is virtually synonymous with deforestation because the forest is cut and burned to plant agricultural crops or to establish pastures for beef cattle (5). Thus forest management techniques that permit use of the tropical forest without destroying the habitat have the potential to integrate conservation and development objectives.

SOCIOECONOMIC PROFILE OF THE PALCAZÚ VALLEY

The Palcazú Valley in Peru's Central Selva region (Fig. 1) offers a classic example of the advancing agricultural frontier threatening to destroy the tropical forests under the guise of rural development. With the return of a civilian government in 1980 after a 12-year hiatus of military rule, newly elected president Fernando Belaunde Terry revived his 30-year dream of a Marginal Highway at the eastern base of the Andes. Belaunde asked development agencies for foreign aid to assist Peru in opening up the Central Selva through agricultural development projects that would turn the region into the breadbasket of Lima (6).

The U.S. Agency for International Development (USAID) chose to focus its development assistance on the small (20 x 70 km) Palcazú Valley as part of the

Figure 1. Map showing Amuesha native communities of the Palcazú Valley in the Central Selva region of Peru.

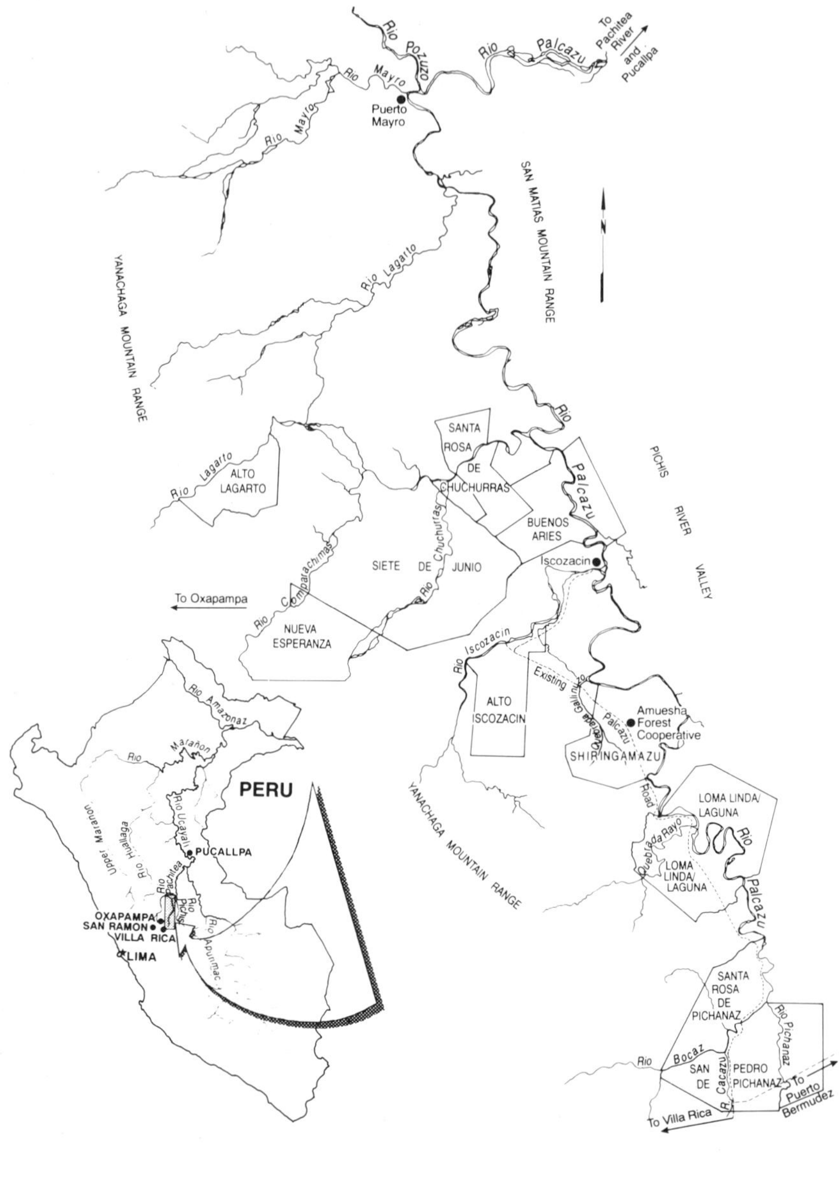

government's supra-ministerial Pichis-Palcazú Special Project (PEPP). In compliance with sections 118 (Appendix 1) and 119 of the U.S. Foreign Assistance Act, a USAID contractor conducted a detailed environmental assessment (7) of PEPP's proposed development projects for the Palcazú Valley. The field assessment concluded that the Palcazú Valley was not only unsuitable for traditional rural development based on agriculture, but also that the Palcazú had a sizable population of Yánesha Indians whose native communities claimed much of the lower southern valley (8). As a consequence of the USAID-required social soundness evaluation of PEPP's proposed development of the Palcazú Valley, USAID included a condition precedent in the loan agreement with the Peruvian government that stipulated that all native community land claims must be officially recognized and legally titled to the Yánesha Indians before any USAID loan disbursements to PEPP could occur (6).

The Yánesha are an Arawakan language group who have been in the Palcazú and nearby valleys for centuries and perhaps millennia (6). Settlement pattern appears to have been large sedentary villages based on floodplain agriculture, supplemented by hunting and fishing. If recent land-use patterns are indicative of earlier agroecological systems, the Yánesha used short-term fallows on recent alluvial terraces to raise beans, maize, peanuts, and squash, while manioc was grown on nearby hilly land using longer rotation shifting cultivation (9). This prehistoric settlement pattern came to an abrupt end in the sixteenth century when the Yánesha population was decimated by foreign-introduced diseases (6). By the time colonists first entered the valley late in the nineteenth century, the Yánesha population was sparsely distributed in the lower southern valley (Fig. 1).

The Amazonian rubber boom attracted the first colonists to the Palcazú late in the nineteenth century. These early entrepreneurs were mostly Swiss, plus a few Austrians and Germans. At least two major colonization attempts failed (on the upper Río Iscozacín and near Puerto Mayro). However, successful colonies of European immigrants and descendents were established just outside the Palcazú Valley in the Chanchamayo, Huancabamba, and Pozuzo regions. During the first half of the twentieth century the Yánesha were in demand to clear forests for cattle operations, to collect rubber for the middlemen, and to migrate seasonally to Villa Rica to pick coffee—all under the control and company store of the economically dominant European families (6).

When the Central Selva Natural Resources Project (CSRM) began in 1981 there were approximately 5,000 inhabitants in the Palcazú Valley, of whom 2,800 were Yánesha. The remainder consisted of a few dozen second- and third-generation European families, some Campa Indians, and mostly ladino immigrants attracted in the 1950s by exploratory oil drilling. The economically dominant European families ran large ranches based on beef cattle (including a small airline company for flying beef out of the valley) along the major rivers in the central and northern parts of the valley. Colonization pressures forced the Yánesha to the wetter and steeper parts of the Palcazú Valley, principally the narrow southern end of the lower valley, and to middle sections of the Iscozacín and Chuchurras tributaries (Fig. 1).

At the start of the CSRM project there were several conflicting claims to land in the valley, hence the importance of the condition precedent to the loan agreement mentioned above. Although there were concerted official and individual attempts to reduce the size of native communities' claims to land, USAID support eventually led to official government recognition and legal title for 11 Yánesha native communities in the Palcazú Valley. The Yánesha communally hold about 80% of the southern half of the lower valley. Furthermore, as of 1984 some 80% of the Yánesha lands were in primary forests. Thus the Yánesha forests would be high-graded for valuable timbers by sawmillers from the adjoining deforested valleys (Villa Rica, Oxapampa) already converted to extensive pastures for beef cattle, or become the base for the sustainable production of timbers.

ECOLOGICAL FOREST MANAGEMENT

Physical Environment of the Palcazú Valley

The Palcazú watershed (190,000 ha) is in the abrupt physiographic transition of the Central Selva (9° 50' to 10° 45' S latitude) from the flat Amazon Basin to the rugged outliers of the Andes Mountains. Over an east-west distance of only 20 to 30 km, the Palcazú Valley is bordered on the east by the low San Matías Range (1,000–1,200 m), on the west by the Yanachaga Cordillera (3,800 m), and to the south by the intermediate and broader San Carlos Range (now traversed by a spur of the Marginal Highway). The namesake Río Palcazú is squeezed against the eastern base of the San Matías Range by the uplifted Yanachaga Cordillera. The valley floor (ca. 20 x 70 km) is separated by the large Río Iscozacín into a wide, relatively flat northern section (base elevation 270 m) and a much narrower, more rugged southern section (base elevation 350 m). Forestry activities are restricted to the lower valley, below 600 m elevation.

USAID and The Nature Conservancy assisted with the creation and consolidation of the 123,000 ha Yanachaga-Chemillén National Park (600 to 4,200 m) that protects the upper western watershed of the Palcazú Valley. The San Carlos and San Matías Ranges are decreed Protection Zones in order to conserve watershed forests. In addition, lands in the southern portion of the valley that are between the National Park and the Yánesha native communities have been declared a Communal Reserve for traditional, noncommercial uses by the Yánesha Indians. Thus, nearly all of the southern part of the Palcazú watershed is in national park, protection zone, communal reserve, or native community land. The few exceptions tend to be privately owned farms along the major rivers.

The southern part of the lower valley is characterized by low, rolling, often steep hills dissected by several east-flowing rivers with active floodplains. Geographically older terraces of gray sandy-loam soil also occur in the lower valley (e.g., in the Shiringamazú production forestry unit). Soils of the hilly landscape are deep, highly weathered, easily eroded, acid (pH 3.8–4.5), high in aluminum, with very low base saturation and cation exchange capacity (10). Land-use capability classification of the entire lower valley (11) gives the following maximum sustainable uses: annual or seasonal crops (8%); pasture (13%); perennial crops (14%); production forestry (46%); and protection forests (19%). Because the Yanachaga Cordillera functions as a front range intercepting westerly moving moist air masses, the valley receives almost 7,000 mm of annual rainfall. January is the rainiest month (>1,000 mm) and June through September is the least rainy.

Most of the lower valley is classified as tropical premontane rain forest in the Holdridge Life Zone Classification System (12). The rolling hills of the lower Palcazú Valley have typical tropical rain forest that is tall (45 to 55 m), species-rich (175 to 200 species per ha >10 cm dbh [diameter at breast height]), with abundant epiphytes, lianas, and vines. Primary forest on the old alluvial terraces differs from the hilly forests by having a higher density of small trees, a lower canopy with huge emergents (especially *Cedrelinga catenaeformis*, Mimosaceae) and an abundance of rubber trees (*Hevea brasiliensis*, Euphorbiaceae). A striking anomaly is the dearth of palms in the Palcazú primary forests, possibly caused by traditionally high demand for palm leaves for thatch. The aguaje (*Mauritia flexuosa*, Palmae) is occasional where drainage is poor; however, extensive palm swamps do not occur in the valley.

Status of Important Species and Genetic Resources

Despite insignificant commercial exploitation of native species in the Palcazú Valley (primarily due to its inaccessibility), there has been a long tradition of subsistence gathering and hunting from the streams and forests of the valley. Fishing is still a popular night and weekend activity, usually with a baited hook on a handline. During low water, small seining nets are used occasionally. In small, clear-water streams, the "carachama" is caught by hand. Road construction crews were prone to use dynamite to kill or stun fish, but that destructive technique is no longer common because dynamite is scarce. Though no recent studies have been conducted on the freshwater fishes of the Palcazú Valley, it is our opinion that populations have not been severely depleted by overfishing or destructive techniques.

Wildlife is quite sparse in the valley, primarily due to hunting. Preferred game include brocket deer, tapir, peccary, paca, agouti, monkeys, turtles, and any bird larger than a toucan—with the notable exception of the hoatzin. Any cat is killed for its pelt. In the forests of the valley floor, one sees very little wildlife, including snakes. Nevertheless, it is unlikely that any native faunal species has been eliminated from the valley's forests because of overhunting. Rather, most of the larger terrestrial or arboreal vertebrates have reduced populations in the accessible forests, while presumably healthier populations still remain in the more remote forests now protected in the National Park and two Protection Zones.

Even prized heavy hardwoods are still present in the lower valley's forests, owing to lack of access. Though valuable timbers such as Spanish cedar (*Cedrela odorata*, Meliaceae) were logged up to 500 m from any stream and floated to Pucallpa 40 to 50 years ago (13), it is easy to still find large trees of this highly sought species. Special nontimber forest products, such as palm leaves for thatch (from a *Geonoma* sp.) and the supple tamshi liana (*Philodendron* sp.), are reported to be scarce near native communities (14). However, many Yánesha now have tin roofing, which greatly reduces the demand for palm thatch; similarly, the cutting of strips in the production forests provides appreciable quantities of the tamshi liana (a pliable woody vine) used by the Yánesha for weaving baskets.

Forest Management Approaches

The record of sustained-yield management of complex tropical forests is notoriously poor. Despite more than a century of efforts and a variety of systems (15) for managing tropical forests, there are very few successes. Some serious constraints to tropical forest management include (1) the low volume of commercial timbers per unit area; (2) expensive extraction costs associated with the common practice of selective logging of only the high quality timber; (3) poor understanding about tropical forest dynamic processes; (4) a general lack of information about the regeneration requirements and silvics of most canopy tree species; (5) government policies (e.g., short-term concessions for large volumes) that discourage sustained yield forest management or make it economically unattractive; (6) national and international agencies that promote agricultural colonization on lands that cannot sustain agriculture or cattle pastures; and (7) agencies responsible for forestry that do not define and protect the national forestry estate, nor promote sustainable forestry practices on their forest lands.

These and other problems led to a growing consensus in the 1970s that it was impossible to manage complex tropical forests (16). Nevertheless, the 1980s saw major changes in that misperception supported by improving economic values of tropical woods, increasing pressure from consumers for sustainable production of tropical woods, and innovative approaches to integrating local people in sustainable forestry practices (17). The Yánesha Forestry Cooperative's effort to manage its production forests on a sustainable basis using a strip-cutting technique (see below) is one of the exciting new initiatives promoting wise use of tropical forests.

Conservation and Development Objectives

Many developing countries have made impressive progress in creating a national system of protected areas, yet few countries have attained the recommended 10% of the national territory in national parks and equivalent reserves. Most forest-rich developing countries are well below the 10% minimum, thus leaving vast areas of tropical forests open for development, either through political schemes like Indonesia's transmigration program, or undirected colonization of the "frontier" as in the Brazilian Amazon (18). Innumerable attempts to colonize tropical forests have failed, often because the soils are not suitable for permanent agriculture. In fact, many ecologists and other tropical specialists contend that there is no more agricultural frontier to colonize.

Integrated conservation and development projects based on the wise use of tropical forests are essential from at least two perspectives. First, ecologically sound management of tropical forests can produce timber as well as nontimber forest products without destroying the forest, while maintaining natural habitats for the great majority of species. Thus, sustainably managed tropical forests could contribute much needed habitat to the conservation objectives of a developing country rich in tropical forests. Second, sustainable forest management in the buffer zone of a protected area is a viable means of involving local people in economically attractive productivity, while the production forests buffer a protected area from more intensive land-use pressures.

Yánesha Forestry Cooperative

The combination of minimal unclaimed forest land in the lower Palcazú Valley and the vast primary forests on land claimed by the Yánesha prompted a major change in focus of CSRM forestry plans. Rather than use production forestry by small-holder colonists as the basis for development activities, the Yánesha Forestry Cooperative (COFYAL) was constituted in 1986 by five native communities and 70 individual Yánesha. The five founding communities are along or near the spur road that enters the southern end of the valley and ends at the Río Iscozacín, thus offering good access to the production forests of these communities. The Shiringamazú community ceded land to COFYAL for construction of timber processing facilities. COFYAL objectives are to (1) provide a source of employment for members of the native communities; (2) manage the communities' natural forests for sustained yield of forest products; and (3) protect the cultural integrity of the Yánesha people.

Strip Cutting Technique

Harvesting of timber is done on long, narrow (30 to 40 m wide) strips on which all trees are cut, hence the designation of this as a strip cutting or strip clear-cut system. Length of a strip (usually 100 to 300 m long) is determined by topography and the logistics of log extraction. The narrow width simulates the approximate diameter of a canopy opening caused by a tree fall (19), thus a strip cut is an elongated gap.

More important, strips are not cut in a sequential manner that would be analogous to the advancing agricultural frontier. All strips that will ever be harvested in a production stand (often 100 to 500 ha) are marked on the ground; then several strips are skipped between the first and second strips to be cut. After the first cycle of strips has been harvested, the second cycle commences by harvesting one strip intermediate between two strips harvested in the preceding cycle. The third cycle harvests a strip intermediate between the first and second cycle strips. Over a 30- to 40-year rotation, there may be four to six cycles of harvesting of strips in a production stand (20). Thus, it may take 5 to 10 years per cycle of strip cuts, e.g., a four-cycle sequence is [1|3|2|4|1|...], while a six-cycle sequence is [1|3|5|2|6|4|1|...]. The number to strips to be harvested each year is determined by dividing the total area of strips in a production stand by the rotation length and then assigning the annual cut to the strips in the predetermined sequence of cutting cycles.

Harvesting the timber on a strip begins with the small vegetation that can be cut with a machete, i.e., stems up to about 5 cm dbh. The felling and extraction proceeds in an orderly way from the smaller trees to the largest, which is the antithesis of traditional logging operations that seek only large trees, or forest clearing that fells the canopy trees onto the smaller ones. When a tree is felled, the log is bucked and extracted, and the crown branches are cut up in order to minimize obstacles to movement on the strip. Every effort is made to directionally fell the trees into the strip, so as to not damage the bordering forest or advanced regeneration.

Strips are oriented more or less perpendicular to an all-weather graveled forest road built by COFYAL to access the maximum number of strips in a production stand. However, in order to avoid soil compaction and ruts, no machinery is allowed on a strip. Log extraction is by animal traction (oxen in the Palcazú, though mules or water buffalo could also be used) to a log yard at the head of a strip. Logs are extracted on a narrow skid trail usually along an edge of a strip. A cable winch mounted on a platform truck is an alternative for strips up to 150 m from the forest road. Logs are transported from the log yard to the processing facilities by a platform truck or tractor and wagon.

IMPACTS OF *IN SITU* CONSERVATION

Ecological Benefits

The strip-cut technique of tropical forest management has several ecological benefits. Strip cuts simulate the gap-phase dynamics of natural forests, thus promoting excellent natural regeneration of hundreds of native tree species. Strip cuts favor the regeneration of shade-intolerant tree species that may comprise about half the tree species in a species-rich tropical forest such as the Palcazú (21). Because the harvested strips are not used for agriculture, nor is the logging slash burned, there is abundant sprouting from the cut stumps. This vegetative regeneration complements the more abundant regeneration from seeds on the harvested strips. Initially, stump sprouts dominate the early regrowth on a strip, principally because the 30- to 50-cm thick cover of organic debris causes a six- to nine-month lag in the establishment of seedlings. By the second year of regrowth, saplings from seed have outgrown the stump sprouts.

Annual censuses of tree regeneration on the two demonstration strips cut in 1985 (19) show the outstanding regenerative capacity of Palcazú forests using the strip cut technique. The first few years of tree regeneration following strip cutting show a rapid accrual of species, e.g., the first demonstration strip (0.15 ha) had the following number of tree species (individuals >50 cm tall): 110 at 11 months; 132 at 16 months; 135 at 23 months; and 209 at 2.5 years. Even at six years since clear cutting, the first demonstration strip has lost only 14 tree species, while the larger, second demonstration strip (0.5 ha) lost 18 species. Most of the tree species lost from these two strips were short-lived, weedy pioneers. Inventory data from the 1991 census of the two demonstration strips are summarized in Table 1.

The richness, diversity, and abundance of tree regeneration on strip clear-cuts offer the opportunity to silviculturally favor particular species and to maximize the growth potential of preferred trees. If no silvicultural interventions occur, the natural regeneration will be subject to normal successional processes, which means that growth of all but a few individuals will be less than maximum and that most of the young trees will die from severe competition. Thus we have the opportunity to adjust the competitive balance among trees and to influence the composition and structure of the regenerating forest for the next rotation. Silvicultural treatments

Table 1. Summary of natural regeneration data for trees >50 cm tall on two demonstration strips in the Palcazú Valley, Peru.

	STRIP 1	STRIP 2
Size	20x75 m	50x100 m
Area	0.15 ha	0.50 ha
Harvesting completed	May 1985	Dec. 1986
Maximum age of regeneration	5.8 yr	5.5 yr
Maximum height	18 m	20 m
Maximum diameter (dbh)	21 cm	24 cm
Sprouted stumps	287	848
Seedlings	885	2,370
Total number of stems	1,172	3,218
Species only from stump sprouts	34	46
Species only from seedlings	85	81
Total number of tree species	182	259
Number of tree species lost	14	18
Number of subplots (10x10 m)	15	50
Average no. stems/subplot	78	64
Average no. species/subplot	45	34

begin in the second year to control climbers and to reduce the number of stump sprouts to one to three, depending on the diameter of the stump. The first thinning occurs in the fourth year and future thinnings will be scheduled based on the growth increments of the preferred trees.

A fascinating aspect of the natural regeneration on the strip clear-cuts is the appearance of rare tree species. For example, there are several juveniles of *Minquartia guianensis* (a highly prized timber of the Olacaceae) on the two demonstration strips—and we have been unable to locate a putative parent tree. Similarly, seedlings of several other rare tree species have been documented in the regeneration on the strip clear-cuts. It appears that the excellent natural regeneration of hundreds of tree species on strip clear-cuts and the use of silvicultural treatments can not only maintain the exceptionally high tree species diversity of the Palcazú forests, but they may even enhance the regeneration success of rare, shade-intolerant tree species.

By simulating the natural gap-phase dynamics of a tropical forest, the strip clear-cut technique maintains the forest structure, which is an exceptional ecological benefit of a forest management system. The dispersed location of strips in any cutting cycle is almost like a surgical insertion of a strip cut; in fact, it is surprisingly difficult to see a recently cut strip from a low-flying light airplane. Maintenance of the structural matrix of the natural forest should ensure provision of ecological services (e.g., watershed protection, minimal erosion) and provide habitat for forest animals and plants. Not all of a production stand is assigned to strips for harvesting because steep slopes, stream edges, and swampy areas are excluded. Detailed land-use capability classification assigned only 46% of the first production stand (ca. 500 ha) in Shiringamazú to strips. The exclusion of forest land requiring protection further enhances maintenance of forest structure.

Economic Returns

The development of timber processing facilities by COFYAL is a key feature of the economic viability of ecological forest management in the Palcazú Valley. On land ceded by the Shiringamazú native community to COFYAL, the Co-op has a small sawmill (circular saw plus a Mighty-Mite mill) for producing sawn wood, a carpentery shop for making rural furniture, a bank of 44 PresCaps for preserving small-dimension (10 to 35 cm butt diameter) roundwood, and a 6 cubic meter portable kiln for charcoal production. This modest array of processing facilities located near the production forests enables the Co-op to use most of the hundreds of timber species harvested from the strips. Furthermore, the sale of wood products by the Co-op provides positive feedback to the Yánesha communities for conserving their tropical forests and managing them for the sustainable production of timber.

Despite astronomically high inflation in Peru and occasional attempts by Shining Path guerrillas to intimidate valley inhabitants—which prompted USAID to cancel financial support and technical assistance to the CSRM project in 1988—COFYAL has managed to survive. Meager sales of forest products within Peru have been supplemented by the sale of an occasional container of premium-quality hardwoods to the Ecological Trading Co. in London and to the California-based Luthiers Mercantile Co. (22). COFYAL exported about $30,000 of sawn wood in 1991 (23). Net returns run about 50% of gross revenues from sales. Even though the profits from national and export wood sales are relatively modest in terms of U.S. dollars, they are appreciable in Peruvian currency.

A comparative analysis of economic returns from two common land-use systems on the low-fertility soils of the lower Palcazú Valley with the COFYAL production forestry system shows that the latter is far superior to pasture operations for beef cattle, but somewhat inferior to robusta coffee (*Coffea canephora*, Rubiaceae) (6). This suggests the need for government incentives and/or donor assistance to encourage local people to adopt sustainable forestry systems over low quality coffee. The projected economic returns from these three options also suggest that controls on land use based on land-use capability classification are probably necessary. It is quite possible that continued diversification of wood products through local processing, as well as integrating nontimber forest products into the forest management system, could well increase the net returns from production forestry to where forestry is more viable economically than even robusta coffee.

SOCIOECONOMIC COSTS AND BENEFITS

Projected Regional Impacts

Production forestry for commercial purposes requires access to markets. Though river transport is sometimes used to move logs to sawmills or plywood plants (24), timber operations are often dependent on roads for extraction and hauling of logs. Local processing by COFYAL is fundamental to using most of the timbers in the Palcazú production forests, hence reliable road access from the forests to the processing facilities and from there to markets is an integral part of the operations. Penetration roads to reach forests for commercial harvesting also facilitate access by colonists, hunters, and others, which is a serious risk in this type of development project. Building the Palcazú spur of the Marginal Highway was an extremely risky venture that could have flooded the valley with colonists. It is highly unusual in tropical America that a penetration road—or even the plans for such a road—did not stimulate spontaneous colonization along the expected route. The legal titling and official recognition of Yánesha native community claims of land were extremely important first steps that preempted entrepreneurial colonists. Once the CSRM project began field activities, strong government control over forest conversion for agriculture and permits for logging also contributed substantially to maintaining the integrity of Yánesha lands.

Formation of the Yánesha Forestry Cooperative and its initial successes with production forestry have had and continue to have significant socio-cultural impacts on the Yánesha. COFYAL channels appreciable portions of its net returns to native community priorities such as housing, health, and education. Training of COFYAL members in all aspects of production forestry is an almost continual process, hence they have become competent foresters, sawmillers, mechanics, drivers, accountants, salesmen, administrators, managers, and so on. Even Yánesha women have become regular members of the Cooperative. The enfrachisement of the Yánesha through the production forestry activities of COFYAL has had a salutary effect on their culture. Many cultural traditions (e.g., dance, identity, oral history, pride, special events) have been revived by the Yánesha, attracting the attention of other indigenous groups of the Central Selva region.

Actual and Potential Development Conflicts

The greatest conflict is over natural resources, particularly land and land use. If COFYAL fails, it is doubtful that the Yánesha and their culture will survive.

Outside pressure from loggers and colonists will continue and possibly increase as land and timber become increasingly scarce. The apparent large land areas under the control of a relatively small population may prompt politicians, wealthy land owners, or squatters to try to reduce the land holdings of the Yánesha. Although extensive pasture lands under single ownership are politically correct in most Latin American countries, large holdings of "undeveloped" forests are usually a political liability. Another persistent conflict over land use is caused by the cattle culture that prefers pasture for beef cattle (25) in spite of information showing it is inferior economically to several other land uses.

Economic viability of COFYAL is absolutely essential not only for survival of the Yánesha, but for the Palcazú forests as well. Even within a few of the Yánesha native communities, some unscrupulous members have attempted to sell valuable timbers to outside sawmillers, rather than participate in COFYAL's production forestry; one native community member was removed from a leadership position by his peers for attempting such a sale. Similarly, the large areas of forest now assigned to the National Park, two Protection Zones, and the Communal Reserve will come under ever increasing pressure from loggers, poachers, colonists, and speculators. Without a viable COFYAL, it is very unlikely that these protected areas will survive.

Transferability

The Yánesha forest management model is eminently transferable to other tropical forest regions with ample natural forests suitable for production forestry and under low population pressure. It is an excellent production forestry model for forest-rich countries, such as those that share the Amazon Basin. Not only is this forest management system appropriate for communally held forests such as those of the Yánesha, but it would also work for sizable forest areas held by a single owner or concessionaire.

We believe it would be a suitable development model for a colonization project based on production forestry rather than agriculture; however, we recognize that coordinating timber extraction and processing from dozens of individual forest owners would be more difficult than from communally or individually held production forests. For a forest colonization project, each family would need to have sufficient production forest for a full rotation of strips and to produce enough timber to meet annual income needs.

Access to markets is a key requisite for the transferability of the Yánesha forest management system. Without access to markets this, or any, production forestry project will fail. Thus, there are both geographic and temporal windows where this type of forest development project is possible. A candidate area must be ahead of the agricultural frontier, i.e., primary forests should dominate the landscape. Preferably, the natural forests should not have been subjected to selective harvesting of the valuable timber. Yet, a candidate area must not be so far ahead of the penetration roads that it precludes access to markets.

CONCLUSIONS AND LESSONS LEARNED

The Palcazú strip-cutting model of forest management is the first attempt to manage complex tropical forests for the sustained production of timber from virtually all tree species. The Palcazú project pioneered the use of long, narrow strip clear-cuts as a management technique for tropical forests. Most significantly, the Palcazú project generates attractive economic benefits for local people.

The most important ecological conclusion from the Palcazú forestry project is that strip cutting promotes excellent natural regeneration of hundreds of native tree species. Furthermore, the vigorous growth of young trees, both from seed as well as from stump sprouts, suggests that the full rotation period may be as short as 30 to 40 years. Silvicultural treatments can enhance survival and growth of preferred species, including rare trees. A well-designed production stand harvested on a series of strip clear-cuts over four to six cycles maintains the structural integrity of the forest and preserves natural habitat.

From an economic perspective, local processing adds value to the wood products and permits utilization of nearly all of the hundreds of tree species occurring in the Palcazú production forests. Production forestry is almost as profitable as robusta coffee and much more economically viable than the traditional production of beef. Though the Palcazú forestry component did not originally include the export of high quality timber, the modest exports have played an important role in the survival of the Yánesha Forestry Cooperative. External technical assistance has been an essential ingredient in getting the Co-op on sound technical, administrative, and financial footing.

The Central Selva Natural Resources Project pioneered some truly revolutionary social changes. For the first time in the Amazon Basin, an Indian tribe received official recognition and legal title to its native lands. The Palcazú project helped establish the first Indian forestry cooperative in the Amazon Basin. And the Yánesha Forestry Cooperative appears to be on the track to profitability. Production forestry is contributing to a renewal of Yánesha cultural identity and traditions. Leaders of the Co-op are transferring the production forestry model to other Indian groups in the western Amazon.

The Palcazú production forestry model illustrates that tropical forestry projects can no longer focus just on the production of biomass. Any development project must be ecologically sound, economically viable, socially responsible, and politically acceptable. If any one of these four essentials is missing or fails, the probability of success decreases greatly. This type of integrated, trans-disciplinary approach to development is especially important for forest-rich tropical countries. Unless ecological forest management of production forests becomes a much more common approach, very few tropical forests will survive well into the twenty-first century.

REFERENCES

1. World Resources Institute. 1990. *World resources 1990–91*. New York: Oxford University Press.
2. Wilson, E.O. 1992. *The diversity of life*. Cambridge, MA: Harvard University Press. (In press).
3. World Resources Institute. 1990. *World resources 1990–91*. New York: Oxford University Press; Sader, S.A., and A.T. Joyce. 1988. *Biotropica* 20:11.
4. MacKinnon, J., K. MacKinnon, G. Child, and J. Thorsell. 1986. *Managing protected areas in the tropics*. Gland, Switzerland: IUCN.
5. Hartshorn, G. S., *et al.* 1982. *Costa Rica country environmental profile: A field study*. San José, Costa Rica: Tropical Science Center; Leonard, H.J. 1987. *Natural resources and economic development in Central America: A regional profile*. Washington DC: IIED.

6. Stocks, A., and G.S. Hartshorn. *J. of Int'l Forestry* 1. (In press).
7. McCaffrey, D., *et al.* 1981. *Central Selva Natural Resources Management Project.* McLean, VA: JRB Associates.
8. Miller, C., and H. Martínez. 1981. *Social soundness analysis: Central Selva Resource Management Project.* Lima, Peru: USAID; Smith, R.C., *Cultural survival.* Occasional paper #8.
9. Salick, J. 1989. *Advances in Economic Bot.* 7:189.
10. Bolaños M., R., and V. Watson C. 1981. In *Central Selva Natural Resources Management Project*, appendix C. D. McCaffrey, ed. McLean, VA: JRB Associates.
11. Tosi, J.A., Jr. 1981. In *Central Selva Natural Resources Management Project*, appendix N. D. McCaffrey, ed. McLean VA: JRB Associates.
12. J.A. Tosi, Jr., personal communication.
13. J. Quijano, personal communication.
14. C. Mateo, personal communication.
15. Baur, G.N. 1968. *The ecological basis of rainforest management.* Sydney: Forestry Commission of New South Wales.
16. Leslie, A.J. 1977. *Unasylva.* 29:2.
17. Hartshorn, G.S., R. Simeone, and J.A. Tosi, Jr. 1987. In *Management of the forests of tropical America: Prospects and technologies*, J.C. Figueroa C., F.H. Wadsworth, and S. Branham, eds., 235–43. Rio Piedras, PR: Institute of Tropical Forestry.
18. Burbridge, P.R., R.B. Norgaard, and G.S. Hartshorn. 1988. *Environmental guidelines for resettlement projects in the humid tropics.* Rome: FAO.
19. Hartshorn, G.S. 1989. *Ecology* 70:567.
20. Tosi, J.A. Jr. 1982. *Sustained yield management of natural forests.* Lima, Peru: USAID.
21. Hartshorn, G.S. 1989. In *Botanical dynamics, speciation and diversity*, L. Holm-Nielsen, I. Nielsen, and H. Balslev, eds., 65–73. London: Academic Press.
22. Landis, S. 1990. *Fine woodworking* 82:75.
23. J. Arce, personal communication.
24. Gentry, A.H., and R. Vasquez. 1988. *Forest Ecology and Mgmt.* 23:73.
25. Leonard, H.J. 1987. *Natural resources and economic development in Central America: A regional profile.* Washington, DC: IIED.

APPENDIX 1. U.S. FOREIGN ASSISTANCE ACT

Section 118. Tropical Forests

(a) Importance of Forests and Tree Cover

In enacting section 103(b)(3) of this Act the Congress recognized the importance of forests and tree cover to the developing countries. The Congress is particularly concerned about the continuing and accelerating alteration, destruction, and loss of tropical forests in developing countries, which pose a serious threat to development and the environment. Tropical forest destruction and loss: (1) Result in shortages of wood, especially wood for fuel; loss of biologically productive wetlands; siltation of lakes, reservoirs, and irrigation systems; floods; destruction of indigenous peoples; extinction of plant and animal species; reduced capacity for food production; and loss of genetic resources; and (2) can result in desertification and destabilization of the earth's climate. Properly managed tropical forests provide a sustained flow of resources essential to the economic growth of developing countries, as well as genetic resources of value to developed and developing countries alike.

(b) Priorities

The concerns expressed in subsection (a) and the recommendations of the United States Interagency Task Force on Tropical Forests shall be given high priority by the President: (1) In formulating and carrying out programs and policies with respect to developing countries, including those relating to bilateral and multilateral assistance and those relating to private sector activities; and (2) in seeking opportunities to coordinate public and private development and investment activities which affect forest in developing countries.

(c) Assistance to Developing Countries

In providing assistance to developing countries, the President shall do the following: (1) Place a high priority on conservation and sustainable management of tropical forests;

(2) To the fullest extent feasible, engage in dialogues and exchanges of information with recipient countries a) which stress the importance of conserving and sustainably managing forest resources for the long-term economic benefit of those countries, as well as the irreversible losses associated with forest destruction, and b) which identify and focus on policies of those countries which directly or indirectly contribute to deforestation;

(3) ..., support projects and activities a) which offer employment and income alternatives to those who otherwise would cause destruction and loss of forests, and b) which help developing countries identify and implement alternatives to colonizing forested areas;

(4) ..., support training programs, educational efforts, and the establishment or strengthening of institutions which increase the capacity of developing countries to formulate forest policies, engage in relevant land-use planning, and otherwise improve the management of their forests;

(5) ..., help end destructive slash-and-burn agriculture by supporting stable and productive practices in areas already cleared or degraded and on lands which inevitably will be settled, with special emphasis on demonstrating the feasibility of agroforestry and other techniques which use technologies and methods suited to the local environment and traditional agricultural techniques and feature close consultation with and involvement of local people;

(6) ..., help conserve forests which have not yet been degraded, by helping to increase production on lands already cleared or degraded through support of reforestation, fuelwood, and other sustainable forestry projects and practices, making sure that local people are involved at all stages of project design and implementation;

(7) ..., support projects and other activities to conserve forested watersheds and rehabilitate those which have been deforested, making sure that local people are involved at all stages of project design and implementation;

(8) ..., support training, research, and other actions which lead to sustainable and more environmentally sound practices for timber harvesting, removal, and processing, including reforestation, soil conservation, and other activities to rehabilitate degraded forest lands;

(9) ..., support research to expand knowledge of tropical forests and identify alternatives which will prevent forest destruction, loss, or degradation, including research in agroforestry, sustainable management of natural forests, small-scale

farms and gardens, small-scale animal husbandry, wider application of adopted traditional practices, and suitable crops and crop combinations;

(10) ..., conserve biological diversity in forest areas by a) supporting and cooperating with United States Government agencies, other donors (both bilateral and multilateral), and other appropriate governmental, intergovernmental, and nongovernmental organizations in efforts to identify, establish, and maintain a representative network of protected tropical forest ecosystems on a worldwide basis; b) whenever appropriate, making the establishment of protected areas a condition of support for activities involving forest clearance or degradation; and c) helping developing countries identify tropical forest ecosystems and species in need of protection and establish and maintain appropriate protected areas;

(11) ..., engage in efforts to increase the awareness of United States Government agencies and other donors, both bilateral and multilateral, of the immediate and long-term value of tropical forests;

(12) ..., utilize the resources and abilities of all relevant United States Government agencies;

(13) Require that any program or project under this chapter significantly affecting tropical forests (including projects involving the planting of exotic plant species a) be based upon careful analysis of the alternatives available to achieve the best sustainable use of the land; and b) take full account of the environmental impacts of the proposed activities on biological diversity, as provided for in the environmental procedures of the Agency for International Development;

(14) Deny assistance under this chapter for a) the procurement or use of logging equipment, unless an environmental assessment indicates that all timber harvesting operations involved will be conducted in an environmentally sound manner which minimizes forest destruction and that the proposed activity will produce positive economic benefits and sustainable forest management systems; and b) actions which significantly degrade national parks or similar protected areas which contain tropical forests or introduce exotic plants or animals into such areas;

(15) Deny assistance under this chapter for the following activities unless an environmental assessment indicates that the proposed activity will contribute significantly and directly to improving the livelihood of the rural poor and will be conducted in an environmentally sound manner which supports sustainable development: a) Activities which would result in the conversion of forest lands to the rearing of livestock; b) the construction, upgrading or maintenance of roads (including temporary haul roads for logging or other extractive industries) which pass through relatively undegraded forest lands; c) the colonization of forest lands; and d) the construction of dams or other water control structures which flood relatively undegraded forest lands.

(d) PVOs and Other Nongovernmental Organizations

Whenever feasible, the President shall accomplish the objectives of this section through projects managed by private and voluntary organizations or international, regional, or national nongovernmental organizations which are active in the region or country where the project is located.

(e) Country Analysis Requirements

Each country development strategy statement or other country plan prepared by the Agency for International Development shall include an analysis of: (1)

The actions necessary in that country to achieve conservation and sustainable management of tropical forests; and (2) the extent to which the actions proposed for support by the Agency meet the needs thus identified.

(f) Annual Report

Each annual report required by section 634(a) of this Act shall include a report on the implementation of this section.

12

Managed Forest Gardens in West Kalimantan, Indonesia

Christine Padoch and Charles Peters

West Kalimantan, the third largest Indonesian province on the island of Borneo, encompasses a total land area of almost 147,000 km^2. Much of this land forms part of the Kapuas River watershed and lies directly on or very near to the equator (latitude 2°N to 3°S). The natural vegetation of the region is characterized by a variety of different forest types including mangroves, peat forest, freshwater swamp forest, heath forest, and lowland or hill mixed Dipterocarp forest. Some of these forests are reputed to be the oldest and most species-rich in all of Southeast Asia (1).

In spite of their extent and inherent diversity, very little is known about the forests of West Kalimantan relative to other areas of Borneo (e.g., East Kalimantan or the East Malaysian states of Sarawak and Sabah). The few floristic studies that have been conducted, however, suggest that the forests of this region are exceptionally rich in edible fruits, rattans, oil seeds, medicinal plants, resins, and other useful plant products. Leighton (2), for example, reports that the forests at Gunung Palung in the Ketapang district contain 21 species of wild mangosteen (*Garcinia* spp.), eight species of rambutan (*Nephelium* spp.), seven species of durian (*Durio* spp.), four species of mango (*Mangifera* spp.), and a host of lesser-known fruits such as rambai (*Baccaurea*; 23 species) and cempedak (*Artocarpus*; 13 species). Small tracts of forest may also exhibit a high abundance of useful plants. A 1.0 hectare plot of hill Dipterocarp forest inventoried in the Sambas district was found to contain three species of illipe nut (*Shorea* spp.), 25 species of edible fruits and nuts, 35 timber species, five species producing *damar* (oleo-resin) or other useful exudates, two species of rattan, three species whose leaves or bark are used medicinally, and one species used locally as a fish poison (3).

Much of the valuable germplasm growing in the forests of West Kalimantan is rapidly disappearing. Although reliable data on forest clearing are difficult to obtain, recent estimates suggest that almost 50% of the original forest land in this province has already been lost (1, 4); the destruction of lowland Dipterocarp forests and mangrove areas has been especially pronounced. Extensive logging, the establishment of large-scale plantations of industrial crops, and the increasing agricultural demands of a growing rural population appear to be the major factors responsible for this alarming reduction in forest area.

Intensive timber extraction in West Kalimantan was initiated soon after the enactment of the Basic Forestry Law in 1967. By 1970, there were approximately 400,000 hectares of forest being logged (5). Currently, there are over 5.8 million hectares of timber concessions in this province; the total area of each concession averages about 92,000 hectares (6). Almost all of the local freshwater swamp and mangrove forests have been designated for logging (7). In many cases, the actual boundaries of these concessions are poorly defined and large tracts of forest reserved specifically for conservation or watershed protection are also logged. The Directorate of Forest Planning (8), for example, estimates that almost 800,000 hectares of forest concession in West Kalimantan are composed of lands that are actually designated as nature reserves, national parks, wildlife sanctuaries, or protection forests. Over the last 10 years, local concessionaires have logged an average of 1.4 million cubic meters of wood per year from the forests of this region, making it the third largest producer of sawlogs in Indonesia (9).

In theory, the Indonesian Selective Felling System (*Tebang Pilih Indonesia*, or TPI) was designed to provide a sustainable flow of wood products with minimal ecological impact on the concession areas being exploited (11). Only timber trees with a minimum diameter of 50 cm can be cut, and at least 25 merchantable trees per hectare must be left to grow until the next cutting cycle. The concessionaire is also required to mark the boundaries of the area, make inventories before and after harvest, conduct felling and skidding such that logging damage is kept to a minimum, replant logged-over stands deficient in regeneration, and submit periodic production reports to the Ministry of Forestry (11). A small, nonrefundable reforestation fee or *dana reboisasi* (approximately U.S. $7.00 for every cubic meter of wood extracted) must also be paid to cover the costs of site rehabilitation.

In practice, the concession system in West Kalimantan, as well as in other parts of Indonesia, is plagued by frequent abuses. As an illustration of this, the Director General of Forest Exploitation recently announced that less than 20% of the 522 concession holders in the country abide by the official regulations governing timber extraction (12). One major problem is that the concessions are usually granted for 20 years, while the TPI specifies a 35-year cutting cycle. Concession holders know that they have only one opportunity to log a particular tract of forest, and there is, therefore, a strong economic incentive to extract as much timber as possible during each harvest operation. Similarly, reforestation activities are frequently viewed as an unnecessary expense because the concessionaire will be unable to harvest the timber when it is mature. Excessive logging damage is also a chronic problem. Unless properly controlled, removing as few as 10 trees/hectare from mixed Dipterocarp forest can damage over 50% of the residual trees, denude up to 30% of the soil surface, and significantly increase the rate of run-off and erosion from the site (13). In spite of the original objective of sustained-yield management, the actual result of much of the logging currently occurring in West Kalimantan has been the degradation and progressive impoverishment of local forests.

The areal extent of both government-sponsored and small-holder plantations of estate crops in West Kalimantan has also increased dramatically in recent years. Official estimates (14) indicate that over 388,000 hectares of rubber and 43,000 hectares of oil palm have been established in this province. Coconut plantations occupy 80,000 hectares, most of these planted on the extensive peat soils found near the estuary of the Kapuas River, and plantations of pepper, cloves, coffee, and assorted citrus contribute an additional 20,000 hectares. Given that many large-scale plantations have been established in areas formerly covered by species-rich forest, the increased production of industrial crops also results in a significant net loss of biodiversity.

The spread of swidden farming (also known as shifting cultivation or slash-and-burn) is frequently cited as the most important source of deforestation in Indonesia, and one recent study (15) estimates that over 2 million hectares of land in West Kalimantan are being farmed in this manner. It is difficult to assess the degree to which this practice is actually the cause of extensive forest degradation because only a small percentage of local swidden plots or *ladang* are created in undisturbed forest. The overwhelming majority of swidden fields are made in previously farmed sites or on lands that have been logged. A small percentage (approximately 170,000 hectares) occurs on lands that have been designated nature reserves or other protected forests, but this area is considerably smaller than the extent of logging concessions in conservation areas.

The rapid expansion of logging concessions and tree plantations has caused a notable reduction in the area available to local farmers for small-scale agriculture. In response to the pressure caused by this diminishing resource base, some swidden cultivators have moved into forested areas zoned for other purposes. This movement is facilitated by the construction of logging roads. Upon arrival at a suitable site, which is frequently logged-over forest, the valuable timber remaining may be felled and sold and the plot then cleared, burned, and planted. The common perception of swidden farming as a practice that necessarily leads to the invasion of Imperata grass and severe land degradation ignores the reality of traditional swidden systems in much of West Kalimantan. While Imperata has indeed become a problem in some areas, other zones have been used by swiddeners for centuries and yet remain relatively free of the noxious weed. Many of the indigenous farmers of inland West Kalimantan employ sophisticated techniques of forest management as part of their swidden cycle. Although frequently overlooked or ignored by government officials and policy makers, these management practices appear to make a significant contribution to the conservation of biodiversity in West Kalimantan.

CASE STUDY: TRADITIONAL SYSTEMS OF FOREST MANAGEMENT

The dozens of indigenous tribal groups found in Kalimantan are known collectively as Dayaks. Each of these groups exhibits fundamental differences in language, customary law, artistic expression, housing architecture, and social organization (16). There is also considerable variation in the traditional pattern and intensity of resource use among groups. Descriptions of Dayak resource use have frequently mentioned the practice of collecting forest products and planting tree crops. Very few studies, however, have focused on local systems for managing forest vegetation (17). An unfortunate result of this oversight is that, in most circles, Dayaks are still perceived as highly mobile, aggressive groups of slash-and-burners who live at low population densities and are, at best, indifferent resource managers. While this stereotype may indeed be valid in some limited areas of Borneo, this is certainly not an accurate portrayal of most of the Dayak groups found in West Kalimantan. Results of ongoing studies in the Sanggau district suggest that Dayaks in this region employ a multitude of techniques to both manage and conserve the limited areas of forest under their control.

One of our study sites and the focus of this chapter lies approximately 25 kilometers north of the Kapuas River in the Balai subdistrict of Sanggau (Fig. 1). As of 1989, the population of this subdistrict was estimated at 21,206 inhabitants (18), most belonging to a previously undescribed Dayak group we have tentatively designated the Tara'n Dayak of Balai. The population density of this region is approximately 54 people per square kilometer, or more than twice the provincial average. The area is characterized by low hills of intermediate to steep topography, separated by several broad valleys and numerous small rivers and streams. The natural vegetation is hill Dipterocarp forest.

The study area of Tae, a village comprising five hamlets, extends over 1,600 hectares and supports about 88 people per square kilometer. Although mature forest is confined largely to steep hillsides and ridgetops unsuitable for cultivation, much of the area appears wooded. Patches of secondary vegetation of varying age testify to the continuing practice of swidden cultivation in the village. However, in the hamlet of Tae (one of the five hamlets that make up the village of the same name), as well as in other parts of the Balai subdistrict, land-extensive swidden cultivation is waning in favor of land-intensive wet-rice or sawah farming.

Figure 1.

Figure 2 (above). The hamlet of Tae surrounded by woodlands. What appear to be forests are actively managed forest gardens.

Figure 3 (right). A mature *tembawang* forest garden along a path between Tara'n Dayak hamlets.

In March 1990, for example, we found that only 19 of the 37 households in Tae hamlet farmed any swiddens at all; about half relied only on irrigated or rain-fed permanent fields for their rice crops, while the others employed various combinations of both sawah and ladang cultivation. As Tara'n Dayak farmers gradually decrease their dependence on swidden cultivation and extend and intensify their use of permanent fields, their need to clear wooded slopes diminishes. Tae hillsides are being increasingly transformed into forest gardens, managed forests, or agroforestry plots. These three principal vegetation types reflect distinct differences in the origin, management priorities, and rights of access and inheritance of each stand. In Indonesian they are known, respectively, as *tembawang*, *tanah adat*, and *tanah usaha*.

Tembawang Forest Gardens

The term tembawang refers specifically to former house sites, and many forest gardens in Tae started out as mixed plantings of fruit trees around longhouses, individual dwellings, or temporary farm huts. Tae residents report that on average they change the site of their village about every 20 years, often moving only a few hundred meters away. According to several informants, the reason for these frequent moves is that the fruit trees become too tall and start to shade the front yard, thereby eliminating the principal function of this multi-purpose area, i.e., drying rice in the sun. Rather than cut or prune the fruit trees, residents prefer to rebuild their houses elsewhere. After several centuries of residence in the Tae area, the Tara'n Dayak of Balai have created a landscape dotted with forest gardens. Contiguous areas of tembawang may be 10 hectares or more in size; the total area of forest gardens in Tae amounts to several hundred hectares.

Older tembawang contain not only planted species, but also a mixture of spontaneous vegetation that has grown up around the fruit trees. This vegetation is further enriched by the occasional planting of rubber, construction timbers, rattans, and medicinal plants. Tara'n Dayak informants cite the most casual of planting techniques as important in the formation of tembawang. For example, they report that many trees sprout from seeds discarded by people eating fruit as they walk through the garden. Selective weeding, particularly around durian and illipe nut trees during harvest season to facilitate fruit collection, is an important management tool. During these weedings, the seedlings and saplings of especially valuable species will be spared while others are cut. In most tembawang, it is extremely difficult to distinguish between those trees that were intentionally planted and those that became established as volunteers and were later spared during weeding operations.

Many of these forest gardens are extremely rich in tree species. One 10 X 200 meter transect (0.2 ha) sampled in one tembawang contained 224 trees [>5.0 cm diameter at breast height (dbh)] or clumps of bamboo pertaining to 44 different species. Thirty of the species produce edible fruits or shoots. In addition to common market fruits such as durian, rambutan, langsat (*Lansium*) and mangosteen, the transect contained numerous examples of the semi-domesticated or wild relatives of these species. Sugar palms (*Arenga*), rubber trees (*Hevea*), illipe nuts (*Shorea* spp.), medicinal plants, and a variety of species useful for construction, e.g., bamboo and ironwood (*Eusideroxylon zwagerii*), were also tallied. The Tara'n Dayak we interviewed in Tae claimed to have planted at least 74 species of fruit trees and they identified more than 100 different species of edible fruits in their managed forests. Although the claim is never directly made by Tae residents, it is difficult to avoid the conclusion that tembawang are consciously managed for diversity.

Many of the useful plant resources contained in these forest gardens are sold. Illipe nuts, an important source of edible oil and a major forest-product export in West Kalimantan, are routinely harvested from tembawang wherever available. Within the last 10 years, Tara'n Dayak villagers have also started marketing a large quantity of durian fruits. During the height of durian season in 1991, more than 10,000 fruits were hauled to market (by motorcycle) from four of Tae's hamlets in a single day. Even at a minimal price of about 15 cents per fruit, the income earned by many families throughout the month-long season was substantial.

Figure 4. In the early morning a motorcycle is loaded with durians for sale in the market town of Batang Tarang.

Tanah Adat Forest Reserves

The second type of managed forest is termed tanah adat, or land preserved by traditional law. While these small areas of forest appear similar to tembawang, they were created in a totally different way. Tanah adat usually represents forest that was set aside many generations ago and has never been cleared. Tae hamlet has four such sites with a total area of about 12 to 13 hectares. Two of the tanah adat sites include village cemeteries.

Figure 5. Tara'n women carrying durians they have harvested in *tembawang* near Tae.

Although traditional law prohibits the clearing of tanah adat, these forest areas are heavily managed by the Tara'n Dayak of Balai. A variety of valuable species such as fruit trees and construction timbers are planted, while less useful trees that are interfering with useful species may be removed. The planting of durian trees to commemorate or assist the dead is a particularly common practice in the plots that serve as cemeteries. Forest composition is further enriched through periodic selective weeding at fruit harvest time. The results from a 20 X 100 meter transect sampled in tanah adat suggest that these areas are especially important as a communal source of timber for house construction. Among the 229 trees and 55 species recorded in the transect, 30 species were pointed out as valuable construction woods. Additional species of interest encountered in the forest reserve include illipe nut, rattan, sugar palm, bamboo, and a variety of different native fruits such as durian, langsat, rambai, and mentawa (*Artocarpus*). The fruits from the durian and illipe nut trees are collected and sold; most of the other plant resources found in tanah adat are destined for household use.

Cyclic Tanah Usaha Plantings

The final category of managed vegetation in the village of Tae is referred to as tanah usaha or commercial plantings. These areas differ from the two previous types in that they are usually part of a cyclic agroforestry system wherein swidden plots planted to annual crops are gradually transformed into stands of trees. Such land management patterns have recently received considerable attention in the Latin American tropics (19). Similar cyclic shifts are apparently a common component of Bornean systems (20). The perennial crops planted by Tae swiddeners, usually rubber mixed with a varied assortment of fruit trees, are planted as the hill rice is ripening. After 30 years or so, when the rubber trees have passed their peak production and the demand for rice intensifies, these plots are cleared and burned to make ladang and the cycle is started again.

Rubber is the most important and common tree species planted in tanah usaha. Hevea seedlings were first acquired by Tara'n Dayak farmers in the late 1930s or early 1940s, and, as in many Bornean communities, this crop has provided an important income supplement to local farmers ever since (21). All of the households in Tae have substantial rubber gardens that produce a minimum of three kilograms of latex per day; carefully tended plots can yield a daily production of up to eight kilograms. It should be noted, however, that rubber is not the only commercial crop planted in tanah usaha, and many villagers have also incorporated cacao and other market fruits in their plots. Areas close to the village are often far more diverse in species, resembling tembawang in terms of the number and variety of fruit trees present. Those too far away to be conveniently harvested contain a smaller number of fruits.

FOREST GARDENS AND THE CONSERVATION OF BIODIVERSITY

Recent literature related to the conservation of biodiversity in Indonesia cites the importance of local community involvement in development (22). This strategy represents a radical departure from the more traditional approach to forest conservation that relies heavily on patrolling, law enforcement, and the total exclusion of local communities from designated protection areas. The exact details of this community involvement have yet to be defined and surely will differ greatly from region to region. There is general agreement, however, that the planning of conservation areas must begin to consider the activities and needs of local people, particularly boundary communities, in a participatory approach to the management and conservation of natural resources. Of particular interest is the development of appropriate management systems for these areas.

The traditional systems of forest management employed by the Tara'n Dayak of Balai would seem to be especially promising prototypes in this regard. Forest gardens are clearly less diverse than most "natural" forests (23), especially in larger animal fauna. They do, however, present significant advantages relative to more prevalent forms of land use practiced in West Kalimantan (e.g., selective logging or total forest conversion). These managed forests provide soil protection, maintain the hydrological functions of local watersheds, and offer habitat to certain, albeit limited, types of fauna. Based on the long history of this practice in Tae, forest gardens also appear to be operated on a sustained-yield basis. Perhaps their greatest importance, however, is that both tembawang and tanah adat are management systems that actually promote the *in situ* conservation of important cultivars and their wild relatives. The maintenance of genetic diversity is a cornerstone of Tara'n Dayak resource management. Instead of planting only one species of illipe nut, they plant nine. They plant commercial rambutan, yet they also favor the regeneration and growth of several wild species of Nephelium. Finally, the fact that managed forest gardens provide food and are an increasingly important source of both income and various subsistence products is a powerful incentive for local communities to protect and care for these areas.

In spite of these obvious ecological and economic benefits, the continued existence of the forest gardens in Tae and other Dayak communities in West Kalimantan is severely threatened. One reason for this is the inherent "invisibility" of these traditional management practices. None of the foresters or agronomists who staff provincial or regional government acknowledge that what Tae residents do should even be called forest management. Tembawang and tanah adat do not looked managed. The plantings are not even-aged or evenly spaced and they appear haphazard. Spontaneous vegetation growth is not only tolerated but encouraged in these stands. Ecologically, this fostering of diversity is perhaps the most attractive feature of forest gardens. From political and economic standpoints, however, it represents tenuous middle ground. Forest gardens are viewed as far too managed for those interested in nature conservation, but not managed enough for those interested in crop production. They are dismissed by many as primitive and are easily encroached upon by development schemes for enhanced commercial production, especially monocultural plantations of oil palm or improved rubber. Given their diversity, the realization of the full commercial potential of managed forests would require the development of a wide range of processing and marketing facilities. The economic value of their genetic riches is also difficult to establish or compare to the worth of other enterprises, such as a palm oil plantation. As long as the management of diversity remains an issue of secondary importance in West Kalimantan, the traditional expertise of the Tara'n Dayak of Balai will probably continue to be overlooked and unappreciated.

Official recognition of the value of tembawang and other managed forests would not guarantee their continued prosperity, but would certainly lessen pressures for the substitution of monocultures in the name of "economic progress." Even more effective, however, might be programs for economically upgrading diverse stands with improved varieties of commercial species that do well under such circumstances. The Tara'n Dayak of Balai are no different from other folk in desiring economic betterment of their region. They have eagerly adopted improvements in irrigation for rice and economically valuable species such as rubber. Agricultural advances that work within their systems of forest vegetation management would surely be embraced by these wise and innovative farmers.

REFERENCES AND NOTES

1. FAO. 1981. *National conservation plan for Indonesia, Vol. V: Kalimantan*, Field Report No. 17, INS/78/061. Bogor: Food and Agricultural Organization.
2. Leighton, M. 1990. Seed dispersal syndromes of Bornean fruits: The dependence of plants on birds and mammals for seed dispersal and forest regeneration. Research Report. Jakarta: Center for Research and Development in Biology-LIPI and Subdirectorate of Nature Conservation (PHPA).
3. Peters, C.M. 1991. Illipe nuts in West Kalimantan: Use, ecology and management potential of an important forest resource. Paper presented at Conference of Interactions of People and Forests in Kalimantan. New York: New York Botanical Garden.
4. Sutter, M. 1989. *Forest resources and land use in Indonesia*. FAO Field Document No. I-1, UTF/INS/065. Jakarta: Food and Agricultural Organization; RePPProt. 1990. *The land resources of Indonesia: An overview*. Regional Physical Planning Program for Transmigration. Land Resources Development Center, ODA, Foreign Commonwealth Office and Direktorat Jenderal Penyiapan Pemukiman, Departemen Transmigrasi, Jakarta.
5. MOF. 1989. *Data Perkembangan Hak Perusahaan Hutan*. Jakarta: Ministry of Forestry.
6. FAO. 1989. *Forest management for sustainable production and conservation in Indonesia*. Field Document No. I-2, UTF/INS/065. Jakarta: Food and Agricultural Organization.
7. Silvius, M.J., *et al.* 1987. *The Indonesian wetland inventory*. Bogor: PHPA,AWB/INTERWADER.
8. Directorate of Forest Planning. 1981. *Report on the forests of Indonesia*. Publication No. 18. Jakarta: Ministry of Forestry, Directorate of Forest Planning.
9. Statistik Kehutanan. 1980–1990. *Statistik Kehutanan Indonesia*. Jakarta: Directorate General of Forest Utilization.
10. In some situations, clearcutting of forest concessions is also permitted in Indonesia. Two variants of this form of management, the Clearcutting with Natural Regeneration system (Tebang Habis Permudaan Alam or THPA) and the Clearcutting with Planting system (Tebang Habis Permudaan Buatan or THPB) are authorized by the Ministry of Forestry. The THPA system is designed for use in forests displaying low species diversity and abundant regeneration of commercial species. Given the rarity with which these conditions occur in Dipterocarp forests, the system has not been applied on any significant scale since its approval in 1972 (6). The Clearcutting with Planting system, on the other hand, is routinely used to convert severely degraded natural forest into industrial plantations (Hutan Tanaman Industri). Both systems inevitably cause a drastic reduction in biodiversity.
11. DGF. 1972. Surat Keputusan No. 35. Jakarta: Director General of Forestry.
12. Indonesian Observer. 1990. "Only twenty percent of forest concessionaires rated as good." November 30, 1990. Jakarta.
13. Hamzah, Z. 1978. *Proc. Symp. on the long-term effects of logging in Southeast Asia*. BIOTROP Special Publ. No. 3, 27–34; Tinal, U., and J.L. Palinewen. *Ibid.*, 91–96; Abdulhadi, R., K. Kartawinata, and S. Sukardjo. 1981. *Malay. For.* 44:407–8.
14. Kantor Statistik. 1990. *Kalimantan Barat Dalam Angka*. Pontianak: BAPPPEDA and Kantor Statistik.

15. RePPProt. 1990. *The land resources of Indonesia: An overview*. Regional Physical Planning Program for Transmigration. Land Resources Development Center, ODA, Foreign Commonwealth Office and Direktorat Jenderal, Penyiapan Pemukiman, Departemen Transmigrasi, Jakarta.
16. Ave, J.B., and V.T. King. 1986. *Borneo: The people of the weeping forest*. Leiden: National Museum of Ethnology.
17. A few that have are Morris, H.S. 1953. *Report on a melanau sago producing community in Sarawak*. London: Her Majesty's Stationery Office; Chin, S.C. 1985. *Agriculture and resource utilization in a lowland rainforest Kenyah Community*. Kuching: Sarawak Museum; Weinstock, J.A. 1983. *Econ. Bot.* 37:58–68.
18. Kantor Statistik. 1989. *Kabupaten Sanggau Dalam Angka* Sanggau: Kantor Statistik.
19. For instance: Denevan, W.M., and C. Padoch, eds. 1988. Swidden-fallow agroforestry in the Peruvian Amazon. *Advances in Economic Bot.* 5. New York: New York Botanical Garden; Padoch, C., and W. de Jong. 1987. In *Agroforestry: Realities, Possibilities, and Potentials*, H.L. Gholz, ed. Dordrecht: Martinus Nijhoff; Posey, D.A. 1982. *Garden* 6(1):18–24; Alcorn, J. 1984. *Huastec Mayan Ethnobotany*. Austin: University of Texas.
20. Padoch, C. 1989. *Expedition* 30(1):18–28; Chin, S.C. 1985. *Agriculture and resource utilization in a lowland rainforest Kenyah Community*. Kuching: Sarawak Museum.
21. For instance, the Kantu of West Kalimantan (Dove, M.R. 1985. *Swidden agriculture in Indonesia: The subsistence strategies of the Kalimantan Kantu*. Berlin: Walter de Gruyter/Mouton.) and the Iban of Sarawak (Padoch, C. 1982. *Migration and its alternatives among the Iban of Sarawak*. Leiden: Instituut voor Taal-, Land-, en Volkenkunde/Martinus Nijhoff).
22. For instance: Petocz, R.G. 1987. *Conservation in Indonesia: Current status and development of an action strategy*. Washington, DC: World Bank; ADB. 1991. *Conservation of tropical forest ecosystems and biodiversity*. Report TA 1430-INO. Asian Development Bank; BAPPENAS. 1991. *Biodiversity action plan for Indonesia*. Jakarta: BAPPENAS.
23. Although the floristic diversity of Bornean forests is quite variable, detailed studies have shown that small tracts of mixed Dipterocarp forest may contain from 200 to 250 tree species per hectare (e.g., Kartawinata, K., R. Abdulhadi, T. Partomihardjo. 1981. *Malay. For.* 44:397–406; Proctor, J., J.M. Anderson, P. Chai, H.W. Vallack. 1983. *J. Ecol.* 71:237–60; Ashton, P.S. 1984. In *Plant systematics*, W.F. Grant, ed. New York: Academic Press; Riswan, S. 1987. In *Proc. Third Round Table on Dipterocarps*, A.J.G.H. Kostermans, ed. Bogor: UNESCO. The heath or kerangas forests found on sandy podzolic soils commonly exhibit 25 to 100 tree species per hectare (S. Riswan, op cit. [1987]; P.S. Ashton, op. cit. [1984]); peat forests are generally reported to be the lowest in terms of species richness (Anderson, J.A.R. 1976. *Soil Res. Inst. Bogor Bull.* 3: 45–55; Whitmore, T.C. 1984. *Tropical forests of the Far East*. Oxford: Clarendon Press. The tembawang and tanah adat surveyed in the present study contained 42 and 51 tree species per 0.2 hectare plot, respectively.

13

The Ancient Forests of the Pacific Northwest: A Case Study in Conservation and Economic Development

David S. Wilcove and Jeffrey T. Olson

Within the last two years, the protection and harvesting of old-growth and ancient coniferous forests in the Pacific Northwest has emerged as one of the most heated environmental battles in the United States (1). The controversy has been portrayed as a choice between economic health and environmental protection—a choice between spotted owls and jobs. Although this simple dichotomy makes for good headlines, it does not convey the complexity and significance of the issue. The ancient forest/spotted owl debate reflects a profound change in the way a growing number of people value the goods and services derived from forest ecosystems. It is also a reflection of changing economic conditions in the Pacific Northwest, as that region moves away from its traditional dependence on timber extraction toward a more diverse, urban economy.

ECOLOGICAL VALUES OF OLD-GROWTH FORESTS

The old-growth coniferous forests of the Pacific Northwest are unrivaled in both the size and longevity of individual trees and in the accumulation of biomass within individual stands (2). They are also rich in numbers of plant and animal species. In western Washington and Oregon, for example, over 200 vertebrate species breed or forage within mature and old-growth stands (3). The number of species of invertebrate animals, vascular plants, fungi, and bacteria found in ancient forests is unknown but undoubtedly enormous. The ground litter and soil of these forests are among the most species-rich sites on earth, with a single square yard holding as many as 200,000 mites from a single suborder, plus tens of thousands of other mites, beetles, centipedes, pseudoscorpions, springtails, and spiders (4). Fungi species in old-growth forests number in the thousands (5).

In western Washington and Oregon, over 45 species of vertebrates are either dependent upon or closely associated with old-growth forests (Table 1). The best known of these is the northern spotted owl (*Strix occidentalis caurina*), now

Table 1. Numbers of species closely tied to ancient forests in select taxa (4). Data are for the Pacific Northwest region only.

Taxon	Number of Species Closely Associated with Old-Growth Forests	Old-Growth Species on Federal List of Endangered Species	Old-Growth Species that are Candidates for Federal Listings
Mammals	12		1
Birds	20	1	1
Amphibians & Reptiles	15		2
Mollusks	43		5
Vascular Plants	48		

protected under the federal Endangered Species Act. The number of species in other taxa that are dependent upon ancient forests is unknown, in large part because the taxonomy and distribution of invertebrate organisms are not known. Forty-three species of mollusks are thought to be closely tied to ancient forests. Many are endemics, restricted to individual creeks and seeps. Most have been described only within the last decade, and many more presumably await discovery. During the spring and summer of 1991, for example, an endemic cluster of five undescribed species in the genus *Fluminicola* was discovered in the Siskiyou Mountains on the California-Oregon border. The five species appear to be restricted to two creeks and associated springs (4).

The ecological value of old-growth forests goes well beyond their role as habitat for rare species. Thirteen species of Pacific Northwest conifers routinely exceed 200 feet in height; among them are some of the world's most important timber species. Temperate, wind-pollinated trees, especially conifers, are among the most genetically variable organisms known (6). Old-growth forests therefore represent an invaluable storehouse of genetic diversity. While seed collections, clonal orchards, and other *ex situ* techniques maintain genetic variation to some extent, only *in situ* protection of natural forests maintains the structure of that diversity. Thus, preservation of old-growth forests protects both the content and context of genetic diversity. This genetic diversity is likely to become an increasingly important response mechanism as greenhouse gases, acidic precipitation, and other waste products of human activities alter the climate and atmosphere over large areas of the earth and as new pests and pathogens spread to the Pacific Northwest. Genotype variation may determine which commercially valuable trees are able to persist in areas subjected to these changes.

Watersheds dominated by old-growth forests provide clean drinking water to communities throughout western Washington and western Oregon, and streams flowing through ancient forests provide spawning habitat for anadromous fish and other aquatic organisms. The commercial harvest of salmonids from spawning grounds in federal forests in Oregon and Washington is conservatively valued at $40 million annually (7). Fish populations are vulnerable to the environmental changes associated with commercial logging. Logging roads can modify natural drainages and accelerate erosion, leading to harmful changes in stream flow, channel configuration, sediment transport and storage, and substrate composition. Timber harvesting can increase sedimentation rates and water temperature, change seasonal flow patterns, and alter the quantity and distribution of woody debris in streams, all of which rebound to the detriment of anadromous fish. Recent surveys indicate that over 760 fish stocks in Washington, Oregon, and northern California are in danger of extinction (4).

Finally, thousands of people are drawn to ancient forests every year to engage in outdoor activities such as hiking, fishing, hunting, and camping. In addition to enhancing the lives of participants, these activities pump money into local communities. Because so many of the values of old-growth forests are not traded in the marketplace, and others remain to be discovered, the full value of this ecosystem cannot be calculated (8, 9). We shall return to this point when we consider the forces driving policy decisions about ancient forests.

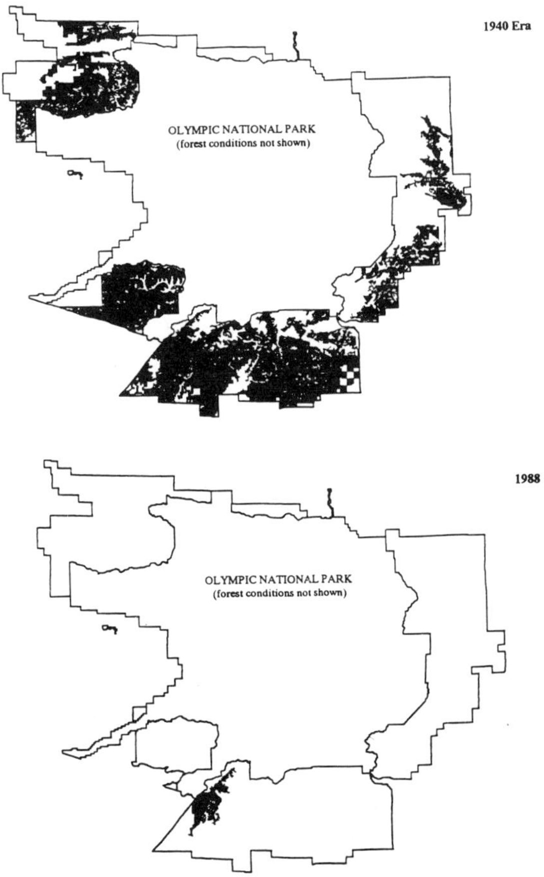

Figure 1. Loss of old-growth forests in Olympic National Forest Between 1940 and 1988. Old-growth forests are indicated in black. The area in the center of the figure is Olympic National Park, where logging is prohibited. Areas on the periphery of the park constitute the national forest, where extensive logging of old growth has occurred. From P. Morrison, (12).

CURRENT STATUS OF OLD-GROWTH FORESTS

No comprehensive inventory of Pacific Northwest old growth exists. The most accurate data on past and present amounts of old growth are available for Washington and Oregon west of the Cascade Mountains. Prior to the arrival of white settlers, this region contained an estimated 15 to 19 million acres of old-growth forest (10, 11). Inventory and mapping data compiled by The Wilderness Society indicate that only about 2.1 million acres now remain, an 86 to 89% decrease from historical levels (12). Most of this loss has occurred since World War II (Fig. 1).

In western Washington and Oregon, virtually all of the remaining old growth is found on federal lands. Approximately 35% occurs within national parks or designated wilderness areas and is therefore protected from logging. The protection vs. harvesting controversy centers around the status of the other 65%, which is largely open to logging. In recent years, over 87,000 acres of mature and old-growth forest have been logged annually on federal lands in Washington and Oregon alone (13).

Timber harvest techniques in the Pacific Northwest fall into two broad categories: uneven-aged and even-aged management (14). In uneven-aged management, trees are harvested singly or in small groups. This system results in a forest of trees of many ages. In even-aged management, the harvest is designed to allow regeneration to occur throughout the site at once, so trees are cut at approximately the same time. In its simplest form, clearcutting removes all of the trees on the site, essentially resetting the successional clock to zero. Regrowth can be accomplished by planting seedlings (a form of artificial regeneration) or by allowing seed stock from adjacent, uncut areas to colonize the logged area (natural regeneration). On federal lands, even-aged techniques are used far more often than uneven-aged management, with clearcutting as the predominant harvest method. Harvest cycles are rarely lengthy enough to permit the development of significant amounts of woody debris, an important structural component of ancient forests.

The pattern of logging is also significant. In national forests, the practice has been to disperse cutting units across the landscape to minimize their visual and (presumably) ecological impact. This has led to extensive fragmentation of old-growth forests, thereby increasing the risk of extinction for certain species (15) and the magnitude of various edge-related phenomena (i.e., blowdown, desiccation) that damage old-growth forests (16).

IMPACT ON GENETIC DIVERSITY

Silviculture

Whether the genetic diversity of tree species is diminishing as a result of national forest management practices is difficult to judge. Few studies have explored this issue, and the data are expensive to obtain. Nonetheless, few biologists would deny the inherent potential of contemporary forestry practices to diminish genetic diversity.

All silvicultural actions are likely to affect the genetic composition of timber stands, although not necessarily in deleterious or irreversible ways. Logging by its very nature is a direct selective force on trees causing the loss of certain individuals or undesirable species from the population. Genetic diversity can also be disrupted by forestry practices that result in the creation of smaller, reproductively isolated populations of trees or that alter the physical environment in which trees grow and reproduce. In the United States, a number of silviculturists have expressed concern that genetic changes due to current forestry practices will diminish the vigor and productivity of domestic forests (20).

With artificial regeneration, changes in the forest gene pool are inevitable. In most cases, in fact, an explicit goal of artificial regeneration is to "improve" the genetic composition of timber stands by replacing existing trees with stock selected to boost timber yields. In addition to intended genetic changes, artificial regeneration can have a number of unintended or ancillary consequences. For example, seeds obtained from tree orchards may suffer from a reduction of the genetic diversity as selection is intensified to collect from trees that tolerate grafting, perform well under orchard conditions, or respond to treatments to stimulate early and frequent seed production.

The Forest Service has developed seed transfer guidelines and planting zones to ensure that sites are replanted with locally adapted trees. These guidelines are an important step in protecting genetic diversity. Unfortunately, anecdotal comments and at least one survey of seed availability for Douglas-fir in California (21) indicated that seeds from the proper zones and elevation ranges often were not available for planting in areas where harvests were planned.

New Forestry—Solution or Illusion?

In recent years, some forest ecologists have advocated an alternative approach to forest management in the Pacific Northwest, one that attempts to better accommodate ecological values while still permitting commodity extraction. Often called "New Forestry," it has been described by its supporters as "a kinder and gentler forestry" (17). The goal of New Forestry is to maintain biological diversity and resilience by mimicking the dynamics of forest ecosystems. Central to this approach is the retention of "biological legacies" on harvest sites; these include fallen and standing coarse woody debris, green trees, and other elements that add structural diversity to forests (17). Timber sales are designed to minimize fragmentation of residual forest stands, often by permitting larger and more closely spaced cutting units. Longer rotations and less destructive methods of site preparation are also components of New Forestry (18).

New Forestry has met with skepticism by a variety of interest groups. Many environmentalists are wary of it, fearing it will be used to justify continued logging of old-growth forests. They view New Forestry as a promising concept for enhancing the biological diversity of existing plantations and other intensively managed stands, but will not accept further losses of old-growth forests. Among some foresters, there is concern that New Forestry techniques will significantly reduce timber yields, increase the costs associated with logging and timber management, and exacerbate problems with insect outbreaks and fires (19). Some of this skepticism surely stems from the vagueness of the concept. Supporters have been slow to advance a quantitative definition (or recipe) for New Forestry, although in fairness to them, the concept itself is evolving in response to new research.

THE COST OF PRESERVING OLD-GROWTH FORESTS

The most thorough assessments of the costs of preserving ancient forests have been made in connection with plans to protect the northern spotted owl. In April 1990, a panel of scientists from the Forest Service, Fish and Wildlife Service, and Bureau of Land Management (BLM) concluded that the owl was "imperiled over significant portions of its range" and that "[c]urrent management strategies are inadequate to ensure its viability" (15). The panel recommended that approximately 3.6 million acres of Forest Service and BLM timberland be set aside (i.e., removed from the commercial timber base) to maintain a viable population of owls in Washington, Oregon, and northern California. Only a portion of this acreage, however, consists of ancient forests (12); much consists of younger stands that could become older forests with the passage of time. The panel's recommendations became the basis for the draft northern spotted owl recovery plan developed by the Fish and Wildlife Service. Since May 1990, at least six studies have evaluated the cost of protecting spotted owl habitat. These costs have been expressed primarily in terms of job losses resulting from reduced supplies of timber; they range from 22,000 direct and indirect jobs (22) to more than 130,000 direct, indirect, and induced jobs (23).

Mead *et al.* (24) estimated the net social cost of preserving old growth for the spotted owl at $26 billion over the next 50 years—primarily as a result of foregone logging opportunities. Job losses would not be evenly distributed across the region. Rather, they would fall disproportionately on communities near the national forests, many of which are small, rural communities with limited ability to replace high-paying jobs in the timber industry with comparable jobs in other sectors. Yet even in rural counties in the Pacific Northwest, employment growth in all sectors other than wood products manufacturing has exceeded the national average (25). Such strong employment growth promises to more than compensate for most reasonable projections of employment loss resulting from old-growth protection.

Of the six studies mentioned earlier, only Mead *et al.* (24) provided any estimate of the benefits of old-growth preservation; most did not even acknowledge that such benefits exist. Even Mead's analysis did not take into account the full range of values associated with old-growth forests, but focused exclusively on recreational benefits.

Although the economic studies have ignored the benefits of protecting old-growth forests, a significant portion of the American public believes these benefits outweigh the costs of preservation. In a paper presented at the 1990 Pacific Northwest Economics Conference, Hagen and Vincent (26) presented estimates of the threshold social value that would make preservation of owl habitat economically efficient. In their most extreme case, Hagen and Vincent found that society would have to receive benefits equivalent to $11.52 per U.S. citizen to equal the costs of habitat preservation; this totals approximately $2.8 billion. Yet Rubin *et al.* (27), using the Contingent Evaluation Method and a 1987 survey of 1,200 randomly selected individuals, estimated that consumers were willing to pay a present value equivalent of $32 billion to ensure the spotted owl's survival. Their estimate exceeds Mead's cost estimate by $5 billion and is significantly greater than the threshold value calculated by Hagen and Vincent.

CHANGING ECONOMY OF THE PACIFIC NORTHWEST

Ongoing changes in the magnitude and structure of the timber industry in the Pacific Northwest, coupled with growth in the regional economy, reduce the economic potential of old-growth forests, thereby lowering the economic cost of preserving them. Recent years have seen a general reduction in the availability of mature timber from private lands in the Pacific Northwest (28), a result of logging at levels that exceed the rate of tree growth (29). Consequently, mills have had to adjust to a diet of smaller diameter logs. Smaller diameter logs carry with them higher processing costs on a board-foot basis. Therefore, timber companies in the Pacific Northwest now face stiff competition from the southeastern United States, where the gentler terrain allows for a more efficient harvest of trees.

In order to remain competitive with the southern states, the Pacific Northwest timber industry has modernized and automated. State employment data for western Washington and western Oregon show that the industry has reduced its work force on the west side of the Cascade Mountains by 24,500 people over the last 10 years while simultaneously boosting lumber and plywood production by 11% (30). Simultaneously, the economic base in the Pacific Northwest has been undergoing a fundamental transition to a more diverse, urban economy. All sectors of the economy have grown rapidly, with the notable exception of the wood products industry (Fig. 2). This growth has been facilitated by the region's perceived "quality of life," location relative to the Pacific Rim, and transportation facilities to

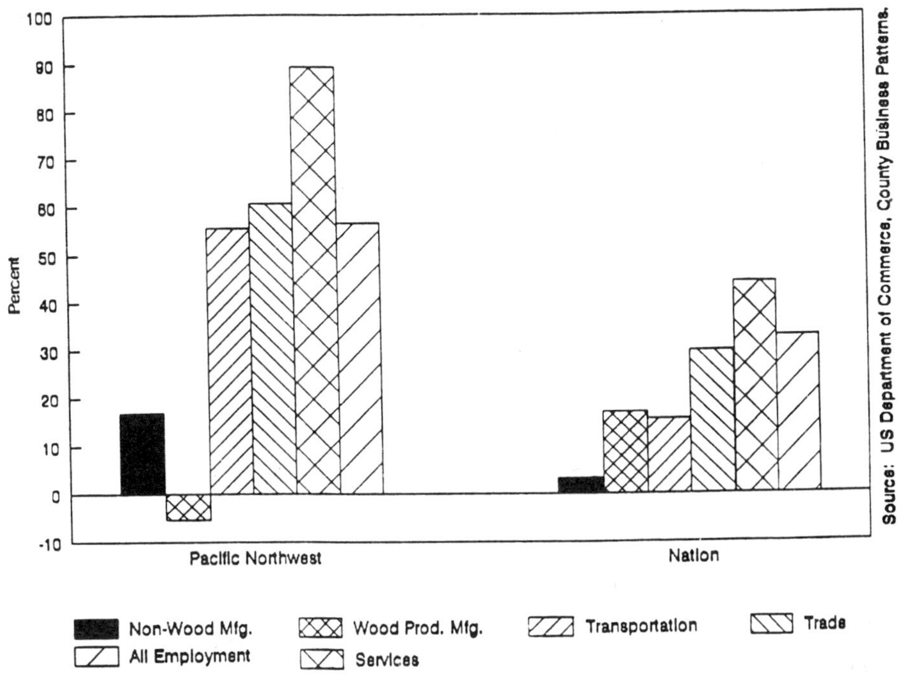

Figure 2. Percent change in employment by sector in the Pacific Northwest from 1975 to 1987. In recent years, the Pacific Northwest has experienced faster economic growth than the nation as a whole. All sectors of the regional economy, with the notable exception of the wood products industry, grew during this period. As a result, the importance of the wood products industry to the economy of the Pacific Northwest has declined.

major domestic markets (31). Growth in the region has been strong, with an average of over 98,000 new jobs created annually since 1986.

As a result of these two trends—automation of the timber industry and growth of the regional economy—the importance of the wood products industry to the regional economy has declined. In 1978, the logging and solid wood products industry accounted for 7% of the gross product of Washington and Oregon. By 1987, that figure had dropped to 3.5% (32).

As timber companies in the Pacific Northwest try to remain competitive with their southern counterparts, they will continue to modernize their mills. This automation process will mean the loss of more jobs, irrespective of the amount of logging that occurs on federal forestlands, and these job losses will far exceed any losses that might result from environmental protection. Nontimber elements of the economy will increase in importance as the timber industry fades, and federal forestlands will be under greater pressure to provide more in the way of nontimber benefits—such as wildlife habitat, drinking water, and opportunities for outdoor recreation—to the people of the Northwest and the nation as a whole. As long as the rural, timber-dependent communities continue to rely on a shrinking timber industry, their future is clouded by the growing competitive disadvantage of that industry. Alternatively, they can attempt to diversify their economic base to take advantage of the opportunities presented by the nontimber benefits of the national forests, which include the remaining old-growth forests. In this respect they have been ill-served by most of the economic analyses produced thus far, which have magnified the costs of preserving old-growth forests while minimizing or even ignoring the benefits. The result has been continued destruction of a scarce and irreplaceable natural resource.

REFERENCES AND NOTES

1. The term "old growth," as used here, follows the definition developed by the USDA Forest Service's Old-Growth Definition Task Group. See Old-Growth Definition Task Group, Interim Definitions for Old-Growth Douglas-fir and Mixed-conifer Forests in the Pacific Northwest and California. Research Note PNW-447. (USDA Forest Service, Pacific Northwest Research Station, Portland, Oregon, 1986). Key components of this definition (for Douglas-fir forests on western hemlock sites) are the presence of at least 8 large trees per acre equal to or greater than 32 inches in diameter at breast height (dbh) or 200 years in age; a deep, multilayered canopy; at least 12 trees of shade-tolerant species per acre equal to or greater than 16 inches dbh; at least 4 snags per acre equal to or greater than 21 inches dbh and 15 feet in height; and at least 15 tons of logs per acre, including 4 pieces per acre equal to or greater than 50 feet in length and 24 inches in diameter. The term "ancient forest" is a nontechnical term referring to both old-growth forests and virgin forests that contain most but not all of the characteristics of old growth.
2. Waring, R.H., and J.F. Franklin. 1979. *Science* 204:1380–86.
3. U.S. Department of Agriculture Forest Service. 1988. Final Supplement to the environmental impact statement for an amendment to the Pacific Northwest regional guide. Spotted owl guidelines. Portland, OR: USDA Forest Service, Pacific Northwest Region.
4. U.S. Department of the Interior. 1992. Draft recovery plan for the northern spotted owl. Washington, DC: USDI.
5. Denison, W.C. 1990. In *Ancient forests of the Pacific Northwest*. E. Norse, ed., 114–24. Washington, DC: Island Press.
6. Rudolph, S.G. 1990. In *Ancient forests of the Pacific Northwest*. E. Norse, ed., 129–32. Washington, DC: Island Press.
7. Frissel, C.A. 1990. Report to The Wilderness Society, Washington, DC: The Wilderness Society.
8. Randall, A. 1988. In *Biodiversity*. E.O. Wilson, ed., 217–23. Washington, DC: National Academy Press.
9. Ehrenfeld, D. 1988. In *Biodiversity*. E.O. Wilson, ed., 212–16. Washington, DC: National Academy Press.
10. Spies, T.A., and J.F. Franklin. 1988. *Natural Areas J.* 8:190.
11. Norse, E. 1990. *Ancient forests of the Pacific Northwest*. Washington, DC: Island Press.
12. P. Morrison, The Wilderness Society, personal communication. See also Morrison, P. 1988. *Old growth in the Pacific Northwest: A status report*. Washington, DC: The Wilderness Society; Morrison, P. 1990. *Ancient forests on the Olympic Peninsula: Analysis from a historical perspective*. Washington, DC: The Wilderness Society; Morrison, P., *et al.* 1991. *Ancient forests in the Pacific Northwest: Analysis and maps of twelve national forests*. Washington, DC: The Wilderness Society.
13. Anderson, D.R., *et al.* 1990. Status review: Northern spotted owl, Strix occidentalis caurina. Washington, DC: USDI Fish and Wildlife Service.
14. Norse, E., *et al.* 1986. *Conserving biological diversity in our national forests*. Washington, DC: The Wilderness Society.
15. Thomas, J.W., *et al.* 1990. A conservation strategy for the northern spotted owl. Report of the Interagency Scientific Committee to address the conservation of the

northern spotted owl. Portland, OR: USDA Forest Service, USDI Bureau of Land Management, USDI Fish and Wildlife Service, and USDI National Park Service.

16. Franklin, J.F., and R.T.T. Forman. 1987. *Landscape Ecology* 1:5.
17. Franklin, J. 1989. *American Forests* 95(11–12):37.
18. Perry, D.A. 1988. *J. of Pesticide Reform* 8(3):8.
19. Atkinson, W.A. 1990, May. Paper presented at the annual meeting of the Oregon Society of American Foresters, May 4, 1990. Eugene, OR.
20. Millar, C. 1987. *Conservation Biology* 1:191; California Gene Resource Program. 1982. Douglas-fir genetic resources: An assessment and plan for California. Berkeley, CA: National Council on Gene Resources; Silen R., and I. Doig. 1976. *Pacific Search* 10(8):7.
21. California Gene Resource Program. 1982. Douglas-fir genetic resources: An assessment and plan for California. Berkeley, CA: National Council on Gene Resources.
22. Hamilton, T., coord. 1990. Economic impact of implementing the recommendations of the Interagency Scientific Committee. Washington, DC: USDA Forest Service and USDI Bureau of Land Management.
23. Beuter, J.H. 1990. Society and economic impacts in Washington, Oregon, and California associated with implementing the conservation strategy for the northern spotted owl: An overview. Washington, DC: American Forest Resource Alliance.
24. Mead, W.J., *et al.* 1990. The economic consequences of preserving old-growth timber for spotted owls in Oregon and Washington. Santa Barbara, CA: Community and Organization Research Institute, University of California.
25. Anderson, H.M., and J.T. Olson. 1991. Federal forests and the economic base of the Pacific Northwest: A study of regional transitions. Washington, DC: The Wilderness Society.
26. Hagen, D.A., and J.W. Vincent. 1990, April. In Proceedings of the twenty-fourth annual Pacific Northwest regional economics conference. Bellingham, WA. In press.
27. Rubin, J., *et al.* 1990. A benefit-cost analysis of the northern spotted owl: Results from a contingent valuation survey. Davis, CA: University of California.
28. Olson, J.T. 1988. *National forests: Policies for the future*. Vol. 4. Pacific Northwest lumber and wood products: An industry in transition. Washington, DC: The Wilderness Society and the National Wildlife Federation.
29. Adams, D.M., and R.W. Haynes. 1989. The future of the Douglas-fir region forest economy: Potential development under changing public policies and private resources. Center for International Trade in Forest Products. Seattle, WA: The University of Washington; Sessions, J. 1990. Timber for Oregon's tomorrow. The 1989 update. Corvallis, OR: Oregon State University, Forest Research Lab, College of Forestry.
30. Warren, D. 1990. Production, prices, employment and trade in Northwest forest industries. USDA Forest Service Pacific Northwest Research Station Resource Bulletin PNW-RB-171. Portland, OR: USDA Forest Service.
31. E. Whitelaw, personal communication.
32. U.S. Department of Commerce. 1990. Regional economic information system. Washington, DC: Bureau of Economic Analysis.

14

Development of Training Programs for Conservation Research and Natural Resource Management in Ecuador

David A. Neill

A large proportion of the earth's biological diversity is concentrated in the developing countries of the tropics. In the face of increasing demographic pressures, habitat destruction, and species loss in most tropical countries, the successful *in situ* conservation of plant and animal resources in these regions requires a large input of "human capital": a cadre of technically trained resource managers and researchers and a scientific institutional base capable of carrying out resource conservation programs.

The greatest share of the world's scientific resources—trained scientists and research institutions—is concentrated in the developed countries of the temperate zones. In recent years, the scientific resource base in some tropical Latin American countries has shown substantial improvement, in part through increased interactions with scientists, universities, and research institutions based in the more developed countries. The development of human capital for natural resource conservation, however, requires a long-term investment, while the need for trained researchers and resource managers in tropical Latin America is extremely urgent. The scarcity of educated technical personnel and the limited institutional infrastructure for research and management are among the principal constraints to effective *in situ* conservation of biological resources in tropical Latin America.

This chapter examines the "human resource base" for the conservation of genetic resources in natural habitats in continental Ecuador (excluding the Galápagos Islands), concentrating on plant resources and plant biologists. In order to alleviate the severe shortage of conservation researchers and natural area managers in Ecuador, a number of programs have been initiated in the last several years to provide professional and technical training for natural resource conservation. Because of the time lag in the educational process, these efforts have, for the most part, not yet resulted in implementation of *in situ* conservation programs.

This is an extremely crucial time for conservation in Ecuador. While the country is beginning to assemble the technical capacity to conserve its natural habitats and biological diversity, these resources continue to be lost at an alarming rate. The prospect facing conservation in Ecuador is literally a race against time. Will newly trained scientists and technicians be able to initiate and carry out effective conservation programs in natural habitats while the resources are still there to be saved?

PLANT RESOURCES IN ECUADOR

With an area of about 283,000 km^2, Ecuador is among the smallest countries in South America. Its complex topography and varied climates, however, create a high diversity of natural habitats. Twenty-five of the 30 tropical Holdridge Life Zones are represented in Ecuador (1); only the cool-dry life zones (in which biological diversity is relatively low) are absent.

Ecuador is very likely the richest country in the world in terms of diversity of plant and animal species per unit area. Estimates of the number of vascular plant species in the country vary from 20,000 (2, 3) to 25,000 (4), including about 1,500 pteridophytes. In fact, due to the inadequacy of floristic inventory, these estimates are very imprecise. An estimated 250,000 flowering plant species exist in the world (5); Ecuador occupies only 0.2% of the earth's land surface, yet contains nearly 10% of all flowering plant species. An estimated 5,000 species of flowering plants, 20% of the total Ecuadorian flora, are endemic to Ecuador.

Ecuador is also located in one the most important centers of genetic diversity for agricultural crops. The Andes of Bolivia-Peru-Ecuador (including adjacent Pacific coastal areas) constitute one of the classic Vavilovian centers of genetic diversity, particularly for tuberous crops and grains (6). More recently, Clements (7) proposed a center of genetic diversity in western Amazonia, including Ecuadorian Amazonia, for domesticated fruit trees and shrubs.

LOSS OF BIODIVERSITY: A HISTORICAL REVIEW

Most of Ecuador's overall plant diversity and endemism is contained in undisturbed, primary forest habitats. Local diversity is generally much lower in secondary vegetation, and most species occurring in secondary habitats are widespread. From the point of view of crop genetic diversity, however, most diversity is contained in the native landraces of cultivated species and in their wild relatives, which are often found in weedy habitats adjacent to cultivated areas.

Continental Ecuador comprises three distinct physiographic regions: Western Ecuador, including the Pacific coastal plain and a range of low coastal mountains, covering 80,000 km^2; the Andes, including the flanks and the highland regions, covering 112,000 km^2; and the Amazon region, covering 81,000 km^2 (9). Nearly the entire territory of Ecuador was once covered by forests, except for 4,500 km^2 of high-altitude Andean páramos and 3,000 km^2 of desert scrub on the southwestern Pacific coast (1).

An estimated 22% of Ecuador's land area, mostly in the inter-Andean valleys and in the tropical dry and moist forest areas of the coast, was deforested and converted to agricultural uses in pre-Columbian times. Up until the mid-twentieth century, the rate of conversion of natural habitats was relatively slow; by 1958, 33% of the land area had been converted to agricultural and other uses (8).

During the next three decades the rate of deforestation accelerated dramatically, particularly in the coastal region where forest cover (excluding the very wet forests in the northwest) declined from 60% of the land area in 1958 to about 2% in 1988 (9). In the Guayas River basin, natural forest cover was almost totally eliminated during the 1960s; the only remnant remaining is the tiny (1 km^2) reserve at the Rio Palenque Science Center. For Ecuador as a whole, only 26% of the land area was estimated to remain in primary forest in 1988, with more than half that in lowland Amazonia.

Currently, estimates of the rate of deforestation in Ecuador range from 2,000 km^2 to 3,400 km^2 per year (9, 10). The upper estimate constitutes an annual deforestation rate of 2.3%, the highest in South America. Various factors contribute to this high rate of deforestation. First, demographic pressures have caused the loss of many natural habitats. Ecuador is the most densely populated country in South America. Its population grew from about 4 million in 1958 to 10.7 million in 1990; its annual growth rate of 2.5% is third highest on the continent, after Paraguay and Bolivia (11). Second, in the coastal region, the conversion of forested land on the relatively fertile soils resulted in the establishment of a lucrative export-oriented agricultural sector. The principal crops include bananas, African oil palm, coffee, cacao, rice, and sugar cane. Third, the discovery of major petroleum reserves in the Amazon basin in the 1960s led to the development of the oil fields in the 1970s. The network of roads built for oil extraction allowed access for settlement by farmers migrating to the region from Andean and coastal Ecuador.

Aggravating these factors are inefficient agricultural practices, which result in low yields and force the utilization of marginal lands. Agricultural yields in Ecuador are low in comparison with neighboring countries; this situation is in part due to the extremely low investment in agricultural research and extension in Ecuador (12). In general, governmental land-use policies continue to favor the expansion of agricultural frontiers into "unoccupied" lands rather than improvement of yields and optimization of land use.

It is difficult to assess with any precision the amount of biological diversity lost due to the destruction, mostly in the last 35 years, of three-fourths of Ecuador's natural vegetation. It is not known how many species are actually threatened with extinction, nor how many have already become extinct. The best possible estimate at the present seems to be Dodson's (4) contention that all of Ecuador's approximately 5,000 endemic plant species are at some degree of risk.

In lowland western Ecuador, where natural habitats have been reduced to 2% of their original extent, the threat of species loss is particularly acute. Many of the species at risk have a real or potential economic value. For example, the "Rio Palenque mahogany" (*Caryodaphnopsis theobromifolia*) was once one of the most valued timber species in the Guayas River basin, used by local inhabitants for construction and cabinetry. The species was not described botanically until 1977. The entire remaining population of *C. theobromifolia* was thought to consist of 12 adult trees at the 1 km^2 Rio Palenque reserve (4); subsequently, a few additional populations have been found in forest remnants on the periphery of the Guayas River basin (13).

NATIONAL PARKS AND OTHER RESERVES IN ECUADOR

The prospects for the *in situ* conservation of plant and animal genetic resources in Ecuador depend to a large extent on the protection of natural habitats within the national system of parks and equivalent reserves, forest reserves, and privately protected conservation areas. Ecuador's system of national parks, ecological reserves, and other conservation units owned and managed by the national government was established in the late 1970s and early 1980s, following the preparation of a preliminary strategy for the conservation of natural areas during the years 1974–1976. This system was designed to include comprehensive coverage of the major habitat types in the country, and includes 17 conservation units ranging in size from Galápagos and Yasuní National Parks, each covering more than 6,500 km^2, down to the Pululahua Geobotanical Reserve (33 km^2).

These 17 national conservation units are managed by the Department of Natural Areas and Wildlife, a division of the Forestry Subsecretariat of the Ministry of Agriculture. Except for the Galápagos Islands, however, the parks have been almost entirely undefended and unmanaged. Considerable loss of natural habitat has already occurred within some officially protected areas such as Machalilla National Park and Cuyabeno Wildlife Reserve as a result of forest removal by settlers and timber harvesters after the establishment of the parks. Podocarpus and Yasuní national parks are threatened, respectively, by future mining and petroleum operations.

A shortage of park management personnel and inadequate technical training for park managers have hampered effective protection of all of Ecuador's national conservation units; this shortage is particularly acute for the more remote parks and reserves. The 9,820 km^2 Yasuní National Park, for example, is attended to by just one park supervisor with an undergraduate degree in biology and two park guards with basic secondary school education.

PRIVATE RESERVES

In the past two decades, a number of privately owned reserves have been established in Ecuador. These *de facto* protected areas are owned and managed by nongovernmental conservations or by private owners. They are small in area but since they are energetically protected from encroachment by their owners, they are effective as *in situ* conservation units. Some of these small private reserves provide protection for particular habitats that are not included in the national park system. Examples include the 1 km^2 Rio Palenque Science Center, which protects the last tiny remnant of primary wet forest in the Guayas River valley, and the 5 km^2 Jatun Sacha Biological Station, which is the only protected area within the extremely diverse tropical wet forest zone in Amazonian Ecuador.

BIOLOGICAL SCIENCES IN ECUADOR

In spite of its wealth of biological diversity, and in contrast to some other tropical Latin American countries such as Costa Rica (14), Ecuador has not had a strong tradition of national institutions devoted to natural history research, biological inventories, and ecological studies. There are 19 public and six private universities in Ecuador; most of these offer degrees in biology and related sciences. The majority of the curricula, however, and particularly at most provincial universities, is designed to produce secondary school science teachers. Few faculty members engage in research, few have obtained postgraduate degrees, and many teach classes part-time while holding down another job. Classroom lecture is the predominant method of instruction, and little current scientific literature is available; even textbooks are frequently lacking. Few opportunities are available within most university curricula for laboratory or field experience.

A few Ecuadorian universities provide a stronger research-oriented education in biological sciences, including the Catholic and Central Universities in Quito and the University of Guayaquil. The two latter universities, which are state-supported, issue doctoral degrees in biology that require two years of schooling beyond the undergraduate degree (*Licenciatura*) and preparation of a thesis. In reality, however, these degree programs offer relatively little in the way of advanced training or research opportunities; the faculties are almost entirely composed of graduates from the same universities. Only a handful of Ecuadorians have gone abroad to earn Ph.D. degrees in biology and related fields, and most of these are in administrative positions rather than being engaged in research. A noteworthy contribution to the research-oriented education of biologists in Ecuador has been that of Aarhus University, Denmark, which has provided visiting professorships for Danish botanists in residence at the Catholic University in Quito since 1979.

Five Ecuadorian universities offer training in forestry and natural resource management, usually within agronomy schools. Only two of the forestry schools, however, have relatively well-established programs that can effectively train professional foresters.

Many public universities, due in part to political pressures and budgets dependent on high enrollment, have an open admissions policy, and no tuition charges, so all applicants are accepted into the first year of university study. At the lower academic levels, classes are very large, but the attrition rate of students is also very high, and typically only 5 to 10% of an incoming class completes the four or five years of instruction.

Most Ecuadorian universities, unlike many U.S. institutions, require a thesis for completion of an undergraduate degree. However, funding and opportunities for thesis research, particularly in field biology and related disciplines, have been very limited until recent years. Many students who have completed their coursework have full-time jobs and are unable to devote time to field research. Lacking the thesis, many Ecuadorian university students fail to obtain a professional degree even after completing a full course of study.

Investment in agricultural and natural resources research is very low in Ecuador, even in relation to other developing countries. Investment in agricultural research declined from 0.85% of agricultural GDP in 1975 to 0.17% in 1988. The world average for developing countries is 0.94% of agricultural GDP, six times the level of investment in research spent by Ecuador (12). The World Bank recommends a 1 to 2% funding level for agricultural research in developing countries. Data are not available on investment in research in forestry and other natural resource management activities in Ecuador, but the level of expenditure is much less than even the modest investment in agricultural research.

Economists contend that Ecuador's lack of investment in a "science base" for agriculture and natural resource management is a major contributing factor to deforestation, habitat loss, and resource depletion in the country. With inadequate agricultural research and extension, crop yields are well below the average for neighboring countries (12). Therefore, the increased demands for food by an expanding population can be met only through expansion of the agricultural frontier (on lands of marginal quality) rather than through improved productivity.

CASE STUDIES: TRAINING PROGRAMS IN FIELD BIOLOGY AND NATURAL RESOURCE MANAGEMENT

In recent years, public awareness of the grave threats to Ecuador's biological diversity has increased considerably, both within the country and internationally.

Ecuadorian conservation organizations such as the Fundación Natura have carried out effective publicity campaigns for conservation at a national level. International aid agencies and conservation organizations have focused increasing attention on Ecuador, especially since the mid-1980s, and a number of *in situ* conservation programs have been initiated, involving collaboration between foreign donor organizations and Ecuadorian governmental and nongovernmental organizations.

In recognition of the constraints placed on conservation efforts in Ecuador by the shortage of technically trained natural resource managers and researchers, some of the new programs supporting *in situ* conservation of biological resources include training programs for Ecuadorian professional and paraprofessional biologists, foresters, and natural area managers. Some of the recently established training programs are described below. This is not intended to be a comprehensive list of such educational programs in Ecuador; but rather it is provided to highlight some programs as examples of the effort being made.

In addition to the training components of *in situ* conservation programs summarized here, some progress is being made in *ex situ* conservation of plant genetic resources in Ecuador. Ecuador's National Institute for Agricultural Research (INIAP), a division of the Ministry of Agriculture, recently established a Department of Plant Genetic Resources, which is actively involved in collecting germplasm of native crop plants and their wild relatives and in maintaining *ex situ* germplasm collections (15).

The Darwin Research Station, Galápagos

The Galápagos Islands, located 1,000 km west of the Pacific coast of Ecuador, have attracted research biologists from around the world for many years; in recent decades, the Charles Darwin Research Station has coordinated research programs on the islands. Since the early 1970s, the Darwin Station has sponsored internship programs in which Ecuadorian biology students participate as research assistants in projects carried out by visiting scientists. Many of the research projects are directly concerned with conservation of native Galápagos plants and animals. Until the mid-1980s, the internships on Galápagos constituted almost the only research opportunities available to young Ecuadorian biologists. The Darwin Station has thus contributed to conservation efforts in continental Ecuador by providing field training to Ecuadorian biologists, who in turn have been able to apply their knowledge to programs on the mainland.

Forestry Sector Development Project

The U.S. Agency for International Development (USAID) supported the Forestry Sector Development Project in Ecuador from 1984 to 1990; it was carried out in collaboration with the Forestry Subsecretariat of the Ministry of Agriculture. This program included training for technical personnel in several subdisciplines of forest management and biological research. An agroforestry extension program in lowland Amazonia provided on-the-job training for foresters and extension agents (16). Several short courses in forest protection (forest entomology, pathology, and fire control) were conducted, and a forest protection unit was established within the Forestry Subsecretariat.

The forestry project also provided substantial support for botanical research in Ecuador: an extensive inventory of tree species in lowland Amazonian Ecuador

(17) conducted by the Missouri Botanical Garden and the Forestry Subsecretariat, and surveys of economic plants by the New York Botanical Garden. Both components included training of Ecuadorians in botanical research methods, with visits to the two U.S.-based institutions for training in herbarium and library research methods for systematic and economic botany.

Field Course in Population Ecology and Conservation Biology

This postgraduate training course is consciously patterned after the tropical ecology "Fundamentals" course presented by the Organization for Tropical Studies (OTS) in Costa Rica since the mid-1960s. The Ecuadorian version, a one-month course that travels to four different field sites, has been taught in 1989, 1991, and 1992, and plans are to repeat it annually. The organizers of the course are U.S. and Costa Rican biologists, graduates of the OTS tropical ecology courses; Ecuadorian professionals also participate as part-time instructors. The students are Ecuadorian graduates of local universities in biology, forestry, and related fields. The curriculum, like the original OTS version, emphasizes the posing of testable ecological hypotheses, research design, and statistical analysis. Each student writes several short research papers during the course, which are compiled into a book distributed to all participants. The course is supported financially by USAID through the World Wildlife Fund and the Sustainable Uses for Biological Resources (SUBIR) project described below.

The Fundación Natura Debt-for-Nature Swap Program

In 1989, Fundación Natura, the leading nongovernmental environmental organization in Ecuador, entered into an agreement with The Nature Conservancy and the World Wildlife Fund–US to acquire from commercial banks, at discounted rates, bonds worth $\not{\$}$10 million of Ecuador's external debt. The government of Ecuador agreed to issue stabilization bonds in Ecuadorian currency to Fundación Natura over a seven-year period; these funds are used for a special conservation program administered by Fundación Natura in collaboration with several Ecuadorian government agencies, including the national park service within the Ministry of Agriculture, and several other nongovernmental conservation organizations.

This conservation program, which is scheduled to continue through 1996, includes several activities designed to assist conservation of natural habitats both within and outside of the national system of protected areas. A major component of the program is the training of managers of natural areas and park rangers through a series of short courses.

Botanical Training and Research: The PROMOBOT Program

Part of the debt swap program carried out by Fundación Natura with the Nature Conservancy includes funds donated by the Missouri Botanical Garden. The Ecuadorian currency bonds generated by this portion of the debt exchange are used to finance botanical research, floristic inventories, and training of Ecuadorian botanists. The program, known as PROMOBOT (Proyecto de Promoción Botánica) is carried out in collaboration with the National Herbarium, a section of the Ecuadorian Museum of Natural Sciences.

A major contribution of the PROMOBOT project is to develop the National Herbarium as a center for information on the flora of Ecuador and to provide services that aid in planning for conservation of protected areas, management of forest resources, and related activities. The project staff of nine Ecuadorian botanists and foresters carry out botanical inventories in several regions of the country as well as curatorial activities at the herbarium in Quito.

The PROMOBOT project, like the botanical research program within the USAID-supported Forestry Support Development Project that preceded it, provides training programs for Ecuadorian botanists. The on-the-job training component for project staff includes periodic visits to the Missouri Botanical Garden for instruction and practical experience in herbarium and library research methods for systematic botany. In Ecuador, the program also includes short courses in plant taxonomy and research for upper-level university students and young professionals.

The Conservation Data Center

Another project of the debt swap program is the establishment of a Conservation Data Center (CDC) in Quito. Sponsored by the Nature Conservancy, the data center is one of 11 centers in Latin America. The principal mission of the CDC is to compile information on the status of threatened plants, animals, and habitats in Ecuador, using a specially designed database system, and to make this information available to public and private institutions and individuals carrying out research, conservation, and management of natural resources in Ecuador. Five young Ecuadorian biologists comprise the staff of the Conservation Data Center. They receive technical support and training from the staff of the Nature Conservancy based in Washington, D.C.

EcoCiencia Foundation

EcoCiencia is a nonprofit Ecuadorian foundation devoted to ecological research, education, and conservation. It was established in 1989 when several young Ecuadorian biologists, who had obtained undergraduate training at the Catholic University in Quito and then went on to obtain Master's degrees in zoology, botany, and conservation biology, returned to Ecuador and found that the existing governmental and academic institutions in the country could not accommodate them in their desire to launch careers as research biologists. Not to be disheartened, they decided to form their own research institution.

The EcoCiencia staff has carried out a series of basic and applied research projects during the first two years of its existence, including the documentation and evaluation of the few remnant areas of natural forest in western Ecuador; surveys of fauna and of hunting practices in the vicinity of Podocarpus National Park and Cayambe-Coca Ecological Reserve; ethnobotanical studies in Amazonian Ecuador in conjunction with the New York Botanical Garden; and environmental impact studies for several planned public works projects.

Research Scholarships

Several of the newly established conservation organizations in Ecuador have initiated programs of research scholarships that allow Ecuadorian biology students to carry out field research for their undergraduate theses. Jatun Sacha Biological

Station in the Amazon region has provided scholarships for research on ecology and systematics of plants and animals at the station with funds from private donors in collaboration with the Missouri Botanical Garden. In a similar program, EcoCiencia provides scholarships for conservation-related field research, with support from Wildlife Conservation International.

Natural Forest Management by Indigenous Groups

Cultural Survival, Inc., has collaborated since 1988 with the federation of indigenous communities in Napo province (FOIN), in Amazonian Ecuador, to carry out a regional resource management project. The project initially was devoted to helping secure legalization of land titles for indigenous Quichua-speaking communities, but later expanded to include a program of natural forest management for communities in the upper Napo region near the eastern base of the Andes. The program includes on-site training in forest management for the indigenous project staff. The instructors have included native Amuesha foresters from the Palcazú valley in Peru (see Hartshorn, this volume) and Kuna from Panama. Cultural Survival has received funds for this program from USAID and the World Wildlife Fund.

A similar natural forest management project involving indigenous communities has been initiated with the Awá ethnic group in the wet forest area of northwestern Ecuador. Like the program with the Napo Quichua, it will include a technical training component for the indigenous forest managers.

The SUBIR Project

The Sustainable Uses for Biological Resources (SUBIR) project is a 10-year program in natural areas conservation and sustainable development in Ecuador, funded by USAID and carried out by a consortium of CARE, the Nature Conservancy, and Wildlife Conservation International. The initial three-year phase of the project is centered on three protected areas in Ecuador and the buffer zones around the reserves: Cotacachi-Cayapas Reserve on the western slopes of the Andes, Cayambe-Coca Reserve on the eastern slopes, and Yasuní National Park in lowland Amazonia.

The SUBIR project activities planned within and around each of these conservation units include (1) strengthening local community organizations to develop their capability to manage biological resources; (2) improving management of the core biological reserve; (3) developing nature-oriented tourism as a source of income for local inhabitants; (4) improving land use and biological resources and developing sustainable resource management strategies; and (5) developing research and monitoring programs to improve basic knowledge of the biological resources present, assisting in preparation of management plans, and monitoring the impact of project activities. Training components are built into all of the planned SUBIR activities, ranging from postgraduate courses in ecological research methods for professional biologists to agroforestry extension programs for local farmers.

Postgraduate Education

Since 1985, opportunities have increased markedly for Ecuadorian biologists to undertake postgraduate study at universities in North America and Europe. In that time, at least six Ecuadorians have obtained Master's degrees in botany, zoology, ecology, and conservation biology at U.S. universities and have returned to their native country to engage in research on natural resource. Several more individuals are now pursuing M.S. and Ph.D. studies in these disciplines. Through an agreement between the botany department of Aarhus University and the Catholic University in Quito, five Ecuadorians are participating in a Ph.D. program in systematic botany at Aarhus.

CONCLUSIONS

In some sense, Ecuador was "discovered" by international conservation organizations only during the 1980s, coinciding with the recent expansion worldwide of public interest in tropical forests and biological diversity. The realization on the part of both international and Ecuadorian organizations of a severe shortage of technically trained people in the country who are able to carry out research and management in conservation biology has led to the extensive program of courses, scholarships, and training programs described above.

Education is a process that requires time, however, and the conservation programs in Ecuador are mostly still in the initial training phase. Up until the present, very little in the way of *in situ* management and conservation of biological resources has been applied to field situations in the country. It is hoped that the new generation of professional biologists and resource manangers will be able to implement effective, on-the-ground conservation programs in Ecuador within the next few years.

The training of professional researchers in conservation biology and managers of natural areas is only one of the steps that must be taken to ensure protection of biological diversity in Ecuador. Fundamental reforms in government macroeconomic and land use policies are also required (10), as is the funding for implementation of conservation measures. Nevertheless, in Ecuador, as in other developing countries in the tropics that possess such a large share of the earth's biodiversity, a well-trained technical corps of biologists, researchers, foresters, and natural area managers is a *sine qua non* for effective protection of *in situ* biological resources. The next few years will tell whether the young group of Ecuadorian conservation biologists will be able to carry out the protection of their country's magnificent wealth of flora, fauna, and diverse ecosystems before those resources are lost forever.

REFERENCES

1. Cañadas, L. 1983. El Mapa Bioclimático y Ecológico del Ecuador. Quito: Ministerio de Agricultura y Ganadería.
2. Gentry, A. H. 1978. Floristic knowledge and needs in Pacific tropical America. *Brittonia* 30:134.
3. Harling, G. 1986. Flora of Ecuador—its present status. In *Current scandinavian botanical research in Ecuador*, B. Ollgaard and U. Molau, eds., 9–10. *Rep. Bot. Inst. Univ. Aarhus* 15.
4. Dodson, C. H. 1989. History and status of forests in Ecuador. In *Assessment of biological diversity in Ecuador*, B. Cabarle, ed., Appendix, 30–54. Washington, DC: World Resources Institute.
5. Raven, P. H. 1988. Our diminishing tropical forests. In *Biodiversity*, E. O. Wilson, ed. Washington, DC: National Academy Press.

6. Vavilov, N. I. 1951. *The origin, variation, immunity and breeding of cultivated plants*. Waltham, MA: Chronica Botanica.

7. Clements, C. R. 1989. A center of crop genetic diversity in western Amazonia. *BioScience* 39:624–31.

8. Dodson, C. H., and A. H. Gentry. 1991. Biological extinction in western Ecuador. *Ann. Missouri Bot. Gard.* 78:273–95.

9. Cabarle, B. J., ed. 1989. *An assessment of biological diversity and tropical forests for Ecuador*. Washington, DC: World Resources Institute.

10. Southgate, D. 1990. Development of Ecuador's renewable natural resources. In *The role of agriculture in Ecuador's economic development: An assessment of Ecuador's agricultural sector*, M. D. Whittaker, D. Colyer, and J. Alzamora, eds., 81–116. Quito: Instituto de Estrategias Agropecuarias.

11. 1991 World Population Data Sheet. 1991. Washington, DC: Population Reference Bureau.

12. Whittaker, M. 1990. The human capital and science base. In *The role of agriculture in Ecuador's economic development: An assessment of Ecuador's agricultural sector*, M. D. Whittaker, D. Colyer, and J. Alzamora, eds., 367–402. Quito: Instituto de Estrategias Agropecuarias.

13. A. Gentry, personal communication.

14. Gómez, L.D., and J. M. Savage. 1983. Searchers on that rich coast: Costa Rican field biology 1400–1980. In *Costa Rican natural history*, D. H. Janzen, ed., 1–11. Chicago: University of Chicago Press.

15. Castillo, R. 1991. Nuevo departamento de recursos fitogenéticos en Ecuador. In *Memoria de la II Reunión Sobre Recursos Fitogenéticos*, R. Castillo, C. Tapia, and J. Estrella, eds., 69–83. Quito: Instituto Nacional de Investigaciones Agropecuarias.

16. Peck, R. B. 1988. Promoting agroforestry practices among small producers: The case of the Coca Agroforestry Project in Amazonian Ecuador. In *Alternatives to deforestation*, A. Anderson, ed., 167–80. New York: Columbia University Press.

17. Neill, D. A., and W. A. Palacios. 1989. *Arboles de la amazonía Ecuatoriana: Lista preliminar de especies*. Quito: Dirección Nacional Forestal.

15

Community and Governmental Experiences Protecting Biodiversity in the Lowland Peruvian Amazon

Miguel Pinedo-Vasquez and Christine Padoch

The valley of the Ucayali river is one of the largest whitewater river basins of the world and is extremely high in biological diversity. It contains several types of swamp, riverine, and upland forests, as well as abundant oxbow lakes. These areas provide habitat for numerous plant, fish, and animal species. The richness of its habitats is among the factors that have made the Ucayali basin one of the most populated regions in the Peruvian Amazon (1). Until recently, the rural people of the basin were accustomed to finding the plant and animal species they hunted, fished, and extracted not only in great abundance but also located conveniently near river banks and in oxbow lakes close to their settlements. However, in the last two decades many plant and animal populations have declined sharply throughout much of the region, even within national reserves (2).

The major cause for the decline in many species and the loss of total biodiversity in the Ucayali basin is increased commercial exploitation rather than the factors cited as paramount in other parts of Amazonia: road construction, colonization projects, mining, or coca production. The Ucayali region, like other areas of the lowland Peruvian Amazon, has suffered a long history of exploitation of both its forests and its peoples. Overhunting, increased commercial fishing, and timber extraction have all affected rural populations. For instance, in a period of only four years, from 1970 to 1974, 1,247,285 skins of 15 wildlife species were exported to the United States and Europe from the surrounding region (Fig. 1). Nine of these 15 species were killed exclusively for their skins, while the other six were used for skins and meat (Table 1). Few Ucayalinos have ever benefited from the extraction and export of these products (3). For example, the value of the animal exports shown in Table 1 was approximately U.S. $2,611,463; almost all of which went to only a few families (4).

When commerce in food species rises rapidly, the effects of overharvesting can be even more serious. In the last 10 years, the supplies of a favored fish, paiche (*Arapaima gigas*), has drastically fallen as demand for dried paiche has increased

Table 1. Wildlife species hunted for their skins and exported to U.S. and European markets, 1970–1974.

COMMON NAME	SCIENTIFIC NAME	SKIN	MEAT
Sajino	*Tayassu tajacu*	X	X
Huangana	*Tayassu pecari*	X	X
Lagarto blanco	*Caiman crocodilus*	X	X
Lagarto negro	*Melanosuchus niger*	X	X
Venado colorado	*Mazama americana*	X	X
Camaleon rojo	*Dracaena guianensis*	X	
Anaconda	*Eunectes murinus*	X	
Ronsoco	*Hydrochoerus hydrochaeris*	X	X
Mantona	*Boa constrictor*	X	
Tigrillo	*Felis pardalis*	X	
Nutria	*Lutra longicauda*	X	
Gato peludo	*Felis colocolo*	X	
Tigre	*Felis concolor*	X	
Huamburusho	*Felis wiedii*	X	
Lobo de rio	*Pteronura brasiliensis*	X	

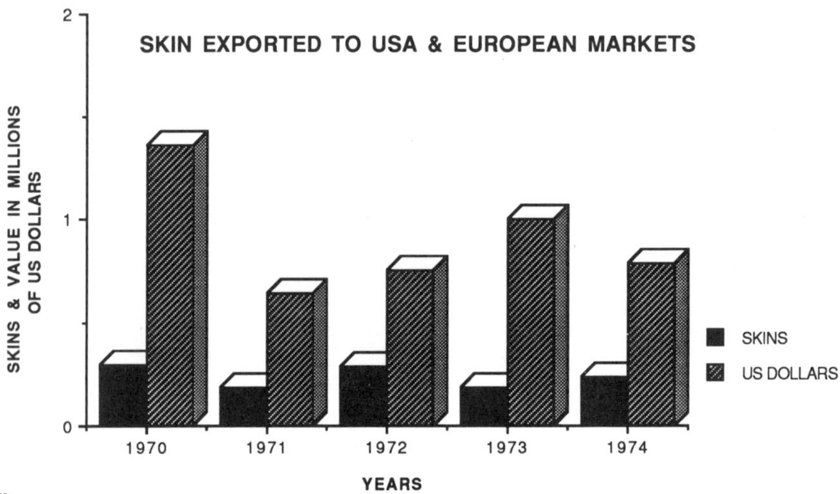

Figure 1.

in Amazonian urban markets. For instance, the weight of dry paiche bought by three middlemen in villages within the Pacaya-Samiria national reserve fell from 15 metric tons in 1985 to six tons in 1991 (Fig. 2). The average price paid by the middlemen in 1985 was 22 cents per kilogram of dry paiche; by 1991 it had risen to $2. A kilogram of dry paiche then sold by the middlemen to rematistas (small wholesalers in urban markets) climbed from 75 cents in 1985 to $4 in 1991 (Fig. 3). Increased urbanization and increased urban demand have fueled overexploitation of the resource. Thus, although little large-scale deforestation has taken place along the lowland Ucayali, many important resources are becoming scarce.

Several species of animals are particularly threatened. These include the *charapa* (*Podocnemis expansa*), the largest Amazonian river turtle, and the manatee (*Trichechus inunguis*). These two species, exploited commercially for centuries, are now almost never seen; and virtually all other large animal species are in decline (2,

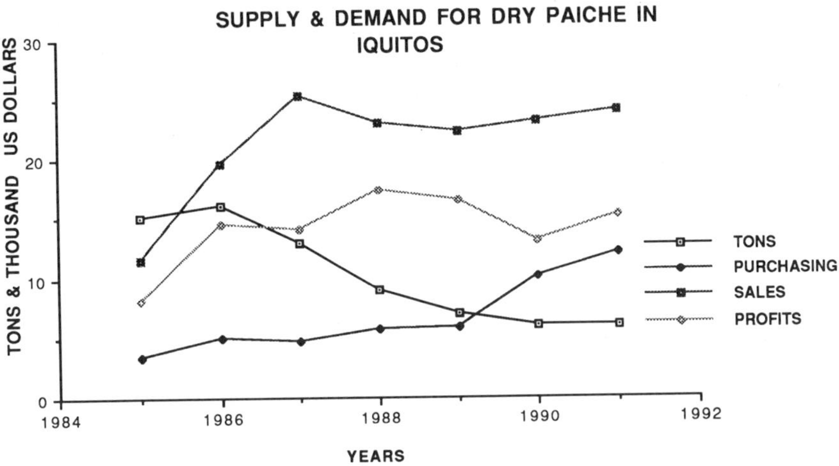

Figure 2.

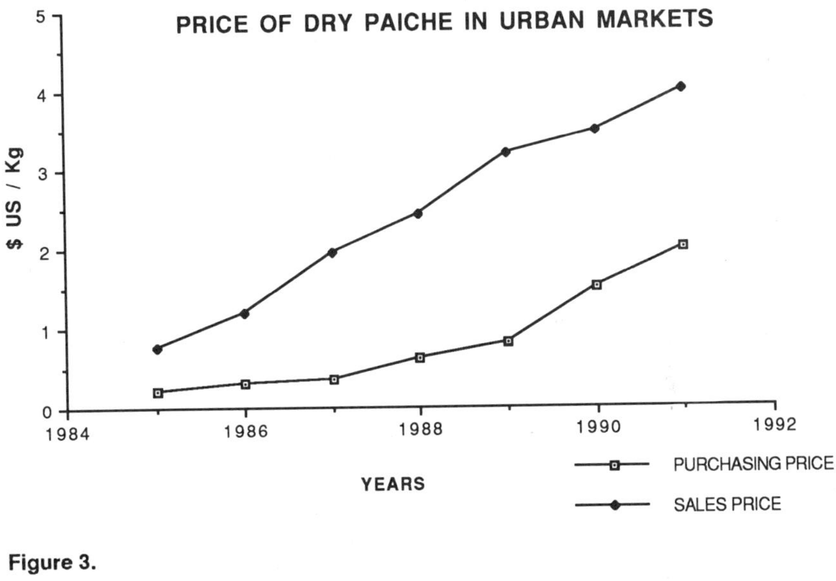

Figure 3.

5). Many oxbow lakes that once were full of fish are now depleted, as commercial fishing boats enter even very remote regions. High-grading of timber species has likewise produced severe scarcity of valuable woods. Mahogany (*Swietenia macrophylla*), tropical cedar (*Cedrela odorata*), and lupuna (*Ceiba* spp.) are especially overexploited.

The majority of rural Ucayalinos are *ribereños*, who are largely the descendents of Cocama and Lamista indigenous Amazonian groups and of migrants who came to the region during the rubber boom at the turn of the century. Ribereño villagers and agencies of the Peruvian government share the goals of protecting biodiversity and promoting sustainable development in the Ucayali region. The ribereños'and the central government's strategies are, however, different. Rural Amazonians have

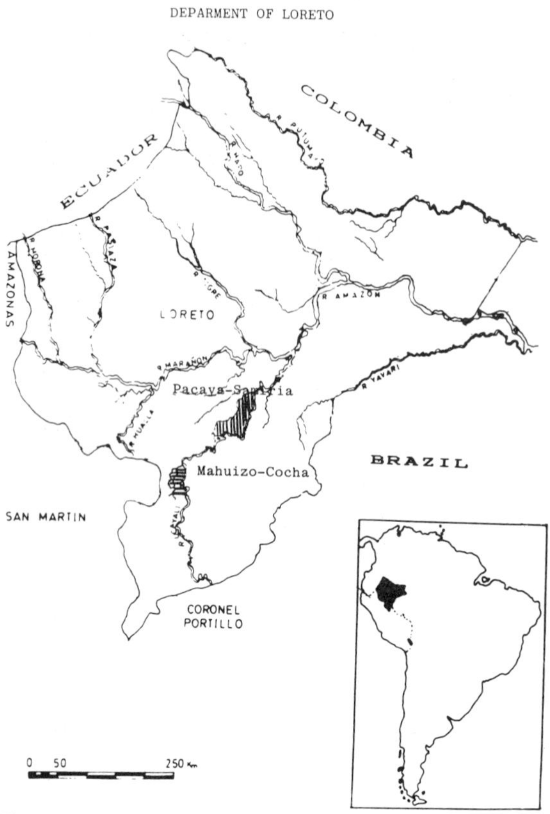

Figure 4.

treated the need to protect biodiversity and the need to improve the livelihoods of rural populations as two closely linked problems. In contrast, governmental institutions have treated these two issues separately. Ribereños and their unions have promoted the diversification of land and resource use within the region as their main strategy for protecting biodiversity and promoting sustainable development. In contrast, government functionaries have promoted agriculture and cattle production to develop the region and the establishment of national parks and reserves to protect biodiversity.

The Pacaya-Samiria national reserve and the Mahuizococha intervillage reserve (Fig. 4) both aim to protect the lower Ucayali's biodiversity. In this brief chapter we compare how this biodiversity has been protected in the two reserves. Common problems that villagers and officials have long faced in regulating resource use are discussed. We then suggest alternatives for controlling exploitation and enforcing laws and rules. Some of the successes of the intervillage reserve are highlighted and presented as suggestions for reforming the structure and administration of the Pacaya-Samiria reserve. The necessity of promoting broad participation of local people, government agencies, and other social groups in the administration and control of these two reserves and other regional conservation units is also discussed. Finally, we touch upon the need for controlling or reducing urban market demand for valuable and threatened resources that are now extracted from the two reserves.

ADMINISTRATION AND CONTROL OF BIODIVERSITY WITHIN THE TWO RESERVES

The Pacaya-Samiria national reserve has an area of 2,080,000 ha and is the largest and most important national reserve in the northeast sector of the Peruvian Amazon (2). The Mahuizococha intervillage reserve of Dos de Mayo, San Cristobal, and Nuevo Dos de Mayo is one of 44 documented village and intervillage forest and lake reserves established by village residents in the region (Table 2). This reserve covers an area of approximately 1,500 ha of forest land and oxbow lake.

Table 2. Native and ribereño villages or intervillage lake and forest reserves in the lowland Peruvian Amazon, Iquitos.*

No.	VILLAGE	P (1985)	L	F	L/F	A (ha)
1	Santa Ursula (N)	180			X	310
2	Santa Rosa (N)	260			X	270
3	Pobre Alegre (N)	145			X	570
4	Comandancia (N)	120			X	160
5	Catalan Vainilla (N)	90		X		60
6	Roca Eterna (R)	245		X		470
7	Vainilla (R)	225		X		180
8	San Pedro de Marupa (R)	385	X			340
9	Marupa (R)	360	X			220
10	Caballocochillo (R)	175	X			210
11	Nueva Union del Boyador	110			X	470
12	Nuevo Horizonte (R)	90			X	160
13	Santa Rosa de Vainilla (R)	75	X			140
14	Portugal (R)	237			X	540
15	Siete de Junio (R)	86	X			140
16	Oran (R)	325	X			120
17	Ahizana (R)	135	X			140
18	Yanashi (R)	1,755	X			870
19	Santa Victoria (R)	275	X			270
20	Santa Tereza (R)	370	X			150
21	Pucallpa (R)	275		X		240
22	Santa Sofia (R)	280		X		370
23	Puerto Rico (R)	220	X			120
24	Santa Elena (R)	215	X			110
25	Sapo Playa (R)	150	X			110
26	Timicuro Grande (R)	325	X			570
27	San Rafael (R)	315		X		800
28	Juan Velasco (R)	210	X			240
29	Negro Urco (N)	180	X			220
30	Urco Miraño (N)	240	X			310
31	Tamanco (R)	610			X	310
32	Puinahua (R)	120		X		240
33	Yarina (R)	715	X			420
34	Nuñez Cocha (R)	215	X			420
35	San Alejandro (R)	225		X		130
36	San Pedro (R)	175		X		120
37	Huanana (R)	290	X			270
38	Francisco de Orellana (R)	650	X			710
39	Puerto Aveja (R)	245		X		170
40	Tapirillo (R)	277	X			120
41	Justicia (R)	375	X			360
42	Porvenir Segunda Zona (R)	240		X		110
43	Tapira Grande (R)	420	X			340
44	Dos de Mayo	1,670			X	1,500
TOTAL		14,050				14,100

* N=Native village, R=Ribereño village, P=Population census 1985, L=Lake (cocha), F=Forest, L/F=Lake and Forest, A=Estimated area of each reserve.

The Pacaya-Samiria reserve has in recent years been under considerable pressure because of a major increase in commercial fishing and hunting within its territory. The high demand for game meat and fish in urban markets presents an economic opportunity for urban people, while the low per capita incomes and the economic dependence on natural resources of the reserve's residents prompt them to become involved in commercial exploitation (6). While the public administration of the reserve is trying to preserve the region's biodiversity, local people and urban-based commercial extractors are exploiting the resources. As a result, confronta-

Table 3. Rules governing resource extraction in the Mahuizococha intervillage reserve of Dos de Mayo, San Cristobal, and Nuevo dos de Mayo.

1. Villagers are allowed to fish, hunt, and collect other resources in the Mahuizococha reserve for their own consumption and for sale within the three villages. It is prohibited to extract resources for selling to congeladores (commercial fish companies) and rematistas (middlemen), and to send to urban markets.
2. People from other villages or other outsiders are allowed to extract resources for their own consumption. These people are required to get a special permit from the local authorities and the committee in order to collect resources from Mahuizococha.
3. It is prohibited to fish in the reserve using the following materials: arrastradoras and hondoneras (big fishing nets), barbasco (fish poison), and dynamite.
4. People who disobey the rules will be notified and their fish, game, and other resources will be confiscated. Local authorities will confiscate the equipment of persistent offenders, who will also be driven out of the three villages.
5. To reduce the pressure on Mahuizococha's resources, villagers from the three villages are encouraged to fish, hunt, and collect plant resources in the river, forest, and other oxbow lakes during the dry season in the months of April, May, June, July, and August.
6. People from the three villages have the right to denounce any authority or villager who is disobeying the rules. The accusations can be made directly to any authority from the three villages or in public meetings.

tions between the public administration and the extractors have been common and depletion rates have been on the rise (7). Recent interest in the reserve, availability of considerably increased funds, and expertise from outside the region and the country have begun the process of changing the administration of the reserve. We aim to add to this process by suggesting that valuable lessons may be learned from the nearby local reserve.

Like Pacaya-Samiria, the Mahuizococha intervillage reserve has been under pressure by local and outside commercial extractors during the last 10 years. These pressures have been alleviated by encouraging the participation of villagers in the administration and control of the reserve. The extraction of resources either for subsistence or the market is regulated by local rules (Table 3). Even though commercial exploitation and biodiversity destruction have been controlled in the Mahuizococha reserve, it has not been legally recognized as a conservation area by the Peruvian government. One reason for this failure is that land and resource tenure in the region is denied to villages that are not officially recognized as Indian communities. Thus, ribereño villages such as the three that participate in the Mahuizococha intervillage reserve have no legal right to regulate land and resource use in their territories. Rural unions that represent ribereño villages believe that a new governmental policy on land and resource tenure is needed.

The Mahuizococha reserve provides most of the fish and game meat that is consumed in the villages of Dos de Mayo, San Cristobal, and Nuevo Dos de Mayo. In the last 10 years a large part of the residents' income has come from selling fish caught in the reserve. In contrast with the largely local importance of Mahuizococha, Pacaya-Samiria provides almost all the game meat and fish that is consumed in the urban areas of the Peruvian Amazon (2). Like Mahuizococha, a large percentage of local income is derived from the extraction of the national reserve's resources. During the last 20 years or so commercial exploitation has had a negative impact on the biodiversity in both the intervillage and national reserves. However, the impact of increased commercial extraction on the populations and reproduction of game animals, fish, and other resources within Pacaya-Samiria has been more severe than in Mahuizococha. In 1986, within the national reserve, three wildlife and two tree species were listed as endangered and approximately 25 species as threatened (Table 4). In addition, because of commercial extraction, populations of 20 species have been drastically reduced in the Pacaya-Samiria reserve (Table 5).

Commercial extraction of endangered and threatened species is prohibited by local rules within Mahuizococha and by national and regional laws in Pacaya-Samiria. Village regulations are enforced by an elected committee of five members (*comite de administracion y control de Mahuizococha*). The five members are

Table 4. Three endangered wildlife species, two endangered tree species, and fifteen threatened wildlife species within the Ucayali Valley and other regions of the Peruvian Amazon.

	COMMON NAME	SCIENTIFIC NAME		STATUS
1)	Nutria	*Lonta longicaudis*	Otter	Endangered
2)	Huapo colorado	*Cacajao calvus*	Monkey	Endangered
3)	Lobo de rio	*Pteronura brasiliensis*	Otter	Endangered
4)	Palo de rosa	*Aniba roseadora*	Tree	Endangered
5)	Leche caspi	*Galaotadendron utilissima*	Tree	Endangered
6)	Maquizapa negra	*Ateles paniscus*	Monkey	Threatened
7)	Vaca marina	*Trichechus inunguis*	Manatee	Threatened
8)	Lagarto negro	*Melanosuchus niger*	Alligator	Threatened
9)	Pava negra	*Aburria aburri*	Bird	Threatened
10)	Charapa	*Podocnemis expansa*	Turtle	Threatened
11)	Taricaya	*Podocnemis unifilis*	Turtle	Threatened
12)	Sacha pato	*Merganetta armata*	Bird	Threatened
13)	Matamata	*Phrinops gibbus*	Turtle	Threatened
14)	Yangunturo	*Priodontes giganteus*	Armadillo	Threatened
15)	Tigrillo	*Felis pardalis*	Cat	Threatened
16)	Huamburushu	*Felis wiedii*	Cat	Threatened
17)	Paujil	*Mitu salvini*	Bird	Threatened
19)	Shihui	*Tamandua tetradactyla*	Anteater	Threatened
20)	Oso hormiguero	*Myrmecophaga tridactyla*	Anteater	Threatened

Table 5. Twenty species that experienced population reduction due to commercial extraction in the Pacaya-Samiria Reserve.

	COMMON NAME	SCIENTIFIC NAME	
1)	Paiche	*Arapaima gigas*	Fish
2)	Arahuana	*Osteoglossum bicirrhosum*	Fish
3)	Gamitana	*Colossoma maeropomum*	Fish
4)	Paco	*Colossoma brachypomum*	Fish
5)	Sabalo	*Brycon melanopterus*	Fish
6)	Tucunare	*Cichla ocellaris*	Fish
7)	Boquichico	*Prochilodus nigricans*	Fish
8)	Majaz	*Agouti paca*	Rodent
9)	Añuje	*Dasyprocta variegata*	Rodent
10)	Carachupa	*Dasypus novemcinctus*	Armadillo
11)	Sajino	*Tayassu tajacu*	Peccary
12)	Huangana	*Tayassu pecari*	Peccary
13)	Choro	*Lagothrix lagothricha*	Monkey
14)	Motelo	*Geochelone denticulata*	Turtle
15)	Lagarto blanco	*Caiman crocodilus*	Alligator
16)	Sacha vaca	*Tapirus terrestris*	Tapir
17)	Venado cenizo	*Mazama gonazoubira*	Deer
18)	Venado colorado	*Mazama americana*	Deer
19)	Cupiso	*Podocnemis sextuberculata*	Turtle
20)	Pucacunga	*Penelope* sp.	Bird

elected from among delegates of district authorities (offices of the governor, Justice of the Peace, and the mayor), Catholic Church delegates (*animadores cristianos*), school teachers, and the parents' school association. The membership on the committee changes every two to three years. Violations of the reserve's code are punished appropriately. The Pacaya-Samiria reserve does not have such a committee and the enforcement of national and regional laws and rules within Pacaya-Samiria has been in the hands of a technical-administrative governmental team.

In the last 10 years the populations of several overexploited species of fish such as paiche (*Arapaima gigas*) and rodents such as majaz (*Agouti paca*) have increased in the Mahuizococha intervillage reserve. The population growth of these species is attributed by local residents to the enforcement of local rules and to active local participation in administration. In contrast, populations of the same species have declined in the same period within the Pacaya-Samiria reserve because of lack of enforcement of laws and rules by the governmental team. The main strategy used by the team to protect biodiversity within the national reserve is to restrict extraction by controlling access to the area. This strategy is ineffective because 22 villages and towns with a total population of 4,371 persons are located within the reserve. The people of these villages and towns depend for their livelihoods largely on the commercial extraction of resources from the territory of the Pacaya-Samiria reserve.

The general purpose of the two reserves is to protect forest land and oxbow lakes for the reproduction of plant, animal, and fish species. However, the main objective of the Mahuizococha reserve clearly is to protect biodiversity resources so that they may provide food, shelter, medicine, and other necessities in perpetuity for the people of Dos de Mayo, San Cristobal, and Nuevo Dos de Mayo. In contrast, the main objective of the Pacaya-Samiria reserve has long been to conserve biodiversity, the purpose of which local people have not always readily understood. To achieve this general aim, any kind of resource extraction within the national reserve has traditionally been prohibited. In contrast, controlled extraction of resources has been allowed in Mahuizococha as a way to protect biodiversity.

In Pacaya-Samiria the management of biodiversity has been in the hands of a team of specialists. This team has traditionally been concerned with conducting studies on the management of fish and animal species of particular interest rather than on the management of habitats or ecosystems. As a result, literature about the taxonomic classification, biology, and reproductive characteristics of fish and wildlife species are relatively abundant (8). However, literature on the sociological, economic, and ecological aspects of the reserve are very sparse and descriptive (2).

The management of the Mahuizococha intervillage reserve depends exclusively on an intervillage committee and local participation. The protection of biodiversity within this reserve is entirely based on traditional management techniques. Based on this expertise the biodiversity of Mahuizococha is protected by managing habitats or ecosystems. For instance, the territory of the intervillage reserve has been divided in two main areas: (1) extraction areas, where plant, animal, and fish species may be extracted; and (2) reproduction areas, where resource extraction is prohibited.

The reproduction areas are locally known as *cochas muertas* (dead oxbow lakes). Cochas muertas are areas where palms such as chotilla (*Bactris riparia*) and aquatic plants such as putu-puto (*Alchornea grassipes*), huama (*Pistia stratiotis*), and piri-piri (*Cyperus* spp.) are dominants. Many Ficus and Bombax species are also abundant in cochas muertas. Because of the abundance and dominance of these plant species, these areas are judged favorable for the reproduction of game animals, fish, and other plants. Several beliefs or myths exist about cochas muertas and serve to explain why they favor the reproduction of species. The most common is that these areas have "mothers" (*madres*), which are mainly magical giant anacondas, cayman, monkeys, or jaguars. These mythical animals protect all fish, animals, and plants within cochas muertas by killing any hunters, fishermen, or people who try to extract resources from these areas. As a result of the protection of cochas muertas, the populations of commercially highly desirable and commonly overexploited species such as paiche (*Arapaima gigas*) and majaz (*Agouti paca*) have increased in the last 10 years.

Despite the effectiveness of local enforcement of regulations for the protection of biodiversity, several technical problems remain unresolved within the intervillage reserve. Among these is a lack of information on matters such as the reproductive biology of particular fish and terrestrial animal species. For example, accurate technical information on the reproductive biology of paiche (*Arapaima gigas*) could help villagers devise more effective management plans, estimating how many years it will take for this overfished species to recover its previous numbers in the Mahuizococha reserve. It also could help villagers to establish annual fishing limits for paiche to ensure its continued healthy reproduction. Since 1985, the number of paiche annually caught within the Mahuizococha reserve has been below 100 (Fig. 4). The majority of villagers want to have more precise information on how the number of fish caught in the period 1985–1991 has affected this species' reproduction in the reserve. Technical information on the adundance and dynamics of many resources, so that they can be extracted without adversely affecting the biodiversity of the area, must be sought through long-term studies in the Pacaya-Samiria and Mahuizococha reserves. The administrative committee of the intervillage reserve has asked regional governmental institutions for technical assistance. These institutions are not able to provide such assistance due to limitations of economic and human resources. However, the committee hopes that Pacaya-Samiria's team of experts will provide information and help the villages that wish to take an active role in protecting their resources.

PROTECTION OF BIODIVERSITY AND SUSTAINABLE DEVELOPMENT

Conservation units such as national parks and reserves in less developed countries like Peru are expected to fulfill many functions. Rather than just conserving wildlife and other natural resources, they are called upon to play an important role in the development of the surrounding area and to provide economic opportunities to people living within and outside these conservation unit's areas (9). In reality, however, most parks and natural reserves in Peru, like Pacaya-Samiria, have traditionally been oriented toward the conservation of wildlife, fish, and a few tree species. In contrast, intervillage reserves focus on the protection of areas with high biodiversity from destruction by commercial extractors so that local people may continue to depend on them (10).

Villagers in the Ucayali region have long known that development should be based on the wise exploitation of a diversity of resources and not only on one or two species. The rubber boom of the past and the present coca-cocaine boom are mentioned by Ucayalinos as examples of unsustainable development models. In addition, the regional expansion of urban populations and their rising consumption are identified as direct threats to biodiversity within national parks, nature reserves, villages, and intervillage reserves. For example, it is estimated that people living in Iquitos consume approximately 9,077 metric tons of fish per year. In contrast, rural people living within and near Pacaya-Samiria consume approximately 442 metric tons of fish per year. Approximately 90% of the fish that are consumed in Iquitos and other urban areas in the region are caught within the Pacaya-Samiria reserve (4). These estimates strongly suggest that the strategies of protecting biodiversity in the Pacaya-Samiria and other national parks must be directly linked to region-wide programs for rural and urban development.

The limitation of human and economic resources is the main reason given by conservation officials for not extending their programs for the protection of biodiversity beyond officially designated parks and reserves. This argument highlights the traditional dependency on outside economic and human resources characteristic of conservation programs in the Ucayali and other regions of the Peruvian Amazon. While it may be necessary to look outside the region for some technical information, the Mahuizococha intervillage reserve illustrates that much valuable biodiversity can be effectively protected in the region without depending on outside economic and human resources. Allocating part of these resources to areas outside of parks and officially designated reserves can help protect biodiversity at low cost while stimulating sustainable development in the region. Resources for conservation should not be used exclusively to build infrastructure within parks and reserves or to train technicians.

Local people know a great deal about the behavioral characteristics of many important wildlife species, and the value and uses of these resources. As a rule, people who rely daily on game, fish, and forest plants are far more sensitive to signs of overexploitation and are far more interested in biodiversity protection than are city-bred technicians and administrators. The problem is that rural people do not have the opportunity to communicate their knowledge, their concerns, or their needs to the authorities. The apparent lack of participation of many ribereños in the protection of biodiversity within the Pacaya-Samiria reserve is not simply a case of insensitivity to the problem. It is, instead, a reaction to the apparent lack of interest on the part of technical-administrative teams in the past to rural welfare and the officials' connections to more powerful commercial extractors. The negative attitudes of both the rural and urban people to official regulations and to the administrators of the Pacaya-Samiria reserve have basically resulted from the perception that no benefits of those regulations will ever reach them.

To change the attitude of people concerning national parks and reserves, it is first necessary to promote effective communication between the regional authorities and rural dwellers. Communication must take place in a manner that rural dwellers feel respects them and their rights. How and why the preservation of biodiversity will ultimately help more than hurt them is an idea that must be communicated honestly. The necessary communication, trust, and cooperation will not take place unless both individual and group rights are recognized. The example of the Mahuizococha intervillage reserve illustrates how ribereños can protect biodiversity in their territory by managing their resources and regulating access to their forest and oxbow lakes. Villages within and near Pacaya-Samiria find themselves in an unusual situation, but they too would be far more eager to participate in the protection of biodiversity if they had some faith that benefits were not destined to go only to urban commercial operators or government officials.

CONCLUSIONS AND RECOMMENDATIONS

We suggest that the modest village-based reserve may offer some examples for the national reserve of ways to preserve biodiversity and to present several ideas that can be used to improve the administration, management, and protection of biodiversity in these two and other reserves in the Peruvian Amazon. Solutions to many of these problems may require technical and economic changes. Other solutions can be found, however, by considering and altering social factors. Based on these assumptions the following alternatives should be considered:

1. The administration of national parks and natural reserves should not continue to be exclusively in the hands of government or extralocal nongovernmental institutions. A new administrative body comprising representatives of local residents such as district mayors, Justices of the Peace, lieutenant governors (*teniente-gobernadores*), animadores cristianos, rural teachers, and union leaders, as well as representatives of the regional government office and nongovernmental institutions, should be responsible for the administration and control of Pacaya-Samiria and other national parks and reserves in the region. These bodies should be responsible for defining policy, obtaining financial support, and keeping society informed about the successes and problems of the reserves.

2. Economic resources that are allocated to national parks and nature reserves should be made available to finance the function of the new administrative bodies.

3. The effective organization of the different social groups that are the beneficiaries of the Pacaya-Samiria reserve can help solve a variety of problems. Through their organizations, these social groups could effectively participate in protecting biodiversity within and without this and other conservation units. Priority should be given to the formation and maintenance of good channels of communication among social groups.

4. Existing grassroots organizations need to participate in political and administrative decisions concerning biodiversity protection and development plans in the region. One such organization is the Federacion Departamental de Campesinos y Nativos de Loreto (FEDECANAL), which is the largest ribereño organization in the region. This grassroots group should have access to all available information about national parks and reserves, including Pacaya-Samiria. Only through access to such information can this and other grassroots groups help effectively in protecting biodiversity in the region.

5. Commercial extraction needs to be limited in Pacaya-Samiria and other reserves, as in the Mahuizococha intervillage reserve. This can be accomplished not only through tightened administrative and political measures, but also through reducing demand in regional and national markets for game, fish, and other resources, and altering the relative attractiveness of extraction as a source of income. These changes must be tied to the development of

alternative food and income sources. Several lines of action can be followed in rural and urban areas to reduce demand. These might include the following: (1) Prices of agroforestry or agricultural products that are produced in the region should be made more competitive with those of extracted resources. Currently, the price of game meat, fish, and turtles gives a far better return than the price of agricultural products. The current attractiveness of paiche prices, for example, means that people living close to or within the Pacaya-Samiria reserve are far more interested in fishing than in producing crops. (2) Middlemen should be given opportunities to invest their capital in small processing industries based on agroforestry or agricultural products. (3) The media should be used more effectively to educate urban consumers about the benefits of preserving biodiversity and the endangered status of some commercial species. These actions must be instituted gradually, with constant participation of local communities and their organizations, and with adequate technical and administrative support.

6. Economic alternatives must be provided to the reserve's rural residents. "Modern" alternatives are difficult to offer in regions like Pacaya-Samiria and Mahuizococha where agriculture is limited and transportation and communication facilities are very poor. Therefore, attention must be given to traditional technologies and knowledge, and plans should be based on existing resources and capabilities.

7. Each village needs to be encouraged to formulate its own regulations and rules concerning management and protection of biodiversity. The Mahuizococha intervillage reserve should be promoted and used as an example in the region. At the very least, villages in the region must be given legal status as regulators of access to land and resources within their territory. Without this basic right, local people will be subject to intrusion by outsiders and biodiversity cannot be protected.

8. Authorities should recognize that the protection of biodiversity and sustainable development in the region must lie primarily in the hands of people who live with and benefit from the resources. With this realization must also come respect for local residents as experts on their own environments and as potentially valuable allies in the struggle to protect biodiversity and promote sustainable development in the Peruvian and other Amazonian regions.

REFERENCES

1. Villarejo, A. *Asi es la selva* 3ra. ed. 1979. Centro de estudios teológicos de la Amazonia. Lima: Editorial Universo S.A.
2. Comite Local de Desarrollo de la Reserva Nacional Pacaya-Samiria. 1986. COREPASA, *Plan maestro de la reserva nacional Pacaya-Samiria*. Loreto, Perú: COREPASA.

3. Padoch, C. 1987. In *Swidden-Fallow Agroforestry in the Peruvian Amazon*, W.M. Denevan and C. Padoch, eds. *Adv. Econ. Bot.* 5. New York: New York Botanical Garden.

4. Barletti, J. personal communication.

5. Soini, P. 1986. Estudio e incubación de nidadas de charapa (*Podocnemis expansa*). Informe de Pacaya No 18. Iquitos, Perú: Ministerio de Agricultura.

6. Aguilar, R., M. Runciman, C. Pezo. 1986. *Diagnóstico socio-económico del área de influencia de la reserva Pacaya-Samiria*. Iquitos, Perú.

7. Ministerio de Agricultura. 1985. *Memorias anuales*. Dirección Forestal y Fauna, RAXXII-L. Iquitos, Perú: Ministerio de Agricultura.

8. Berger, C. 1972. Estudio del factor de condición del paiche (*Arapaima gigas* Cuvier) en la zona reservada del río Pacaya, entre los años 1959–1969. Tesis Bach. Cienc. Biol. Lima: UNMSM; Fachin, A. 1982. Contribución al conocimiento de la bio-ecologia de la taricaya (*Podocnemis unifilis* Troschel) en el río Samiria, Loreto. Tesis Bach. Cienc. Biol. Iquitos, Perú; Ridgely, R. 1985. *Birds of Pacaya-Samiria National Reserve, Department of Loreto, Peru*. Philadelphia: The Academy of Natural Sciences of Philadelphia.

9. Ranjitsingh, M.K. 1980. In *Tropical ecology and development: Proceedings of the Vth International Symposium of Tropical Ecology*, J.I.Furtado, ed. April 16–21, 1979. Kuala Lumpur, Malaysia.

10. Pinedo-Vasquez, M., D. Zarin, P. Jipp, J. Chota-Inuma. 1990. *Conserv. Biol.* 4:405–16.

16

An Ecosystem Perspective on Threats to Biodiversity in Eastern Amazonia, Pará State

Christopher Uhl, Oswaldo Bezerra, and Adriana Martini

Amazonia is brimming with life. But human activities are increasingly in conflict with the health and longevity of this natural life. Tensions began centuries ago with the arrival of merchants and fortune seekers and continue today as colonists, ranchers, loggers, miners, and large hydroelectric projects occupy Amazonia. The species-rich forests and aquatic environments of Amazonia are sources of livelihood for some and wealth and power for others.

Amazon exploitation and settlement activities are concentrated in the southern and eastern flanks of the Basin, the fronts of exploitation having extended into Amazonia from Brazil's Northeast and Central-South regions. Nowhere is this human advance on Amazonia more apparent than in Pará State in Eastern Amazonia. While Pará occupies one-third of the land area of the Brazilian Amazon, it accounts for over 50% of the timber, cattle, and mineral production of the region (1).

Within the State of Pará, land-based economic activities (e.g., ranching and logging) are associated with road systems (Fig. 1) while aquatic-based activities (e.g., gold mining and dam building) are associated with the tributaries flowing out of the Guianian Shield to the north and the Brazilian Shield to the south of the Amazon River. Pará is also noteworthy for its geologic, topographic, and climatic diversity, housing an array of vegetation types and animal communities (2). Upland evergreen forest occupies significant areas below the Amazon River down through the central part of Pará. Semi-deciduous forests, palm forests, and vine forests are concentrated in regions of transition where tall forest grades into "Cerrado" vegetation. Isolated formations of low hill forests occur in the region of the Carajas iron-ore deposits. Savannas are common to the north of the Amazon River, in the eastern half of Marajo Island and in the lower reaches of the Tocantins River (3). Meanwhile, the Lower Amazon and its estuary contain a rich mix of aquatic ecosystems, including forests that are flooded seasonally or even daily (tidal influence), natural grasslands, and varzea lakes.

In 30 years of aggressive development, one thing has be come clear—the

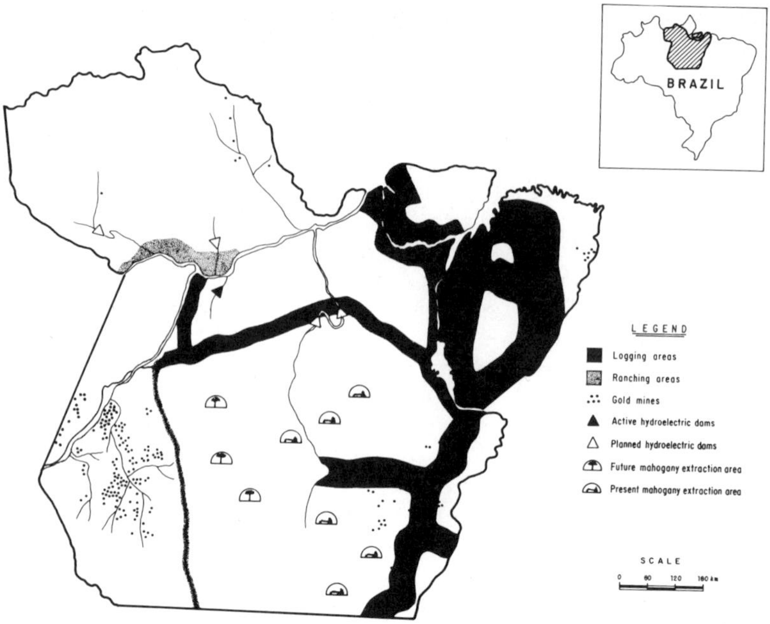

Figure 1. The State of Pará in the Eastern Amazon showing the locations of ranching, logging, gold mining, and dam-building activities.

Eastern Amazon is rich in resources and humans are determined to exploit these resources. The purpose of this chapter is to discuss the impacts of this resource use on the plant and animal biodiversity of the Eastern Amazon. We will do this by considering the most important human activities that are affecting both terrestrial and aquatic ecosystems of Pará. In the terrestrial case, we focus on selective logging, which represents a significant alteration of forest structure, and on forest cutting to establish cattle pastures, which represents the complete removal of forest ecosystem structure. Likewise, in the aquatic case, we focus on a significant alteration of aquatic ecosystem structure, the processing of river sediments to extract gold, and the complete removal of aquatic ecosystem structure—river damming to establish hydroelectric plants. In our analysis of the impacts of these activities on biodiversity, we consider both direct impacts on individual species at the site of the activity as well as biological impacts that might extend to more distant pristine ecosystems via air or water. We conclude by considering what might be done to reconcile development and conservation objectives in the Eastern Amazon.

IMPACTS OF TERRESTRIAL-BASED ECONOMIC ACTIVITIES ON BIODIVERSITY

A Significant Alteration of Terrestrial Ecosystem Structure—The Case of Logging

Logging has been practiced on a small scale for several centuries in Amazonia in areas of forest accessible by water, but a quantum leap in timber exploitation occurred in the late 1970s, once a reliable network of roads had been established in eastern Pará. Roads, such as the Belém- Brasília Highway (Fig. 1) linked an immense expanse of upland forest to Brazil's densely populated and wood-starved Northeast and to Brazil's industrialized Central-South. The advance on Amazonia occurred just as Brazil's extra-Amazonian roundwood stocks were nearing exhaustion.

At present there are four distinct patterns of logging practiced in the Eastern Amazon, each with its own rationale, economic benefits, and potential impact on regional biodiversity. In new frontier regions, where roads have recently been established (e.g., along Pará Highway 150) logging is highly selective. Only about a dozen species are extracted and only the largest and best formed individuals of each species are removed (4). In older frontier regions (e.g., along the Belém-Brasília Highway), logging is more intensive. For example, in the logging town of Paragominas, approximately 115 sawmills process over 100 different tree species. Logging is done with heavy machinery and impacts on the forest are large, with two cubic meters of wood being severely damaged for every cubic meter harvested (5).

Both selective and intensive logging activities also occur in the floodplain forests of the Eastern Amazon. Selective logging is done in more remote areas of the Lower Amazon. The principal species of interest is *Virola surinamensis*. The impacts of this type of logging are slight because generally only one to three individuals are extracted per hectare, and extraction is done by hand. Meanwhile, with the recent establishment of more than a thousand small, family-run sawmills in the Lower Amazonia, a more intensive system of floodplain logging has been initiated. These family-run mills provide lumber to the regional construction industry and specialize in sawing small boles (15 to 60 cm in diameter). Forests that have been intensively logged to supply these mills are left in an open state with little wood, save that which could serve for firewood.

Logging impacts on local biodiversity

Recently, we tallied approximately 350 tree species that are being harvested in the Brazilian Amazon for wood uses. Knowing the ecological characteristics of these species, we have been able to distinguish among those species that are apt to be threatened by logging, as well as those that might be favored by this activity (Fig. 2). For example, a species that is limited geographically to the Eastern Amazon, is poorly represented in the forest regeneration, does not regenerate well in light gaps, does not sprout, has thin bark (i.e., fire sensitive), and has valuable wood is likely to be threatened by logging. By contrast, a species that occurs throughout Amazonia, is well represented in the forest regeneration, grows rapidly, sprouts readily after cutting, and has thick bark (fire resistant) is likely to withstand logging pressure and might even experience a population increase in response to logging-type disturbances.

Based on this rationale, we have developed a simple point system to rank the present-day timber flora of Amazonia, distinguishing among those species that are most likely and least likely to experience population reductions as a result of logging pressure (6) (Fig. 2). In this system, ecological characteristics that would be disadvantageous (e.g., sparse population, poor regeneration, slow growth, high fire sensitivity, etc.) have a point value of "one," whereas characteristics that might confer an advantage have a point value of "three," and intermediate characteristics receive a value of "two." Summing the point values for the various characteristics for a given species provides an indication of that species' sensitivity to logging pressure. Using this system, we have found that some 30 of

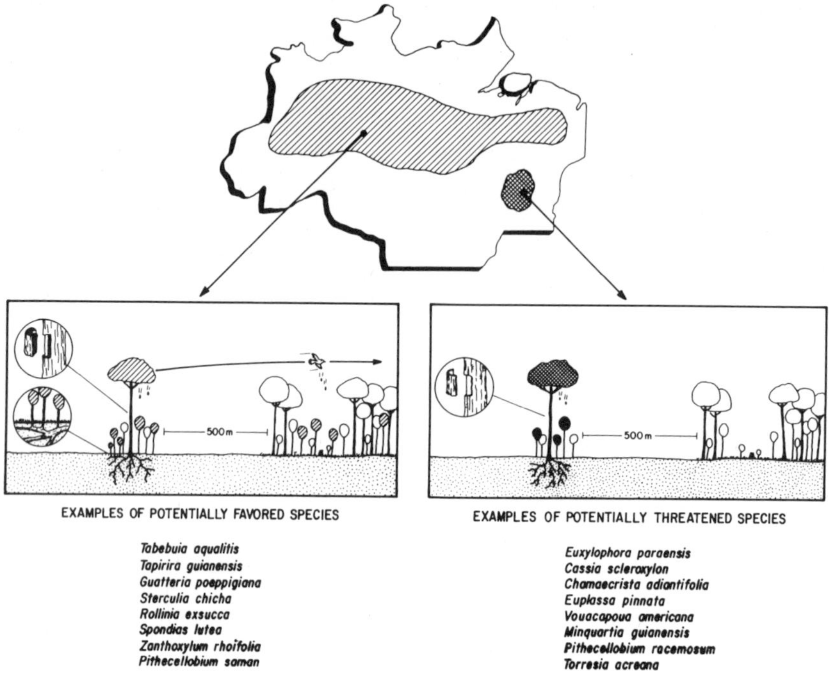

Figure 2. Species that might experience population reductions ("species threatened by logging") as a result of logging activities in Amazonia have characteristics such as a limited geographic range, thin bark (i.e., fire sensitive), inability to sprout, and poor representation in the forest regeneration (e.g., in treefall gaps). Species that might experience population growth ("species favored by logging") as a result of logging activities have opposing characteristics.

the 350 timber species currently being harvested in Amazonia have biological characteristics that may leave them poorly equipped to maintain their populations in the face of intensive logging pressure.

This exercise is important, first, because it shows that adequate information is now available in collections and in the minds of natural historians to begin to predict which plant and animal species might be less likely to resist human intrusion. The same exercise could be done in the case of mining, dam building, and ranching. Second, the list is a pragmatic way to begin to develop population biology research strategy for tree species of the Eastern Amazon. In our view, applied research on tree population biology should be directed to those timber species having characteristics that might render them vulnerable to logging pressure.

Logging might also affect other organisms aside from timber species. Indeed, many bird and mammal species depend on timber species for food. We estimate that 42% of the timber flora of the Brazilian Amazon is fed on by mammals and 41% by birds. It is possible that animal populations that depend heavily on timber species will experience population reductions and concommitant genetic erosion as the timber species that they rely on for food are eliminated. In addition, animal groups with poor temperature regulatory ability, such as amphibians, might avoid logged forests because of the hot and dry conditions on these sites. Other animal groups might actually benefit by the new conditions present after logging. For example, because of the abundance of dead wood on logged sites, decomposers, such as wood-boring beetles, might experience population growth.

Logging impacts on regional biodiversity

Logging is having more far-reaching effects on the Eastern Amazon landscape as well. In frontier areas, selective logging is a catalyst that frequently leads to complete deforestation. For example, logging teams in the environs of the Xingu River (Fig. 1) open up hundreds of kilometers of logging roads each year in the extraction of just one species, *Swietenia macrophylla* (mahogany). These logging roads are avenues into the forest: selective logging of mahogany frequently precedes large-scale deforestation by ranchers.

Even when logging is not linked to deforestation, it might, particularly in its more intensive manifestations, affect regional ecosystems. Preliminary results from Paragominas, in eastern Pará, indicate that evapotranspiration from recently logged stands may be significantly less than from intact forest (7). And given that a significant part of Amazonia's rainfall is derived from local sources (8), forest logging could contribute to reductions in regional rainfall with as yet uncharted effects on regional biodiversity.

In addition, intensively logged stands, because of their open nature, are flammable. Based on interviews with forest landholders, we estimated that approximately half of the logged forest area in Paragominas experiences ground fires. Because many tree species are thin-barked, mortality and, therefore, carbon loss can be high in these fires. This carbon loss is one of several sources that contribute to global warming with possible effects on biodiversity.

The Complete Removal of Terrestrial Ecosystem Structure—The Case of Forest Conversion to Cattle Pastures

In the short interval from 1965 to 1990 about 13% of Pará was deforested. Much of this land was converted to pasture. Although forest conversion to pasture has been routinely criticized as uneconomical and ecologically ruinous, the cattle herd in the Eastern Amazon continues to grow. And as the decade of the nineties proceeds, ranching is increasing in floodplain ecosystems as well as in the uplands; it is common in both new and old settlement areas, and it is the preferred land use both among large and small land holders. All indications are that ranching will increase in importance in Eastern Amazonia in coming years.

Ranching impacts on local biodiversity

When a hectare of rain forest is converted to pasture, the entire plant-animal-air-soil-water matrix is transformed. A simplified ecosystem results. Instead of thousands of plant and animal species composing several hundred tons of biomass spread intricately over 40 meters of vertical space, a field is created with a few dozen species containing some 10 t of biomass/ha compacted into one meter of vertical space. Physical conditions in pastures differ dramatically from those in the forest. The low, open structure of pastures results in higher peak air and soil temperatures and higher vapor-pressure deficits compared to the forest understorey (9).

Some hearty forest species do persist in pastures. A study of 13 abandoned pastures near Paragominas, Pará, Uhl, Buschbacher, and Serrão (10) found 94 tree species present as sprouts. Many of these species were present in the original forest and resprouted in the pastures. Some animal species also occur in both open pastures and forests. For example, 14 of 387 bird species and 10 of 22 bat species found in the Paragominas area have been observed moving between forest and open field environments. Some limited habitat overlap also exists for insects, reptiles, large mammals, and small rodents. However, only a small minority of the many plant and animal species present in the forest can tolerate the open conditions prevalent in deforested sites.

This subset of tolerant species, together with a group of hardy exotic species, will make up the future fauna and flora of Amazonia. In the case of plants, the future flora will be composed of those species that persist in cleared areas and/or those that are able to establish in these areas. Hence, species that sprout readily, resist fire, have seeds and foliage that are unattractive to predators, can withstand or avoid drought, and so on, may dominate in these altered landscapes. In the case of animals, it is the species with extremely flexible feeding and habitat requirements that may thrive. Such a future biota, composed of hardy, generalist-type species, would encompass only a very small fraction of the biodiversity now present.

Ranching Impacts on Regional Biodiversity

When considering the impacts of forest conversion to pasture on biodiversity, it is tempting to focus just on the plot of ground that has been denuded, with regard to the difference in species richness between the virgin forest and the cleared land as the loss of biodiversity. But forest conversion to pasture in the Eastern Amazon could also be affecting "pristine" ecosystems far removed from the pastures. For example, forest conversion to pasture leads to the loss of some 1,600 kg of nutrients (NPK) (11). These nutrients eventually move into nearby streams and rivers and could increase productivity leading to the dominance of certain aquatic groups and the elimination of others.

Second, the carbon embodied in pasture ecosystems after a decade of use is only about 10% of that found in the original forest (9). Hence, some 160 tons of carbon are liberated to the atmosphere when forests are converted to pastures. This carbon release could contribute to global warming with possible attendant effects on biodiversity. Note: this carbon release estimate is conservative because reductions of soil carbon stocks during ranching have yet to be carefully assessed, and these reductions may be large.

Finally, conversion of forest to pasture might affect regional rainfall. Root systems in pastures are shallower than those in forests and, hence, pasture ecosystems may have less access to soil water and a lower annual evapotranspiration than forest ecosystems (12). This, in turn, could lead to declines in regional rainfall (8). And reduction in regional rainfall may lead to extended droughts, increased fire incidence, and ultimately to vegetation changes. In addition, insofar as evapotranspiration is reduced in pastures, runoff increases and so does the probability of regional flooding.

Taken together it is easy to imagine, at least, how these changes in terrestrial-ecosystem function, provoked by the conversion of forest to pasture, could represent a more general threat to regional biodiversity. Nutrients lost from pastures may move into surrounding streams increasing primary productivity and simplifying trophic relationships; carbon loss and reduced rates of evapotranspiration in pastures may lead slowly but inexorably to a warmer, drier regional climate.

Linkages Between Ranching and Logging

We have illustrated how ranching and logging activities can affect local as well as regional biodiversity. It is clear that the ecological effects of ranching are much more severe than those of logging. What is less apparent are the linkages between these two activities.

Most of the land in the Eastern Amazon, forested and cleared alike, is owned by ranchers. A typical ranch might have one-third cleared land and two-thirds forest. Ranchers sell logging rights to sawmills. Hence, logging and ranching occur side-by-side (Fig. 1). Ranching activities can directly affect neighboring forested tracts. For example, ranchers routinely use fire to discourage weedy dicot growth and favor grasses in pastures. However, during particularly dry periods, these fires move into nearby logged forests causing much damage, particularly to fire-sensitive plant and animal species (13).

Consideration of the use of capital generated by logging activities reveals a second important interaction between ranching and logging. Ranchers are using the profits realized in the sale of logging rights to rejuvenate badly degraded pastures (5). This pasture restoration process is costly: the rancher must sell the logging rights to approximately four hectares of forest to have enough capital ($260) to reform one hectare of pasture.

This is the third time that ranching has been subsidized in the Eastern Amazon. The first subsidy came with the initial forest felling when the nutrients embodied in the forest biomass were liquidated to help pasture grasses establish (11). The second subsidy came from the government in the form of capital for ranch infrastructure establishment (14). This third subsidy, to restore degraded pastures, like the first subsidy, comes from nature. Insofar as careless logging practices lead to the genetic impoverishment of both local and regional plant and animal populations, biodiversity loss could be a hidden cost in this pasture reformation process.

IMPACTS OF AQUATIC-BASED ACTIVITIES ON BIODIVERSITY

A Significant Alteration of Aquatic Ecosystem Structure—The Case of Gold Mining

The eastern Amazon is a miner's dream, containing rich deposits of bauxite, cassiterite, iron ore, manganese, diamonds, nickel, chromite, copper, and gold. Of these, gold has been the most sought after. The Amazon gold boom started in the 1980s and was the result of the convergence of five elements: (1) high gold prices on the world market; (2) development of an infrastructure in Amazonia including roads, air strips, and radio communication; (3) a ready source of cheap labor from Brazil's Northeast; (4) availability of inexpensive mining equipment; and (5) official sanctioning of mining (15).

In the period 1980–1988, gold production in Pará was 60% of national production (16). Silva (17), considering both official and unofficial sources, estimated that 725 tons of gold, with a value of $10.5 billion, were produced in Pará over this eight-year period. At decade's end there were some 1,000 active mining sites in Pará with some 700 airstrips, and 300,000 persons involved (15).

Gold mining effects on local biodiversity

Gold is present as very fine particles in soils and river sediments (18). Most gold mining activities are concentrated in the small streams flowing out of the Guianian and Brazilian Shield areas flanking the Amazon River. Mining usually involves the extracting and washing of sedimentary deposits with hoses equipped with jet beaks. River morphology can be severely altered by peeling banks away and creating labyrinths of sediment-clogged pools flanking river channels. Abandoned mining sites resemble moonscapes.

There are, on average, 2.0 grams of gold/cubic meter of sediments. Hence, the production of 100 tons of gold in Pará in one year (current production is probably in this vicinity) would introduce some 50 billion cubic meters of sediments into streams. These sediments could affect local biodiversity. Sediments increase water turbidity, reducing light availability and water temperature, which thereby reduces primary productivity. Sediments and suspended particulates also affect respiratory processes, first by smothering bottom channel benthic organisms and second by settling on gill surfaces and interfering with gas exchange. Water turbidity also decreases visibility and might thereby affect predator-prey relationships. This would be particularly true for fish such as Ciclids with yellow eye filters that require high light levels to hunt effectively (19). Finally, insofar as mining is frequently concentrated in headwater streams, fish species that rely on this habitat for spawning or rearing of young might experience setbacks.

Gold mining activities also have potentially severe effects on forest wildlife. Miners live in the forest during much of the year and engage regularly in subsistence hunting of monkeys, peccaries, deer, armadillos, pacas, capybara, toucans, macaws, and other forest animals. Redford (20) estimates that as many as 57 million animals might be killed in subsistence hunting activities in all of Amazonia each year. Many of these hunted species are large frugivores (21). These species play important roles in seed dispersal, seed predation, herbivory, and pollination. Redford notes that as the population size of a game species is reduced, the species reaches a point where it no longer performs its "ecological functions." This point might occur before the population would be regarded as threatened, based on considerations of population size alone.

Finally, gold mining is affecting the biological and cultural diversity of our own species. Tens of thousands of gold miners have entered Indian Territories in the southern Pará, provoking disease outbreaks (malaria, small-pox, and tuberculosis), strife, and cultural impoverishment.

Gold mining effects on regional biodiversity

In aquatic ecosystems, the dispersion of disturbance effects beyond the point of impact is also a problem. The most serious regional impact of gold mining on biodiversity is mercury pollution [See Lacerda and Salomons (18) for an excellent review]. Mercury is used to separate gold from other materials after gravimetric sifting. Gold binds to the mercury and this amalgam is then burned to release the mercury as vapor, leaving gold as the residue. Estimates for mercury loss in gold production vary, but, in general, the amount of mercury added exceeds the amount of gold produced. For example, Pfeiffer and Lacerda (22) estimate that 1.32 kg of mercury are lost in the production of 1 kg of gold. Some 55% of this mercury is lost to the atmosphere with the remainder lost to nearby rivers (23).

Humans who mine gold take few precautions to avoid contamination, and a variety of studies confirm that Amazon miners have elevated mercury levels in their body tissues (17, 24, 25). Mercury also contaminates aquatic organisms. In the river environment, mercury may be transformed into the lipid and water-soluble compound monomethylmercury, which can then enter the food chain. Through the process of bioaccumulation, mercury reaches high concentrations in predators (23, 24). Predatory fish occurring in small streams that are most likely to be affected by mercury contamination are "trairas" (*Hoplias malabaricus*, Erythrinidae) and the "pirapucú" (*Boulengerella* spp, Ctenoluciidae). In the larger river channels fish predators that are sensitive to mercury contamination are the "surubins" and "capararis" (*Pseudoplatystoma* spp, Pimelodidae), "douradas" (*Brachyplatystoma* ssp., Pimelodidae) and "Jaus" (*Paulicea lutkeni*, Pimelodidae), while in Amazonian lakes the "tucunares" (*Cichla* spp, Cichlidae) are predisposed to mercury contamination because of their predatory habits (26). Meanwhile, humans, in so far as they in turn prey on these fish carnivores, run the risk of mercury contamination. At particular risk are the thousands of families that carry on subsistence fishing and forest extraction activities in the floodplain environment. These native residents of the floodplain consume, on average, some 200 g of fish per day (27). While we are aware of no cases of mercury pollution driving a species to extinction, mercury intoxication could lead to significant population reductions (through death or outmigrations) of many species, including human beings.

Overall, gold mining is a disturbance that might be reducing local as well as regional biodiversity. Miners are aggressive hunters capable of severely reducing populations of large mammal and bird species. The sediments liberated during mining are like a cloudy curtain that descends and settles over river networks affecting the physical characteristics of the water stream bottom and, probably, simplifying the biological structure of headwater streams. Meanwhile, mercury spreads up food chains concentrating in carnivores and eventually haunting far-flung human populations.

The Complete Removal of Aquatic Ecosystem Structure—The Case of Hydroelectric Dams

The rivers flowing out of the shield regions to the north and south of the Amazon River are ideally suited for hydroelectric dam construction because they carry large volumes of water, flow over hard rock basements, and drop significantly in their passage to the Amazon River. The favored sites for hydroelectric development in the eastern Amazon are located along the middle and upper reaches of the Tocantins, Xingu, Araguaia, Tapajos, and Trombetas rivers. The complete list of planned hydroelectric dams in Amazonia totals 80 (28). The construction of these dams would flood some 100,000 km^2.

In Pará one major dam has already been constructed and 32 are planned. Pará's big dam, Tucurui, has a reservoir flooding an area of 2,430 km^2 and produces 2,310 megawatts of energy. Forty percent of this production is consumed within Pará. The remainder is shunted to the Northeast.

Dam effects on local biodiversity

The establishment of hydroelectric dam reservoirs is similar to the establishment of pastures in that the preexisting ecosystem is completely eradicated. In the construction of the Tucurui dam, an immense area of forest was flooded to create the reservoir. The floodplain and island communities upstream of the reservoir have disappeared. Insofar as dams eliminate these habitats, they affect biodiversity.

Dams also affect local biodiversity by barring the migrational movements of fish, interrupting upstream migrations of spawning adults and down-river movements of fish fry. The species most affected are the large migratory fish such as curimatá (*Prochilodus nigricans)*, piramutaba and dourada (*Brachyplatystoma* spp), and mapare (*Hypophthalmus marginalus*) (26, 29). These species are of great commercial importance. The use of techniques such as fish ladders to aid migrational movements of fish may not be possible, given the projected size of Amazonian dams. For example, the Tucurui Dam is 70 m tall. Given the height, slope, and the force of the water, fish ladders are not an option at Tucurui (27).

The effects of damming on biodiversity in dam reservoirs are still being contemplated. Water chemistry of the reservoir lake is very different from that of the preexisting river. Millions of tons of organic matter (forests) are under water. The release of sulfur oxides and organic acids in decomposition processes, the formation of nitrates, and the low concentrations of dissolved oxygen make the reservoir bottoms difficult places to inhabit for a riverine fauna. Petrere (30) has shown marked shifts in species abundance relations in reservoirs, as once common river species such as curimatá, jau, and surubim experience population decline. Meanwhile, species that are predators of the main river channel and varzea lakes, such as pescada and tucunares, might increase their numbers in reservoir settings (30).

It is too soon to predict the full effects of the damming of Amazonia's rivers because reservoir environments require a period of years to stabilize. In these early years, the submerged forest is acting as a substrate for many kinds of macrophytes and invertebrates, as well as a shelter for many fish species. In addition, the submerged forest acts as a kind of slow-release fertilizer, with decomposition processes gradually liberating nutrients to the water. As this biomass disappears, nutrient availability will decline and productivity, species composition, and water chemistry will change. As a new reservoir community develops, there is the potential of invasion by exotic species. Although there are no documented invasions at present, it might be easy for a fish such as Tilapia to invade the reservoir and further affect biotic composition (27).

Dam effects on regional biodiversity

As with other activities, the effects of hydroelectric dam construction extend beyond the boundaries of the dam reservoir itself. In a typical river ecosystem, the downstream environments are nourished by the water and suspended solids that pass through them. Dams interrupt this flow. As rivers approach a dam, the water velocity slows and the load of suspended sediments settles. As much as 90% of the suspended sediment load of the Tocantins River may fall as sediment in the Tucurui Dam reservoir (27). The water that passes down river is relatively sterile. This is particularly true during the rainy season when a large fraction of the water that is passed downstream comes from the clear surface waters of the reservoir. As this water spreads out over the varzea of the lower Tocantins River, it no longer drops nutrient-rich suspended sediments.

Fish species and farmers have depended on this annual nutrient bonanza. Many fish species move laterally from the main river channel to the varzea at high water to feed and spawn (31). When the waters recede, this land is planted to food crops. With the interruption of these cycles, both fishermen and farmers who inhabit the lower portion of the Tocantins River note that fishing and farming productivity have declined (27).

Because water is a medium that facilitates movement and population interchange, it might be imagined that the impacts of human activities on biodiversity are less severe in aquatic environments than in terrestrial ones. In the case of the main channel of the Amazon River this might be true, as many vertebrate species are distributed from the Amazon estuary through the Middle Amazon, all the way to the Upper Amazon in Perú. However, isolation and speciation appear to be common processes in the Amazon tributaries that flow out of the shield areas to the north and south of the main channel. The limnology of the main channel of the Amazon River is distinct from that of these shield tributaries acting as a possible barrier to faunal interchanges. Although much more research is needed, it seems at least plausible that some, and perhaps many, vertebrate species may be unique to individual shield basins. And it is precisely in these shield tributaries that human disturbances to aquatic ecosystems (e.g., gold mining and dam building) are concentrated.

Links Between Mining and Dams

As noted in the terrestrial case, human activities might also influence one another in the aquatic realm. The same river systems targeted for dams are frequently the sites of active gold mining operations. The sediments and mercury released in mining can severely compromise the effectiveness of dam projects. Stream sediment loads in regions of gold mining can increase significantly. If this occurs, operational time of hydroelectric dams could be greatly shortened.

Meanwhile, the mercury released in mining could also contaminate dam reservoirs. There are two pathways for this. First, mercury lost directly to rivers can be transported long distances in association with suspended particles (18). Second, an important fraction (>50%) of the mercury used in mining is evaporated in refining processes. This fraction can be oxidized to Hg^{2+} and later removed from the atmosphere by rains. The work of Lacerda *et al.* (18) provides an example of such long-range dispersal and deposition of mercury in Mato Grosso State, Brazil.

Dam reservoirs provide ideal environments for the conversion of these mercury inputs into highly toxic methyl-mercury. This transformation occurs via bacteria and is favored by the warm, slightly acidic, organic-rich reservoir environment (18). These possible interactions between mercury pollution in gold mining and reservoir contamination may compromise the economic value of reservoir fisheries. Overall, this example reinforces the need to look at the interactions among anthropogenic activities in assessments of environment effects on biodiversity.

THE FUTURE OF THE EASTERN AMAZON

If asked to single out one landscape use that is most likely to threaten the biodiversity of Amazonia in coming years, we would choose ranching. Much of the wealth being extracted from the landscape as wood and gold is channeled to cattle ranching. Loggers use their profits to establish ranches. Ranchers use profits from timber sales to restore productivity to degraded pastures. Miners and mining town merchants use gold profits to buy ranches and to improve pastures in their home states in the Eastern Amazon. Meanwhile, small farmers struggle to accumulate capital to buy their first cow and then transform their fallowed agricultural lands into pastures. All these actors use ranching as a way to lay claim to land and to accumulate and safeguard capital. Ranching appears to be an end state that much of the Eastern Amazon landscape will settle into, with disastrous consequences for regional biodiversity.

While cattle ranching is profitable in the Eastern Amazon, there exist values, in addition to traditional economic considerations, that should be considered when evaluating the wisdom of converting rain forest to pasture (Fig. 3). Forests contain many timber species that could supply wood indefinitely, but when forests are removed, this option is forfeited. The timber present at the time of forest conversion has a value of $2800/ha after processing (5). Generally, this wood is not harvested but burned. Meanwhile, during the pasture-use period, a significant portion of the nutrients embodied in the cut forest biomass is leached from the ecosystem (valued at $3480/ha, given current market prices for NPK fertilizers in Brazil) (32). In addition, some 170 tons of carbon are liberated to the atmosphere during burning and decomposition. This carbon release contributes to global warming. Nordhaus (33) has developed an empirical rationale to account for the societal costs of this carbon release. He considers the degree to which CO_2 emissions could reduce planetary GNP and, in a rather conservative set of projections, concludes that each ton of carbon liberated to the atmosphere would account for a $7.33 reduction in planetary GNP. Using this value, the cost of forest conversion to pasture, directly linked to carbon emissions,is approximately $1250 (170 t X $7.33). Just considering these three values (wood, nutrients, and carbon), directly or indirectly, approximately $7000 of value is lost each time a hectare of rain forest is converted to pasture. Considering that the net profits from pastures are about $10 to $25/ha/year (34), it would take a minimum of 300 years for pastures to pay off these environmental costs.

And there are at least three other, no less important, difficult to quantify, costs implicit in the conversion of forest to pasture (Fig. 3). First, forests, with their leafy canopies high above the ground and their cool understories, provide a humid microenvironment that insulates against wildfires. The value of this service has not yet been calculated. Second, the deep root systems of forests are considered to be an important stablizing factor of regional climates (12). Again, no cost data are available for this service. Finally, forests house species and these species have value. As forest regions are converted to pasture, species that could serve society or that might perform irreplacable ecological functions will be lost. These phenomena are too poorly understood to allow the assignment of firm monetary values, but the costs may not be trivial (Fig. 3).

The conversion of forest to pasture is reversible. Our research at Paragominas (9, 12) reveals that pastures, abandoned after long periods of use, slowly grow back to forest in the absence of fire, although the species composition of this regrowth is somewhat different from mature forest. We have also found that certain hardy economic species grow reasonably well in severely degraded pastures. For example, in a weed-infested pasture that had been subjected to grazing for 16 years, introduced seedlings of cashew (*Anacardium occidentalis*), urucu (*Bixia orelana*), and muruci (*Byrsonima crassifolia*) began producing fruit within two years of planting. Seedlings of the timber species taxi (*Sclerolobium paraensis*) and

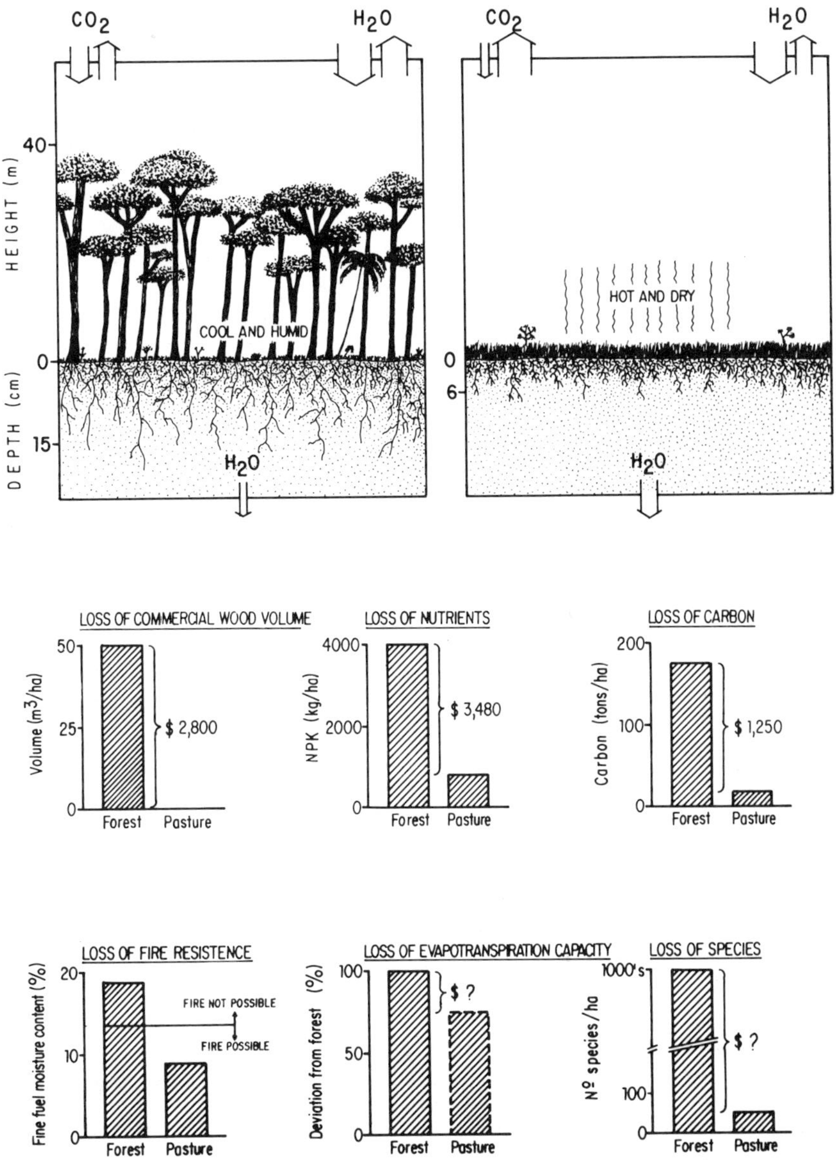

Figure 3. Top. In forest ecosystems of the Eastern Amazon, the fixation and liberation of carbon are in equilibrium and much of the water entering as rain leaves via evapotranspiration. When forest is converted to pasture, there is a large loss of carbon to the atmosphere and less water is recycled. **Bottom.** A first attempt at assigning economic values to the services that forest perform as producers of wood, cyclers of water, barriers to fire spread, and depositories of nutrients, carbon, and species.

mahogany (*Switenia macrophyllum*) are growing 2 to 4 m in height each year in this same setting. No fertilizers were necessary to achieve these results, but plantings were weeded three times each year. However, given the abundance of resources in Amazonia and the uncertainties associated with perennial tree cropping, the restoration of degraded pastures with trees is not an economically attractive investment at present.

MANAGING THE EASTERN AMAZON TO RECONCILE DEVELOPMENT AND CONSERVATIONIST AIMS

Four Basic Steps

What could be done to reconcile development and conservation objectives in the Eastern Amazon? First, it must be acknowledged that Brazil, at present, has no coherent policy with regard to Amazonia. It is true that there are some laws designed to protect the forest, but many of these are studded with loopholes and are seemingly unenforceable. It is also true that there are federal government entities to promote Amazon development [Superintendency for the Development of Amazonia (SUDAM)], to discipline land use [Brazilian Institute of the Environment (IBAMA)], to protect Indian rights [National Indian Foundation (FUNAI)], to survey colonization areas and distribute land titles [National Institute for Colonization and Agrarian Reform (INCRA)], to plan energy use [Northern Brazil Electricity Board (ELETRONORTE)], and to plan government-orchestrated bank programs [e.g., through the Bank of Amazonia (BASA)] to encourage certain economic activities. Yet, none of these entities function in an effective way. None has clear, long-term goals or the means to move toward them. Furthermore, there is very little communication between these entities and, therefore, few coordinated actions. Hence, the starting point in any analysis of how to remedy the disorderly and destructive occupation of Amazonia is to simply admit that Brazil has no coherent policy with regard to Amazonia.

Careless and destructive exploitation will continue in Amazonia unless the Brazilian government takes an active role in the region. Hence, the second step is for the government to move to effectively control the large fragment (perhaps 80%) of Brazilian Amazon territory that is unoccupied. Activities on these lands should only be permitted through government contracts (e.g., concessions for logging, leases for grazing on natural grasslands, long-term occupancy arrangements for forest extractors, etc.).

The tools to determine where the government could legitimately reestablish control are recent wall-to-wall Landsat coverage of the Basin together with INCRA, FUNAI, and IBAMA records of land titling, Indian reserves, and national parks. All lands that have no clear signs of human occupation (based on Landsat analysis), that are untitled (INCRA records), and that have not been designated as Indian Reserves (FUNAI records) or national parks (IBAMA records) could be developing into working forests under state and federal jurisdiction. These lands could be used to benefit the common good.

The third step in reconciling development and conservationist aims in Amazonia is to develop land-use zoning criteria for Amazonia. Researchers have repeatedly emphasized that Amazonia is not a homogeneous expanse of forest (35) and that Amazonia's aquatic areas are not merely waterways (36). Even the designation of dozens of forest types, plant associations, or aquatic subsystems is a gross simplification of the landscape complexity of Amazonia. The goal of zoning is to understand this complexity and, thereby, determine where individual land-use activities should be permitted. The information base to do zoning is at hand (i.e., soil maps, climatic information, vegetation cover data, and crop and cattle performance information). Critical to this zoning work is the demarkation of regions that are rich in biodiversity. The "Workshop 90" meeting (37), with its mission of providing a map of biological richness in Amazonia, was a first step in this process.

Once the government has control of Amazonia and has a clear vision of how this landscape should be used (based on the zoning work), the fourth step is to take concrete measures to promote desirable uses. In general, these favored uses should be intensive rather than extensive. In extensive approaches, yields per unit area are low and production is only possible for a limited time, whereas with intensive approaches production per unit area is elevated and the use period continues indefinitely. Intensive approaches to agriculture in the Eastern Amazon, involving the planting of perennial fruit and nut trees, are many times more profitable than traditional slash-and-burn approaches to farming, although they do require significant capital inputs. Intensive ranching methods, involving better adapted forages and more careful attention to animal movements and animal health, provide yields and profits more than two times those realized in typical extensive ranching activities (34). And forest management for timber production, involving the planning of extraction activities and postharvest silvicultural interventions to promote growth, could reduce the period between cutting cycles by half and guarantee the long-term health of the forest (5). There is adequate technical knowledge to assure intensive, sustainable uses of Amazon land resources.

Final Ingredient for the Wise Management of Amazonia—An Informed Public

The activities that we have focused on either generate capital—logging, ranching, gold mining—or attract capital (e.g., dam building via loans). In many cases, this capital is highly concentrated. Profits of sawmills can run to several hundred thousand dollars per year (5). Profits of gold mine operators may be even greater. And ranchers realize substantial profits, in part because of the grand size of their properties and herds, and, secondarily, through land speculation. These profits are either removed from the region, used to acquire land or to gain more extraction power (e.g., buying trucks and bulldozers in the case of loggers). Increasingly, Amazonia's new monied elite uses its influence to extend control over new Amazonian frontiers while investing in ranches in old frontiers. Indeed, the Amazon occupation process has taken on a life of its own, with the extraction of Amazonia's wealth occurring largely outside the realm of federal and state authority. Amazonian nature is not well served by this unequal development model and neither are the people. Environmental degradation and loss of biodiversity are linked to human greed (Fig. 4).

Given this state of affairs, the simple 1-2-3-4 prescription that we provide will come to naught unless it is accompanied by strong pressure from Brazil's citizenry. Fortunately Brazilians are increasingly aware of environmental issues and the links between development decisions and environmental well being. With the emergence of hundreds of nongovernmental organizations (NGOs) with strong environmental agendas and the realization that public opinion can serve as a strong political tool, it may now be possible to begin to prod the government to assume its responsibilities in promoting rational development in Amazonia.

Meanwhile, capital will remain a major tool for directing rational development in Amazonia. To the extent the Brazilian government is successful in formulating a clear Amazon policy that includes planning of where development

THE AMAZON EXPERIMENT: 1960–2000

PRECONDITIONS: ABUNDANT NATURAL RESOURCES AND ABSENCE OF A COHERENT GOVERNMENT POLICY

PHASE I	INFRASTRUCTURE DEVELOPMENT:	Government Roads, Logging Roads, Airstrips, etc.
PHASE II	ESTABLISHMENT OF ECONOMIC ACTIVITIES:	Cattle Raising, Timber Extraction, Gold Mining, etc.
PHASE III	GENERATION OF CAPITAL:	Capital Invested in Extraction Equipment, Land, Cattle, etc.
PHASE IV	CONCENTRATION OF CAPITAL **LEADS TO** UNEQUAL DEVELOPMENT MODEL:	emergence of a new elite

PHASE V FACING CONSEQUENCES:

	Social Consequences	Ecological Consequences
Timber extraction:	–disabling accidents	–population reductions of species sensitive to logging
	–sickness	–increased incidence of fire
	–poverty	–increased deforestation due to road access
Cattle ranching:	–violence	–elimination of most rainforest species
	–rural poverty	–invasion by aggressive, generalist-type species
	–urban migration	–regional changes in rainfall and temperature
Gold mining:	–genocide	–loss of stream habitat
	–poverty	–sediment effects on reproduction, respiration, and feeding
	–spread of disease	–mercury pollution

OUTCOME: IMPOVERISHED SOCIETY INHABITING A BIOLOGICALLY IMPOVERISHED LANDSCAPE???

Figure 4. The present chaotic occupation of Amazonia is not surprising considering that the Amazon is rich in land, wood, and minerals and that the government has taken a passive role in the region. The result has been an unequal development model with loggers, ranchers, miners, and urban entrepreneurs accumulating riches. The severe distortions in capital accumulation have had social and ecological consequences and may ultimately lead to an impoverished society residing in an impoverished landscape.

should occur (zoning) and what kinds of development are most appropriate in each zone, capital could be applied to stimulate desired ends. The role of the international community could be large both in providing the capital necessary to help Brazil develop a solid Amazon policy and in helping her implement that policy. Brazilian NGOs could fulfill a key role as watchdogs and implementers of new initiatives, insofar as these groups are generally well informed and work close to local problems.

But a lion's share of the challenge rests with the Brazilian people. Brazilian citizens can no longer play a passive, accommodating role. Instead they must educate themselves on environment/development issues and bring pressure to bear on political processes, all the while seeking leaders of integrity and seriousness.

CONCLUSIONS

We wish to stress four conclusions. First, there is adequate knowledge of the ecological characteristics of Amazonian plant and animal species to start to predict which of these species are apt to be threatened by different human resource-use activities. Using this approach, we have identified some 30 timber species that might experience population declines if subjected to severe logging pressure. This consideration of ecological characteristics of plant and animal species also allows us to begin to contemplate what the future flora and fauna of Amazonia might look like if aggressive land transformation activities continue. Such a biota would be composed of hardy, generalist-type species containing only a very small fraction of Amazonia's present-day biodiversity.

Second, we stress the need to include a consideration of secondary or "downstream" effects in the analysis of the effects of human activities on biodiversity. In the Eastern Amazon the effects of forest alteration readily spread beyond ecosystem borders, insofar as changes in ecosystem structure (e.g., loss of carbon or nutrients) and function (decrease in evapotranspiration) in one region can influence productivity and climate in surrounding regions.

Third, in assessing present and future threats to biodiversity in the Eastern Amazon, an understanding of the interactions among individual resource-use activities is fundamental. Practices associated with one resource-use activity can spill into adjacent activities with deleterious effects (e.g., fires set to kill weeds in pastures can escape and severely damage logged forests; mercury and sediments move from gold mining sites to dam reservoirs, possibly compromising reservoir fisheries). Likewise, in the economic realm, capital generated in mining and logging can be applied to a third activity, ranching, with negative effects on regional biodiversity.

Fourth, the progressive expansion of Amazon ranching, with the attendant elimination of Amazonia's biodiversity, is likely to continue until Brazil develops a coherent Amazon policy based on land-use zoning and including programs to foster intensive, sustainable land uses (i.e., will continue until other activities become more attractive than ranching as investment alternatives). As a necessary first step, the state will need to regain its claim on the Amazon landscape. If this were done, access to Amazonian lands could be permitted through government contracts (concessions for logging, leases for grazing on natural grasslands, etc.). In this way, the state would be able to serve as an effective buffer between human greed and landscape uses that both contribute to the common good and protect biodiversity.

We wish to thank Flavio Figueiredo for creating the graphics. In developing the ideas expressed in this paper we benefited greatly from discussions with Dan Nepstad, David McGrath, Adalberto Verissimo, Rui Rocha, Oriana Almeida, Marli Mattos, Nelson Rosa, Patricia Shanley, Peter Jipp, Gordon MacMillan, Ronaldo Barthem, Miguel Petrere, and Luiz Drude de Lacerda. C. Uhl acknowledges the John Simon Guggenheim Memorial Foundation for support during the writing of this chapter.

REFERENCES AND NOTES

1. FIBGE. 1987. Anuário Estatístico do Brasil. Vol. 47. Rio de Janeiro: Fundação Instituto Brasileiro de Geografia e Estatística.

2. Daly, D.C., and G.T. Prance. 1989. Brazilian Amazon. In *Floristic inventory of tropical countries*, B.G. Campbell and H.D. Hammond, eds. New York: New York Botanical Garden.
3. Prance, G.T., and K.S. Brown, Jr. 1987. The principal vegetation types of the Brazilian Amazon. In *Biogeography and quaternary history in tropical America*, T.C. Whitmore and G.T. Prance, eds. Oxford: Clarendon Press.
4. Uhl, C., A. Veríssimo, M.M. Matos, Z. Brandino, and I.C.G. Vieira. 1991. Social, economic and ecological consequences of selective logging in an Amazonian frontier: The case of Tailândia. *Forest Ecology and Mgmt.* 46:243–73.
5. Veríssimo, A., P. Barreto, M.M. Mattos, R. Tarifa, C. Uhl. Logging impacts and prospects for sustainable forest management in an old Amazonian frontier: the case of Paragominas. *Forest Ecology and Mgmt.* (in press).
6. Martini, A., N. Rosa, and C. Uhl. A first attempt to predict Amazonian tree species threatened by logging pressures. Institute do Homem e do Melo Ambiente de Amazonia. Belem. (unpubl.).
7. D., Nepstad, and P. Jipp, personal communication.
8. Salati, E. 1987. The forest and the hydrological cycle, 273–96. In *The geophysiology of Amazonia: Vegetation and climate interactions*, R.E. Dickinson, ed. New York: Wiley.
9. Nepstad, D. 1989. Forest regrowth in abandoned Amazonian pastures of Eastern Amazonia: Limitation to tree seedling survival and growth. Ph.D. Thesis, Yale University, New Haven.
10. Uhl, C., R. Buschbacher, and E.A.S. Serrão. 1988. Abandoned pastures in eastern Amazonia. I. Patterns of plant succession. *J. Ecology* 75:663–81.
11. Buschbacher, R., C. Uhl, and E.A.S. Serrão. 1988. Abandoned pastures in eastern Amazonia. II. Nutrient stocks in the soil and vegetation. *J. Ecology* 76:682–99.
12. Nepstad, D., C. Uhl, and E.A.S. Serrão. 1991. Recuperation of a degraded Amazonian landscape: Forest recovery and agricultural restoration. *Ambio* 20:248–55.
13. Uhl, C., and R. Buschbacher. 1985. A disturbing synergism between cattle ranching burning practices and selective tree harvesting in the eastern Amazon. *Biotropica* 17:265–68.
14. Browder J. 1988. The social costs of rainforest destruction. *Interciencia* 13:115–20.
15. Dourojeanni, M., and Jorge Padua, M.T. 1991. *Goldrush and environment in the Brazilian Amazon*. Brasilia: Funatura. (unpubl.).
16. Ramos, C.R., and A.R. Silva. 1990. Uma visão da macroeconomia Paraense para os anos 90. Para Desenvolvimento. *Instituto de Desenvolvimento Economico e Social do Pará* 26:65–83.
17. Silva, A.R.B. 1988. Ouro do Pará: perdas técnicas, degradação ambiental e a marginalização do Estado na expansão industrial brasileira. Annais do Congreso Brasileira de Geologia. Vol. 5, 2014–22. Belem, Pará.
18. Lacerda, L.D., and W. Salomons. 1991. Mercury in the Amazon: a chemical time bomb? A report sponsored by the Dutch Ministry of Housing, Physical Planning and Environment. Chemical Time Bomb Project.
19. Huntz, W.R.A. 1981. Observações sobre os pigmentos visuais e filtros amarelos nos olhos de peixes Amazonicos. *Acta Amazonica* 11:113–23.
20. Redford, K.H. 1992. The empty forest. *Bioscience* 42(6):412–22.
21. Robinson, J.G., and K.H. Redford. 1986. Body size, diet and population density of neotropical forest mammals. *American Naturalist* 128:665–80.
22. Pfeiffer, W.C., and L.D. Lacerda. 1988. Mercury inputs into the Amazon region, Brazil. *Environmental Technology*. Letters 9:325–30.
23. Malm, O., W.C. Pfeiffer, C.M.M. Souza, and R. Reuther. 1990. Mercury pollution due to gold mining in the Madeira River Basin, Brazil. *Ambio* 19:11–15.

24. Martinelli, L.A., J.R. Ferreira, B.R. Forsberg, and R.L. Victoria. 1988. Mercury contamination in the Amazon: A gold rush consequence. *Ambio* 17:252–54.

25. Fernandes, S.R., F.A. Guimaraes, E.D. Bidone, L.D. Lacerda, and C.W. Pfeiffer. 1990. Monitoramento do mercurio na area do Projeto Carajas. Seminario nacional sobre riscos e consequencias do uso do mercurio. Rio de Janeiro. (manuscript.)

26. R.B. Barthem, personal communication.

27. M. Petrere, personal communication.

28. Brazil Eletronorte. 1985. Polit-kit Ano II, No. 3. O novo perfil da Amazonia. Brasilia: Centrais Electricas do Norte do Brasil S. A. (Eletronorte).

29. Barthem, R.B., M.C.L. de Brito Ribeiro, M. Petrere, Jr. 1991. Life strategies of some long-distance migratory catfish in relation to hydroelectric dams in the Amazon Basin. *Biological Conservation* 55:339–45.

30. Petrere, Jr. M. 1989. Rivers fisheries in Brazil: A review. *Regulated Rivers: Research and Mgmt.* 4:1–16.

31. Junk, W.J. 1989. The use of Amazonian floodplains under an ecological perspective. *Interciencia* 14:317–22.

32. We do not wish to suggest that this amount of fertilizer would be necessary to restore fertility to degraded pastures. The amount necessary is much less than this. Nor do we wish to imply that pastures are closed systems with regard to nutrients—clearly, nutrients enter these ecosystems via atmospheric deposition and weathering. Our point is simply that measurable quantities of nutrients are lost in the conversion of forest to pasture, and given present prices the value of these nutrients is approximately $3500. Although this is not a wholly satisfactory estimate, it represents a starting point for thinking about how to assign values to nutrient loss in the conversion of forest to pasture.

33. Nordhaus, W.D. 1991. To slow or not to slow: The economics of the greenhouse effect. *The Economic J.* 101:920–37.

34. M. Mattos, personal communication.

35. Moran, E.F. 1989. Government-directed settlement in the 1970s: An assessment of Transamazon Highway colonization. In *The human ecology of tropical land settlement in Latin America*. D.A. Schumann and W.L. Partridge, eds., 172–98. Boulder, CO: Westview Press.

36. Goulding, M.G. 1989. *Amazon—The flooded forest*. London: BBC Books.

37. Rylands, A.B. 1990. Um mapa que protege a Amazônia. *Ciencia Hoje* 1:6–7.

Synthesis

The Potential of Biodiversity

Christopher S. Potter and Joel I. Cohen

GENETIC POTENTIAL REVISITED

As emphasized in the introduction to this book, one of the major challenges of sustainable development is realizing the intrinsic value of protecting and using biodiversity wherever it is found. The editors have advocated the idea that conservation of biodiversity enhances the potential for economic development. Conversely, the loss of genes, species, and ecosystems increasingly constrains our options for transforming society and the biosphere in a manner that helps fulfill immediate human requirements, without compromising the potential options of future generations. There are numerous explanations for why we feel this to be so. They include the need for permanent stability of food and fiber production against an ever-changing environmental and socioeconomic backdrop, the continued availability of a wide variety of natural habitats for human occupation and cultural evolution, and the fostering of a global biota with high potential for adaptation to human and geophysical perturbations. Preserving species for their own sake is an admirable and desirable objective; however, within the context of development, biodiversity may be nothing less than the prime human food bank, toolbox, medicine cabinet, and thermostat (Ehrlich and Wilson).

With this premise in mind, the editors have searched for case studies that consider genetic resource conservation as important among the various levels of biodiversity. Moreover, it made sense to initiate an investigation by focusing on an ostensibly compelling facet of the subject. In the case of biodiversity, the study of economic, scientific, and societal values of heritable traits is one such facet. A focus on genetic resources *in situ* (1) provided both an evolutionary perspective advocated by biologists (Erwin) and a comparatively concrete socioeconomic context for conservation (Soulé). Viewed from a developing country's standpoint, genes and populations may indeed be the most heavily affected components on the biospatial spectrum, owing to the relatively high number of species, small geographic ranges, and rapid rate of habitat fragmentation in tropical life zones (Soulé).

While much of the evidence remains anecdotal or merely suggestive at the moment, several case studies from the tropics testify to the notion that the potential economic exploitation of genetic resources in their natural habitats should not be underestimated. For example, the reviews by both primary authors Gliessman and Hartshorn clearly demonstrate that, through detailed research and careful management, biologically diverse ecosystems can meet many of the development requirements of local communities. By extending the idea of genetic resources to include wildlife and fisheries, the work of both Weber and Twilley *et al.* illustrates that certain species conserved in situ can sustain an entire local or even a national economy. While these cases may be viewed as exceptional, their potential is undeniable.

Turning briefly to economics, we concur that genetic resources should be viewed as a public good that is not easily subjected to market valuation (Morowitz). The concept of genetic potential implies a value well beyond what society is willing to pay in the present for conservation of a species and its ecosystem. Case study examples of protecting and using genetic potential *in situ* can help shift the policy debate from the level of singular species to the broader habitat level (Williams, Wilkes), while maintaining a focus on values on a "human" scale.

In the remainder of this chapter, we will elaborate on our major premise and attempt to evaluate three important questions, while drawing upon the themes and lessons brought forward in the other chapters of the book. Insofar as data and information were available, all case study authors examined three common issues: (1) documentation of species loss and genetic erosion in selected natural habitats; (2) presentation of management approaches that sought to meet the needs of both biodiversity conservation and sustainable development, with emphasis wherever possible on genetic resources; and (3) evaluation of potential applicability of a given management approach in similar environments outside the area of immediate interest.

Question 1. Can case studies help document species loss and genetic erosion in selected natural habitats?

While it should be acknowledged that few systematic case studies have been conducted on causes, rates, and implications of species extinction and genetic erosion, current global estimates fall into a general range from 10 to as many as 100 species lost per day (Ehrlich and Wilson). One case study cited that tropical deforestation alone is responsible for the loss of 74 species per day (Hartshorn and Pariona). These numbers are understandably a topic of much debate within and beyond the scientific community (2). However, regardless of which end of the spectrum is more accurate, it is noteworthy that enough information has been collected in some critical cases to make reliable estimates of which species are most threatened with extinction under present land conversion trends (Uhl *et al.*). In other words, if the concept of "threatened species characterization" is valid extensively, a solid biological foundation is developing upon which to base certain fundamental tactics for conservation (Soulé).

Case studies have documented a number of specific instances of species loss and genetic erosion. For example, deforestation in Mexico alone has resulted in 17 documented plant extinctions and over 470 endangered species, representing 17% of the known national plant species richness (Gliessman). There is strong evidence that two close relatives of maize have suffered considerable geographic contraction in Guatemala (Wilkes). Over 20 species of wild potatoes have been lost (Altieri and Montecinos).

In an Indonesian case study, deforestation for the purposes of commercial logging has lead to destruction of approximately 50% of forest lands, with untold loss of associated species (Padoch and Peters). At least 30 major timber species are at risk of serious population declines in Eastern Amazonia (Uhl *et al.*). Commercial extraction has contributed to the status of at least three wildlife and two tree species listed as endangered in the Peruvian Amazon; almost 50 others are threatened (Pinedo and Padoch).

With further regard to wildlife, data exist to show that large primates have experienced dramatic population declines in African forest habitats, chiefly due to deforestation and poaching (Weber). On a global scale, overfishing has contributed to the extinction or near extermination of several aquatic species (Diana). In addition, introduction of tilapia and their subsequent escape into natural aquatic habitats will continue to adversely affect populations of many native species (Diana, Uhl *et al.*). The take of nontarget species in seed fishing to support mariculture is notable (Twilley *et al.*). Moreover, loss of mangrove ecosystems and their genetic resources to shrimp pond development, the area of which has at least doubled since the mid-1980s (Diana), may be influencing the entire fisheries food web in areas where coastal forests play a major role in oceanographic dynamics.

Old-growth forests of the northwestern United States have disappeared at an alarming rate since the 1940s, and with the loss of these unique ecosystems will go over 200 species of vertebrates and many more plant and invertebrate species (Wilcove and Olson). Moreover, these forests support invaluable environmental services, such as fishing, watershed hydrologic regulation, and carbon storage.

While it is difficult to speculate on rates and common causes, one trend that was highlighted by several authors (Altieri and Montecinos, Pinedo and Padoch, Weber, Twilley *et al.*) is that species depletion is occurring fastest in areas nearer to urban centers and their markets. Demand appears high, whereas supply is severely limited and pricing rarely reflects full value. Soulé outlines related causes of biotic degradation that can be identified in many of the case studies in this book.

Unqualified extrapolation of site-specific losses is not valid at the present time, but it appears that even given the relatively modest number of studies conducted to date, dramatic examples of genetic resource depletion exist. "Typical or not?" remains a pressing question.

Question 2. Which *in situ* management techniques appear to be successful in meeting the needs of both genetic resource conservation and sustainable development?

In many instances, natural ecosystems must be managed to achieve the goals of development (Morowitz). Under the subheading of *agroecosystems*, several case studies have documented biologically diverse cultivated areas that combine native and modern (high-yielding) crop varieties. The characteristics of these systems include crop mixtures, rotations, intercropping, biological N fixation, low-cost (ecological) weed and pest control, and a focus on long-term sustainability of crop yields. In contrast to "modern" systems, local, traditional, or indigenous knowledge-based management systems have the advantage of an ecological history by which conservation approaches can be evaluated for conditions of sustainability.

At the bottom line, although yields of *individual* crops grown in biodiverse systems may be diminished, *total* yields of all crops together can be increased relative to monocultures (Gliessman). When deciding what crop varieties to use and where, the cultivator's main choice appears to lie between increasing production and reducing risks, with little distinction made between the source of genetic material (improved varieties or landraces; Altieri and Montecinos). Judging from an albeit limited comparison (Padoch and Peters), this tendency to mix varieties from different sources may be a widespread pattern among agriculturists. The need for international cooperation to assure conservation of plant genetic resources becomes even more urgent in this regard (Wilkes, Williams).

There seems to be high potential for food production in properly managed natural wetland systems that can sustain crop yields at many times the level of mechanized systems, with substantially lower input costs and higher biodiversity conserved. Maintaining ecosystem-function benefits from periodic inundation and high natural organic residue input is the key to continued high yields (Gliessman). On the more intensive management side, biologically diverse, fish-rice polyculture ecosystems result in high grain yields relative to single crop systems, while protecting native fish species (Diana).

Other wetland systems should be protected for the services provided to adjacent agroecosystems. Mangroves must be conserved and managed to sustain production of postlarval fishing activities. These forests are important in preventing eutrophication and maintaining favorable nutrient cycling dynamics in coastal environments. Silviculture, therefore, is not viewed as a practical management option for mangrove ecosystems because of lack of research on techniques and the critical ecosystem services that would be lost if mismanaged (Twilley *et al.*).

For the purposes of forest management, strip-cutting of "gap-openings," which are similar to natural tree falls, followed by selective thinning for favored species, can be an ecologically sustainable management technique for tropical forests (Hartshorn and Pariona). Considerable biodiversity is conserved while growth of certain relatively rare forest species populations is actually promoted by this technique. On the other end of a continuum, managed forest gardens and reserves employ techniques of enrichment planting with native species followed by selective weeding and microclimate alteration for favored species (Padoch and Peters). The result is a highly productive, seemingly sustainable system that yields marketable fruit and wood products. Although the species diversity of managed forest gardens may be somewhat lower than that of an unmanipulated forest, many more species (over 100 in certain circumstances) can be conserved and exploited than in most modern agricultural areas. In most cases, it was stressed that forest management should rely heavily on local knowledge and human resources, with assurances of clear and permanent land ownership by indigenous peoples (Hartshorn and Pariona; Padoch and Peters; Pinedo and Padoch). Socio-cultural lessons of sustainable management for Pacific northwest, old-growth forests (Wilcove and Olson) might be gleaned from successful forest management projects outlined in the book's other case studies.

Restoration of degraded pasture land in Amazonia tends to be economically unwise, despite the observation that potentially valuable timber species will eventually regenerate with proper management (Uhl *et al.*). Intensification of existing agricultural and forest lands is recommended as the best alternative management for these regions. The need for a regional scale approach for Amazonia was highlighted by the studies of both Pinedo and Padoch and Uhl *et al.* The former authors recommend management that seeks to include needs and abilities of local inhabitants in the larger political context of conservation plans. These indigenous communities occupy forested landscapes and lakes and have never been involved in large-scale cattle ranching operations typical of other parts of Amazonia. Theirs is a history, including land reform and resource owership issues, very different from the experience of those offered from the Eastern Amazon. In the Pará State of Brazil, *more* government control over land use and *less* influence of local land owners is viewed as the critical missing factor to protect biodiversity (Uhl *et al.*). This combination of experiences helps the reader to appreciate the imposing diversity of social and natural history throughout Amazonia that compounds any attempt to generalize about saving tropical forests.

Ecologically sensitive tourism can be an attractive means to protect biodiversity and raise impressive revenues, provided several criteria are met: (1) numbers and duration of daily visits are tightly controlled (so as to not adversely affect the subject of interest); (2) training of local guides and curators is given priority; (3) benefits in the form of revenue sharing from ecotourism are extended to communities adjacent to reserves (Weber). Political stability remains a necessary element to the long-term success of ecotourism.

Question 3. Is it possible to generalize about the broader applicability of site-specific habitat management approaches for conserving biodiversity and promoting development?

Every tactic for conservation of biodiversity has its limitations—socioeconomic, cultural, and ecological (Soulé). It is fairly easy to point out differences on a site-by-site basis; it is more useful, however, to seek out common threads in the global biodiversity puzzle. The answer to this third question can be derived only from the most general case study lessons, keeping in mind the risks and problems related to transfer of management approaches from one location to another.

Common constraints cited in almost all studies include lack of (1) financial resources for local initiatives, (2) trained support personnel, (3) access to markets for biological goods and services, and (4) consensus on (and enforcement of) conservation aims from the local to the regional level. Conservation at any level will be difficult to implement where institutional arrangements, government policies, and research programs are biased toward exclusion of local people, extensive replacement of indigenous with modern varieties, and ignorance of prevailing ecological constraints (Altieri and Montecinos). Recognition of local knowledge and human resources, including assurances of clear and permanent land ownership by indigenous peoples, is often a missing component (Hartshorn and Pariona; Padoch and Peters; Pinedo and Padoch).

Several points are worth mentioning with regard to specific habitat and genetic resources uses.

- Transfer of aquaculture success stories from Asia to Latin America will require more research on the potential of fish species endemic to the Americas as the means to sustainable, marketable yields with minimal local species loss (Diana).
- Although lowland tropical forests harbor a higher level of biodiversity, their potential for revenue generation from ecotourism is limited relative to highland forests. This is prima-

rily due to forest structure and visibility considerations. Consequently, alternative means to conserving biodiversity must be sought for these areas (Weber).

- Economic returns of ecologically based forest management in the tropics, compared to perennial cash crops such as coffee, may not yet be competitive. Nonetheless, this type of timber extraction would likely exceed the profitability of beef cattle operations over extensive areas of Latin America (Hartshorn and Pariona). A more potentially severe constraint to extended strip-cutting is increasing population pressure on intact tropical forests.
- Sustainable, intensive use of natural habitat services outside of forest ecosystems is an important strategy that should not be underestimated in meeting basic human needs. Diana cited fish-rice systems; Gliessman highlighted seasonally flooded agroecosystems; Uhl *et al.* cited examples throughout Eastern Amazonia.

CONCLUSION

Obviously, there is a pressing need for more integrated case studies, more data on population genetics and biogeography in the tropics, better appreciation of the contribution of individual species to services that are manifested at the ecosystem level, and a broader definition of economic costs and benefits in conservation policy. Beyond the realm of research, however, it is high time for a serious attempt to dispel the notion that *in situ* conservation of genetic resources is synonymous with park and reserve management. The potential of biodiversity will begin to be realized when development options include conservation of many species in agro-silvi-pastoral systems outside of protected areas. This means rethinking development as a process of environmental transformation that perpetuates adaptation, builds variation, and maximizes potential over sheer yield. Case studies reviewed in this book illustrate that it can be done. They, and others to come, open a window on the possibilities for more sustainable development in a diverse biosphere.

REFERENCES AND NOTES

1. For the sake of simplicity, we take the conservation scope of *in situ* to include all natural or semi-natural areas; hence we have lumped the first six paths for conservation (Soulé) into a single broad category.
2. Various perspectives on this debate were reviewed in *Science*, at 253:739 (1991).

Index